D0152918

ANALYTICAL MASS SPECTROMETRY

ANALYTICAL MASS SPECTROMETRY

Strategies for Environmental and Related Applications

WILLIAM L. BUDDE

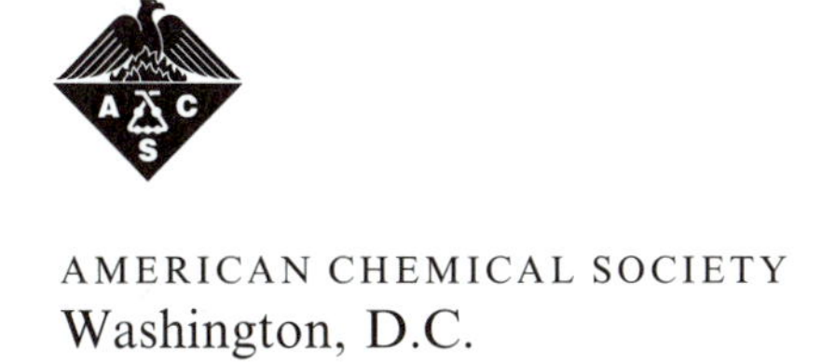

2001

OXFORD
UNIVERSITY PRESS

Oxford New York
Athens Auckland Bangkok Bogotá Buenos Aires Calcutta
Cape Town Chennai Dar es Salaam Delhi Florence Hong Kong Istanbul
Karachi Kuala Lumpur Madrid Melbourne Mexico City Mumbai
Nairobi Paris São Paulo Shanghai Singapore Taipei Tokyo Toronto Warsaw

and associated companies in
Berlin Ibadan

Copublished 2001 by American Chemical Society,
1155 Sixteenth Street, NW,
Washington, DC 20036, and
Oxford University Press, Inc.,
198 Madison Avenue,
New York, NY 10016

Developed and distributed in partnership by
American Chemical Society and Oxford University Press

Library of Congress Cataloging-in-Publication Data
Budde, William L.
Analytical mass spectrometry:
strategies for environmental and related applications/
William L. Budde.
p. cm.
Includes bibliographic references and index.
ISBN 0-8412-3664-X
1. Mass spectrometry. I. Title.
QD96.M3.B84 2000
543′/0873-dc21 00-36186

1 3 5 7 9 8 6 4 2

Printed in the United States of America
on acid-free paper

To

Eileen Elizabeth

Lois Lynn

Michael Xavier

Mark Vincent

Sharon Sue

John Paul

Analysis too often is a secondary activity; unwarranted assumptions are made of the simplicity of the measurement and the reliability of the method . . . . The result in part has been masses of data that are frequently useless for the intended purposes.

Cleaning Our Environment: A Chemical Perspective. A report by the Committee on Environmental Improvement, American Chemical Society, Washington, D.C., 1979

PREFACE

Mass spectrometry has emerged as the single most powerful technique available for the identification and measurement, with great sensitivity, of the elements and essentially any organic or inorganic compound or ion of any molecular weight in any sample matrix. Mass spectrometry is no longer narrow in scope, as it was 30 years ago, or limited to elemental analyses and a relatively small class of low molecular weight, thermally stable, and volatile compounds that form positive ions in the gas phase when bombarded with electrons. Within the broad scope of analytical chemistry, and its sometimes bewildering array of spectroscopies and specialized laboratory procedures, mass spectrometry has been called "The Handsome 600-Pound Gorilla" of analytical techniques (R. W. Murray, *Analytical Chemistry*, April 1, 1993).

Although new analytical concepts and techniques are constantly being generated by a creative scientific research community, mass spectrometry has outpaced most, if not all, other techniques with numerous major discoveries and instrumentation developments. Significant advances have occurred in sample introduction, interfaces with on-line separations, ionization, resolving power, sensitivity, exact mass measurements, quantitative analysis, data processing, interpretation, and a myriad of other capabilities. These developments and the availability of a broad range of high-quality commercial instrumentation have led to the widespread application of mass spectrometry in essentially every scientific discipline and field of technology. Mass spectrometry has dominated the science and business of environmental analyses and several other important areas of application. The cost of mass spectrometric equipment has fallen dramatically over the last 20 years, and compact bench-top or portable instruments are now available from several manufacturers. The discipline of mass spectrometry is now so large and diverse that a complete, detailed description of all fundamentals and applications would require multiple large volumes. The success and widespread applications of mass spectrometry have led to an accelerating pace of fundamental and applied research, and this will undoubtedly lead to still further advances in capabilities and instrumentation.

The purpose of this book is to describe and explain those mass spectrometric and closely related analytical strategies and techniques that are most important and useful for the separation, identification, and measurement of organic and inorganic compounds and ions in environmental and related media. Environmental media are usually considered as the air, water, and earth. Related media, to which most of the strategies and techniques discussed in this book are applicable, include beverages, biological fluids, some consumer products, foods, some industrial materials, and tissue. Substances of environmental interest are invariably used as examples in the descriptions and explanations, but the strategies and techniques considered are usually general and applicable to a variety of compounds and ions of interest in agricultural and food chemistry, forensic investigations, industrial processes, illegal drug determinations, petroleum exploration, pharmaceutical analyses, toxicological studies, and other areas. On the other hand, mass spectrometric strategies and techniques for macromolecules, including biological molecules and synthetic polymers, are beyond the scope of this book.

The approach and organization used in this book are probably somewhat different than in most mass spectrometry books. One approach is to organize chapters by fundamentals such as the theory and principles of mass spectrometers, sample introduction systems, ionization techniques, and the interpretation of mass spectra. The organization of this book is according to analytical and mass spectrometric strategies used to solve problems and according to several broad classes of substances that are central to many problems. The introduction contains a brief historical account of the parallel and mutually supportive developments of mass spectrometric and other analytical instrumentation and the environmental movement in the United States. Chapters 2, 5, and 8 concentrate on general strategies, and Chapters 3, 4, 6, and 7 on broad classes of substances and useful strategies and techniques. Strategic decisions must be made in order to apply mass spectrometry to environmental and related areas of investigation, and decisions often depend on the goals and objectives of the analyses. Routine monitoring of specific substances for government regulatory compliance may call for a different analytical strategy than a science project designed to discover the cause of some effect or the mechanism of some process. Hopefully, this organization and presentation will provide a rationale and understanding of why some techniques are more important than others, why U.S. Environmental Protection Agency regulatory analytical methods were developed as they were, and why some techniques are not often used.

Some of the techniques discussed in this book are relatively new, that is, they were discovered or developed in the last 20 years or less, and some are quite mature. A few of the techniques considered are no longer as important as they were at one time, but they are described in some detail because they illustrate the analytical problems encountered and why different technical approaches were used to address these problems. In order to keep the book to a reasonable size, a number of basic and extremely important topics are only briefly summarized or not covered at all. It is assumed that the reader either has knowledge of, or can acquire elsewhere, the theory and basic principles of operation of the various types of mass spectrometers. Accordingly, there are no detailed descriptions or explanations of any mass spectrometer designs, operating principles, vacuum systems, or general capabilities. References to various classical and new books or reviews of various mass spectrometer designs are provided. The directly connected separation techniques that are so widely and successfully used with mass spectrometry are described briefly, and references to specialized books and reviews are provided. It is assumed the reader either has, or can acquire elsewhere, detailed information about these extremely important separation techniques. A number of interfaces for connecting separation techniques or sampling devices to the mass spectrometer have been developed in the last 20 years. Some of these are described in some detail because there are few places where all these techniques are discussed in the same terms and to the same extent. Since the presence of ions in the gas phase is an essential prerequisite to mass analysis, and since there are now many ways of producing ions for mass spectrometry, ionization techniques are presented and discussed in appropriate detail.

Computer hardware and software play an indispensable role in the operations of all contemporary mass spectrometry systems. It is assumed that the reader has a basic knowledge of standard computer system components, terminology, and of how analytical instrument computer systems generally operate. This book does not include descriptions or explanations of any of the commercial mass spectrometer computer data systems. A few standard types of data processing or display algorithms are described and used frequently. It is also assumed that the reader either has knowledge of, or can acquire elsewhere, the principles of the interpretation of the electron ionization mass spectra of representative organic compounds. A review of this topic would be a large book and is beyond the scope here. References to various classical and new books or reviews on interpretation of mass spectra and fragmentation mechanisms are provided. However, the characteristic fragmentations of the molecular and other ions of some molecules are discussed, as are the collision-induced fragmentations observed with tandem mass spectrometry. A few very important related topics, such as the appearance and use of characteristic patterns of ions caused by naturally occurring isotopes, and the stability of molecular ions, are briefly summarized.

The content, organization, and detail of this book are intended to meet the needs of a broad range of analysts, analytical chemists, environmental engineers, environmental scientists, general chemists, mass spectrometer instrument operators, laboratory managers, and anyone else who is interested in or who needs to know about the analytical applications of mass spectrometry. The professional mass spectroscopist, who has a vast array of more detailed and complete reference material available, will find this book a useful introduction to the background and original literature of analytical and environmental mass spectrometry. Specialists in nonenvironmental but related areas of investigation should be able to use, or adapt as appropriate, the general strategies and techniques with their particular compounds or ions and media.

In 1979, before the widespread application of mass spectrometry to environmental analyses, Ann Arbor Science Publishers published *Organics Analysis Using Gas Chromatography/Mass Spectrometry: A Techniques Procedures Manual* by William L. Budde and James W. Eichelberger. The purpose of that book was "to organize in one place a variety of information that is needed to identify and measure organic compounds in environmental samples using computerized gas chromatography–mass spectrometry (GC/MS)." That purpose was fairly easy to accomplish in a small book during the late 1970s because most of the magnificent capabilities that characterize current analytical instrumentation did not exist. At that time, open tubular or capillary gas chromatography was in its infancy (fused silica columns had not yet been invented), instrument minicomputers had far fewer capabilities than a palm-sized computer has today, many remarkable developments in mass spectrometry instrumentation had not yet occurred, and there was very little experience in environmental analyses using mass spectrometry. Much of the 1979 book was devoted to basic instrument operations, sample preparation procedures, overcoming instrumental problems, quality control, and the use of a rudimentary minicomputer data system to acquire, reduce, and present mass spectrometry data.

This book had its beginning as a one-day short course presented at the Pittsburgh Conferences on Analytical Chemistry and Applied Spectroscopy in New Orleans in 1995, Chicago in 1996, New Orleans in 1998, and Orlando in 1999. The short course was developed at the suggestion of one of the pioneers in mass spectrometry, Andrew G. (Jack) Sharkey of the University of Pittsburgh. The author wishes to express appreciation to Jack Sharkey and to the literally thousands of other dedicated scientists and engineers who contributed in many different ways to make mass spectrometry a magnificent instrumental technique. Special recognition and thanks are due to many former and present U.S. Environmental Protection Agency (USEPA) colleagues, especially Ann Alford-Stevens, Tom Behymer, Tom Bellar, Jack Creed, the late B. F. (Dude) Dudenbostel, Jim Eichelberger, Steve Heller, the late Jim Ho, Doug Kuehl, Ted Martin, Bob Lewis, and Craig Shew. Other friends and associates who were particularly helpful to the author are Bob Finnigan (Finnigan Corporation), Jack Henion (Cornell University), Ron Hites (Indiana University), Larry Slivon (Battelle-Columbus Laboratories), David Sparkman (several mass spectrometer manufacturers and private consultant), and Steve Stein (National Institute of Standards and Technology). The author, who is not a formally trained mass spectrometrist, was first attracted to mass spectrometry after attending the nineteenth annual E. C. Franklin Memorial Lecture at the University of Kansas on 19 April 1967. The distinguished lecturer was Fred McLafferty, then of Purdue University, whose lecture was entitled *Molecular Mass Spectrometry*.

Cincinnati, Ohio
December 2000

W. L. B.

CONTENTS

ANALYTICAL MASS SPECTROMETRY

1

Introduction

The tremendous success of mass spectrometry (MS) as an analytical technique in the environmental and other sciences is based on a number of significant scientific and technical advances during the last 35 years. These advances provided the instrumentation, experimental techniques, and fundamental knowledge needed for the important applications of MS. In addition to these scientific and technical advances, major environmental issues emerged in the United States during the late 1960s, and national concern for environmental protection accelerated during the next two decades. Mass spectrometry, and especially combined gas chromatography/mass spectrometry (GC/MS), were recognized early in this movement as powerful tools to determine the nature and extent of environmental contamination. As federal environmental laws were passed, environmental regulations implemented, and environmental research emphasized, government spending on environmental analytical chemistry increased significantly, and MS was a major beneficiary. Funding of various environmental programs hastened the development of commercial instrumentation and the routine application of MS to a variety of environmental and other analytical problems.

By the end of the 1960s, and after the publication of four pioneering and important books,[1–4] it was clear that MS had become one of the most powerful tools available for the acquisition of information about the composition and structure of organic compounds. However, the mass spectrometers of the late 1960s were invariably rather large, heavy, and expensive instruments which often required 100 square feet or more of air-cooled laboratory space and a steady stream of supplies and replacement parts. Sample introduction systems could accommodate only volatile and thermally stable compounds. Most data processing, including the assignment of mass-to-charge ratios (m/z) to peaks on light-sensitive oscillographic recorder paper, was done manually. Only a small number of government, industrial, and university laboratories could afford the capital and operating expenses of such an instrument.

The late 1960s in the United States was also a time of great public interest in the rapidly degrading state of the environment. The book *Silent Spring* by Rachel Carson, which was published in 1962, is often cited as the beginning of the era of public awareness of the condition of the environment.[5] The 1963 agreement between the United States and the Soviet Union to ban nuclear weapons testing in the atmosphere was driven by environmental health concerns. Some efforts to control smog in the Los Angeles basin began in the 1950s, but were losing ground to an ever-increasing number of automobiles by the 1960s. Although some chemicals had been known for many years to be causes of cancer in humans, chemists were just beginning to recognize this relationship.[6] By the late 1960s and early 1970s environmental crises were regularly reported on the front pages of most newspapers and described in feature articles in news magazines. Television news programs capitalized on the public interest in reports of environmental pollutants and their impact on wildlife and human health. News stories on environmental pollution became so common that people began to expect the *crisis of the month*. Some of the widely reported environmental crises were:

- Unsightly detergent foam at municipal wastewater discharges into rivers
- Failures in reproduction of wildlife at the top of the food chain
- Chlorinated hydrocarbon pesticide residues discovered in birds and fish
- Massive fish kills on the Mississippi and other rivers
- Eye irritating and visibility reducing smog in many cities
- High levels of lead in the blood of some children
- Accelerating eutrophication of Lake Erie and other lakes
- Massive use of dioxin contaminated defoliating agents in Vietnam
- Widespread dumping of raw sewage and industrial wastes into streams and rivers
- High levels of methylmercury in fish taken from streams and lakes near chlorine–sodium hydroxide production facilities
- Pollution of ocean waters and damage to wildlife from offshore oil production and oil tanker accidents
- An infamous fire on the Cuyahoga River in Cleveland
- The discovery of chlorinated and brominated disinfection by-products in drinking water
- Depletion of stratospheric ozone by chlorofluorocarbons
- Mounting evidence of adverse human health effects from a variety of specific organic and inorganic substances

These reports caused negative public reactions to chemicals and demands for government action and controls. The automobile, chemical, energy-producing, and petroleum industries were viewed as major sources of environmental pollution. Prestigious scientific journals published reviews of environmental issues and editorials warning of strong government intervention if voluntary efforts by industry to clean up its act were not successful.[7–9] The American Chemical Society began publication of a new journal, *Environmental Science and Technology*, in 1967 to accommodate the increasing interest in news and scientific studies of the environment. General public unrest, including student protests at several universities, led to the creation of the first National Earth Day on 22 April 1970.

During the latter half of the 1950s and the 1960s many new separation and measurement techniques were developed and commercial instrumentation was introduced on a scale not seen previously. In particular, GC experienced very rapid growth with the development of injection techniques, carrier gases, solid supports, stationary phases, packed columns, temperature programming, general and selective detectors, coated open tubular columns, commercial instrumentation, and even precision syringes.[10] Gas chromatography with various selective and sensitive detectors was an incredibly important development that made possible the separation and detection of organic environmental pollutants at very low environmental concentrations. The discoveries of pesticide residues in birds and fish, pesticides as the causes of fish kills, disinfection by-products in drinking water, and chlorofluorocarbons as causes of ozone depletion in the stratosphere would have been unlikely without GC. Similarly, the development of very sensitive and selective techniques and commercial instruments for the determination of metals in the environment, particularly atomic absorption spectrometry, made possible the discovery of high levels of lead in the blood of some children and the presence of methylmercury in fish. Advances in instrumental techniques, however crude by today's standards, were indispensable in the discovery of environmental pollutants that were linked to adverse environmental effects. Commercial instrumentation played a major role in defining the nature and extent of environmental pollution.

Although a variety of selective and sensitive GC detectors were invented and widely used during this period, deficiencies became apparent, especially in the identification of unknown substances eluting from the GC column. The deficiencies of standard GC detectors for compound identification are discussed in detail in Chapter 2 in the section "Alternatives to Conventional GC/MS." To resolve the identification problem, a time-of-flight mass spectrometer was first used as a GC detector in 1955 by Roland Gohlke and Fred McLafferty at the Dow Chemical Company.[11] The mass spectrometer provided not only a technique for identifying compounds by their characteristic mass spectra, but also a very sensitive detector that is applicable to all classes of compounds amenable to separation by GC. This innovation proved very successful and was followed by the development of many home-made GC/MS systems over the next 10 years.

In 1967 Robert Finnigan and several associates founded the Finnigan Instrument Corporation for the specific purpose of developing and manufacturing a combined GC/MS instrument.[12] This was an important initiative because the mass spectrometer used in the Finnigan GC/MS system was the newly developed quadrupole mass spectrometer. The quadrupole is much smaller, lighter, and less costly than the heavy iron magnets used in traditional magnetic deflection mass spectrometers. The quadrupole also has the ability to scan a mass spectrum at speeds compatible with fast eluting GC peaks and to operate at higher pressures, caused by GC carrier gases, than typical magnetic deflection instruments. Although Finnigan delivered some quadrupole GC/MS systems in the late 1960s, it would be perhaps 10 years or more before this technology was widely accepted by the MS community. In 1971 the Hewlett-Packard Company introduced a GC/MS system containing a dodecapole mass spectrometer, which is a quadrupole with additional rods to allow fine tuning and improved resolution.

During the late 1960s another very important technological development, the moderately priced minicomputer, came on the market and had a profound impact on the practice of MS. During GC/MS of a complex mixture of compounds, literally hundreds of mass spectra can be acquired in a 30-min GC separation. The processing of all these data from a single sample was a formidable task which could occupy an investigator for months if done manually. Therefore, during the late 1960s and the 1970s there was great interest in automating the acquisition and reduction of data from GC/MS instruments. Klaus Biemann and Ron Hites conducted pioneering research in MS automation at The Massachusetts Institute of Technology using more expensive International Business Machines computers.[13–15] During this period the Digital Equipment Corporation introduced a series of small and comparatively inexpensive programmed data processor-8 (PDP-8) minicomputers which opened the way for the development of low-cost and dedicated analytical instrument data systems. The PDP-8 had a significant impact on instrument automation and eventually became one of the most popular minicomputers of the period. By today's standards the PDP-8, a 12-bit processor with typically only 4096 or 8192 words (12 bits each) of ferrite core main memory, was incredibly modest, but the results obtained with this computer were extremely encouraging.

An early partner of the Finnigan Corp., Systems Industries, Inc., used a PDP-8 to develop an integrated computer control, data acquisition, and data reduction system for the Finnigan quadrupole GC/MS instrument. This effort was guided by research at Stanford University on computer control and automation of the quadrupole mass spectrometer.[16] The research at Stanford demonstrated that the scan function of the quadrupole mass spectrometer is readily amenable to computer control, which provided greatly improved m/z calibration accuracy and efficiency of operation. Vast improvements in mass spectrometer automation and data processing capabilities have followed the semiconductor and microcomputer revolutions of the 1980s and 1990s. The computerized GC/MS system made practical the broad spectrum analysis strategy described in Chapter 2. Some of the techniques used in acquiring and processing GC/MS data are also described in Chapter 2 in the section "Data Acquisition Strategies."

In 1969, under intense public pressure to do something about the environment, the Congress of the United States passed, and President Richard Nixon signed, the National Environmental Policy Act. This law defined a general policy on protection of the environment and implemented several requirements including the environmental impact statement for major federal actions that could affect the environment. Under continuing public pressure President Nixon created, by executive order, the U.S. Environmental Protection Agency (USEPA) in December of 1970. The USEPA was formed from units of other federal government departments and agencies which had various responsibilities for parts of the environment. These organizations operated under weak environmental laws and were primarily engaged in research to determine the causes of environmental problems and effects of various pollutants.

During the 1960s the strategies used to determine air quality were mainly measurements of the inorganic gases CO, NO_x, and SO_2. Hydrocarbons of petroleum origin and particulate matter were also important air quality parameters. Water quality was judged mainly by measurements of general and method-defined parameters such as biological oxygen demand, chemical oxygen demand, suspended solids, total dissolved solids, turbidity, pH, total organic carbon, oil and grease, color, hardness, acidity, alkalinity, and specific conductance.[17] Measurements of specific chemical substances in water were usually limited to metal and inorganic ions by colorimetric or spectroscopic methods. Specific organic compounds were generally not determined in air or water although some general classes of organic compounds, such as total phenols by a colorimetric procedure, were estimated. As the number of environmental problems increased, the causes of some of these were linked to specific organic compounds such as chlorinated hydrocarbon pesticides. Some environmental scientists and managers of government laboratories began to realize that analytical methods for specific organic compounds would be needed to evaluate and solve the problems of the environment.

In early 1971 the managers of several USEPA laboratories obtained funding to substantially increase the new agency's capabilities for the identification of organic pollutants in water and related environmental matrices. These managers, led by Bill Donaldson of the USEPA's Southeast Water Laboratory in Athens, GA, formed a team to evaluate commercial GC/MS/computer systems and select one of them for use in several USEPA laboratories. A survey of available equipment was conducted and the brief report of this project contains comments on instruments manufactured by Associated Electrical Industries (AEI), DuPont, Finnigan, Hewlett-Packard, LKB, Nuclide, and Varian Associates.[18] Except for the Finnigan and Hewlett-Packard systems, these mass spectrometers were single- or double-focusing magnetic deflection instruments. While several of these were newly introduced systems that were not sufficiently developed for routine use, the Finnigan model 1015 and the Varian CH-7 had acceptable and similar performance characteristics. The Finnigan 1015 was selected because of its significantly lower cost and the availability of a simplified purchasing procedure through a U.S. Government General Services Administration contract.

Reports of the identification of organic contaminants by GC/MS in various environmental matrices began to appear in the scientific literature during the late 1960s and continued to appear at an accelerated pace throughout the 1970s.[19]

These and other reports of environmental problems stimulated environmental interest groups who, along with the news media, kept environmental issues before the public and Congress. For example, the CBS news program *Sixty Minutes* reported in some detail, and with dramatic impact, the USEPA's discovery of numerous organic compounds in the New Orleans drinking water supply. The Congress of the United States initiated an unprecedented era of environmental action and passed, with presidential approval, a number of major pieces of legislation during the 1970s:

- Clean Air Act Amendments of 1970
- Federal Water Pollution Control Act Amendments of 1972
- Marine Protection, Research, and Sanctuaries Act of 1972
- Noise Control Act of 1972
- Federal Insecticide, Fungicide, and Rodenticide Act Amendments of 1972
- Safe Drinking Water Act of 1974
- Toxic Substances Control Act of 1976
- Resource Conservation and Recovery Act of 1976
- Clean Water Act of 1977
- Clean Air Act of 1977

The importance of this legislation to environmental protection, analytical chemistry, and MS cannot be overstated. In addition to setting federal government goals, policies, and other requirements, these laws specified the development of waste discharge and other regulations and generally charged the USEPA to accomplish the objectives of the Acts. Analytical chemistry was recognized as an essential element in environmental protection. Chemical analyses are needed to determine the nature and extent of environmental contamination and compliance with regulations designed to control environmental pollution.

The Federal Water Pollution Control Act Amendments of 1972 contain the statement in Section 304g that "The Administrator (of EPA) shall ... promulgate guidelines establishing test procedures for the analysis of pollutants ...." This statement is repeated in Section 304h of the Clean Water Act of 1977 which amended the 1972 law. Similarly, the Safe Drinking Water Act of 1974 specifies that primary drinking water regulations include "quality control and testing procedures to insure compliance with such (maximum contaminant) levels and to insure proper operation and maintenance" of treatment systems. These requirements and the other environmental legislation justified increased federal spending for research, development, and demonstration in many areas of the environmental sciences and engineering. The initiatives included research and development of analytical test methods, acquisition of analytical equipment, and the conduct of environmental monitoring studies.

The USEPA's initial purchase in 1971 of six model 1015 GC/MS systems from the Finnigan Instrument Corp. came at a critical time for the small company that was in a precarious financial condition.[12] Additional sales followed and the cash flow from these strengthened Finnigan and some other instrument companies who were able to use the new resources to improve the capabilities and quality of their products. Finnigan was particularly aggressive and introduced several new models of GC/MS systems in the 1970s, obtained a more advanced data system through the acquisition of the Incos Co. in 1976, and commercialized the triple quadrupole mass spectrometer after its initial development (his Chapter 5, "Tandem Mass Spectrometry"). Finnigan also supported internal research, which led to the moving belt liquid chromatography (LC)/MS interface (his Chapter 6 "Mass Spectrometry Interfaces") and the ion-trap mass spectrometer. The developments of the 1970s set the stage for a massive expansion of the application of MS technology to environmental chemical analyses in the 1980s.

Priority Pollutants, Fused Silica Capillary Columns, and the Love Canal

Several major environmental and technical developments occurred near the end of the 1970s that had a huge impact on the future applications of MS. The USEPA was sued by several environmental interest groups for failing to implement certain provisions of the 1972 Federal Water Pollution Control Act Amendments.[20] An agreement to settle the law suits, known as a consent decree, was approved in federal court in 1976. Under the consent decree the USEPA agreed to develop regulations to control the discharge of toxic organic and other substances into rivers and other bodies of water. A list of 129 priority pollutants was prepared as part of the settlement, and considerable USEPA resources were subsequently devoted to the identification and measurement of these substances by GC/MS and other techniques in a variety of industrial wastewater discharges.

On 3 December 1979 the USEPA proposed Guidelines Establishing Test Procedures for the Analysis of Pollutants Under the Clean Water Act.[21] This proposal included the full text of 15 detailed analytical methods for the determination of 113 organic priority pollutants in wastewater (Chapter 3, "Target Analytes—The Priority Pollutants"). Eight of these pollutants were, at the time, commercial products, which are complex mixtures of dozens or hundreds of individual compounds (Chapter 4). Three of the analytical methods, which are designated USEPA Methods 613, 624, and 625, used GC/MS and all 15 methods included fully integrated sample preparation and quality control procedures. This appears to be the first time that strong quality control provisions were integrated into a detailed analytical method description. Method 613 is a single-analyte method for 2,3,7,8-tetrachlorodibenzo-*p*-dioxin that employs an open tubular glass capillary GC column and high resolution MS (Chapter 4, "Dibenzo-*p*-dioxins and Dibenzofurans"). Methods 624 and 625 are multianalyte methods that specify packed-column GC/MS (Chapter 3, "Volatile Organic Compounds in Water", and Chapter 4, "Semivolatile Organic Compounds in Wastewater", respectively).

This regulatory proposal caused considerable comment and debate over the cost of using GC/MS for routine chemical analyses and whether the country had the analytical capacity for the widespread utilization of this technology. In 1979 GC/MS was widely considered to be a research tool for government, industrial, and university laboratories. The 3 December 1979 USEPA proposal was a signal to many that the federal government considered GC/MS to be a practical tool for routine analyses of environmental matrices. The manufacturers of GC/MS equipment celebrated and planned for increased production and sales. However, the controversy did not end and various groups continued to discuss and debate the issue (Chapter 3, "Regulatory Adoption of Purge and Trap GC/MS").

Although open tubular GC columns were invented in 1956, high-resolution glass capillary columns were used by a small number of laboratories in the United States in the late 1970s.[22] Glass columns are fragile and it is difficult to straighten the column ends and fit them into GC injection ports, detectors, and interfaces to mass spectrometers. Furthermore, commercial glass capillary columns were not widely available in the United States until the Perkin-Elmer patent expired in 1977. High-resolution glass columns were more popular in Europe where column coating, injection, and other techniques were largely perfected.[23] Packed GC columns dominated most GC applications in the United States until the early 1980s. In 1979, Hewlett-Packard introduced coated open tubular fused silica capillary GC columns.[24] The narrow-bore fused silica tubes, which were being developed for optical applications, were strong, flexible, and easily manipulated in the small spaces of GC column ovens, GC/MS interfaces, and mass spectrometer ion sources. Fused silica GC column technology developed rapidly amid competition among several companies in the United States. By the end of the 1980s few, if any, environmental analyses were conducted with glass capillary or packed column GC/MS.

At about the same time, near the end of the 1970s, an environmental problem emerged that was to dominate much of the USEPA's resources during the 1980s.[25] During the 1940s and early 1950s waste products from chemical manufacturing operations in the Niagara Falls, NY area were dumped into an abandoned canal.[26] The canal was named after William T. Love who began construction in the late nineteenth century to provide hydroelectric power for a planned industrial development in the area. The canal project failed with only a small portion completed and the ditch was later used for disposal of the chemical production waste products. In 1955 an elementary school was opened very close to the waste site, which eventually was covered with soil and many new homes were constructed in the general area. Several severe winters in 1975–1976 with abundant snow and freeze–thaw cycles caused the soil to subside and created surface ponds contaminated with chemical waste which infiltrated the basements of some homes. In 1978 an emergency was declared, the school closed, 239 homes evacuated, and action taken to install a drainage system, a leachate treatment plant, and a water-resistant clay cap on the site. The USEPA had major responsibilities in this remediation effort and was charged with conducting many studies including exhaustive environmental monitoring to determine the extent of contamination of the air, surface water, ground water, sediments, and soil. This effort in 1980 was the first major application of the new fused silica capillary GC column technology with MS. Although some of the conclusions from some of these studies were criticized at the time, they appear to have withstood the test of time.

The Golden Era of Environmental Analytical Chemistry

The 1980s and early 1990s were the golden era of environmental analytical chemistry in the United States. In 1980 Congress passed and the President signed the Comprehensive Environmental Response, Compensation, and Liability Act (CERCLA) which placed a tax on petroleum and chemical products. The proceeds from this tax are placed in a special fund called the *Superfund* which is used for the investigation and remediation of abandoned hazardous waste dump sites such as the Love Canal. By contrast, regulations under the Resource Conservation and Recovery Act (RCRA) of 1976 are intended to control the future disposal of solid and hazardous wastes. The Superfund program opened a major area of investigation, and thousands of chemical waste disposal sites were located. In a massive application of GC/MS and other analytical techniques, the contents of these waste sites were determined and their impacts on the surrounding environment studied in detail. The extent of the investigation was so large that USEPA laboratories could not begin to handle the sample load. A major external effort, which became known as the Contract Laboratory Program (CLP), was developed by the USEPA to provide private laboratory support for Superfund investigations and remediation. At about the same time, RCRA regulations were implemented and laboratories were also in demand for GC/MS analyses of waste samples bound for controlled solid and hazardous waste disposal sites.

The requirements of the CLP and RCRA were not lost on the analytical instrument industry which continued to improve its products and increase production to meet the demand for analytical instruments. The Hewlett-Packard Company became a major supplier of GC/MS systems with the development of the compact, comparatively inexpensive, bench-top 597X series of GC/MS systems. These had great appeal for the many small industrial and start-up independent testing laboratories. Other instrument companies responded with their versions of bench-top GC/MS equipment, and the competition for sales was a great benefit to the customers and the industry. The Finnigan Corporation introduced the ion-trap mass spectrometer as a GC detector in 1983 and, after a slow beginning, established it as an important and valuable tool in MS. Elemental analyses were also required by the Superfund and RCRA programs which added an environmental stimulus to the developing inductively coupled argon plasma MS technology in the late 1980s (Chapter 7). During this period some commercial laboratories became major environmental data producers with dozens of GC/MS and other analytical instruments operating during two or three work shifts per day to satisfy the demand for chemical analyses.[27] The demand for skilled technicians and analytical chemists was so strong that the Clean Water Act, RCRA, and CERCLA were sometimes referred to as the Analytical Chemists and Mass Spectrometrists Employment Acts.

The Guidelines Establishing Test Procedures for the Analysis of Pollutants, which were proposed under the Clean Water Act on 3 December 1979,[21] did not become final regulations until they were published again in the 26 October, 1984 issue of the Federal Register.[28] This nearly 5-year pause was necessary to consider objections from several industries that the analytical methods were not validated, that is, they had not been tested in multiple laboratories, and multilaboratory precision and accuracy had not been determined. In 1983 the American Chemical Society's Committee on Environmental Improvement published a report titled *Principles of Environmental Analysis* which addressed several controversial analytical methods issues including quality control and the need for method validation. During the 5-year period multilaboratory studies of the proposed methods were conducted and the objections from the industries were satisfied. These studies reinforced the notion that GC/MS was an acceptable qualitative and quantitative technique for organic compounds in environmental matrices. It was also clear that the regulated community would demand validated standard analytical test methods for use in regulatory analyses.

In 1986 the Congress of the United States passed, and the President signed, the Superfund Amendments and Reauthorization Act and remarkable amendments to the Safe Drinking Water Act (SDWA). Among many other requirements, the SDWA amendments mandated the USEPA to establish national standards within 3 years for 83 contaminants that were named in the Act. The law also mandated national standards for 25 additional contaminants every 3 years beginning in 1990. These requirements unleashed a new round of research to develop analytical methods for substances in drinking water at typically much lower concentrations than found in industrial wastewater or at Superfund waste sites. This research produced significant additions to the USEPA's 500 series of analytical methods for organic analytes including the multianalyte GC/MS Methods 524 and 525.[29–31] Unlike their wastewater predecessors of the late 1970s, Methods 524 and 525 utilize fused silica capillary GC columns and Method 525 specifies a liquid–solid extraction technique (Chapter 3, "Volatile Organic Compounds in Water", and Chapter 4, "Semivolatile Organic Compounds in Drinking Water", respectively). These and other USEPA methods were included in SDWA regulations during the second half of the 1980s and the early 1990s. The SDWA regulations also required elemental analyses, and USEPA Method 200.8 for elemental analysis by inductively coupled argon plasma MS (Chapter 7) was incorporated into these regulations.[32–33] Following the changes to the SDWA, the Clean Water Act was amended in 1987, strengthening requirements to safeguard the nation's aquatic resources.

Another major piece of environmental legislation was enacted in 1990 with the passage of the Clean Air Act Amendments. As in other environmental laws, these amendments establish a number of goals and requirements including measures to control CO, NO_x, SO_x, Pb, and particulate matter emissions. The law requires action to control the formation of tropospheric ozone and to phase out chlorofluorocarbons and other substances that cause depletion of the stratospheric ozone layer. It also requires a significant reduction in emissions of organic air pollutants and contains a list of 189 hazardous air pollutants.

Approximately 160 of these are specific organic compounds and the remaining 29 are either commercial products that are complex mixtures of dozens or hundreds of individual compounds or broad families of substances such as all compounds of 11 metals.

During the1980s the USEPA had developed a series of analytical methods for the determination of toxic organic (TO) compounds in ambient air.[34] The TO series of methods was used to determine volatile and other organic compounds in ambient air (Chapter 3, "Volatile Organic Compounds in Air", and Chapter 4, "Semivolatile Organic Compounds in Air, Sediments, Soil, Solid Wastes, and Tissue"). The requirements of The Clean Air Act Amendments of 1990 greatly accelerated research to improve air sampling techniques and develop GC/MS methods for the determination of the hazardous air pollutants. New or expanded and improved versions of some of the TO methods, including several GC/MS methods, were developed, and are used for assessments of toxic organics in ambient air.[35]

Since about the mid-1990s there has been a noticeable reduction in some environmental activities in the United States. While investments have continued to implement existing regulations and studies conducted to evaluate the state of the environment, Congress has not passed additional major environmental legislation and only a few new federal regulations have been adopted. Consequently, the development of new applications of MS in environmental analyses in the United States has slowed considerably, and many analytical chemists and mass spectrometrists have turned their attention to other areas of research. However, interest in improving the state of the environment is strong in Europe and increasing in Asia and Latin America. This period of evaluation and change is an appropriate time to document the current status of analytical and environmental MS and to prepare for new developments in the twenty-first century.

Mass Spectrometry, Chemists, and Other Spectrometries

While there is little doubt that the environmental protection movement was a major driving force in the development and widespread acceptance of GC/MS and several other mass spectrometric techniques, other factors also contributed to the success and acceptance of this technology. One major factor is that mass spectra are relatively easy for chemists to understand and interpret, but other kinds of spectra are not as easy to understand. Nearly all spectrometries measure the absorption or emission of electromagnetic energy by nuclei, atoms, or molecules. These spectra are displayed as a series of absorption or emission maxima as a function of absorbed or emitted energy. Ultraviolet–visible spectrometry, infrared spectrometry, fluorescence spectrometry, and nuclear magnetic resonance spectrometry are examples of spectrometries based on the absorption and emission of electromagnetic energy. While these techniques are valuable tools for the chemist, their spectra can be difficult to interpret because the frequencies of the energy absorbed or emitted are not clearly and directly related to the composition and structure of molecules. The chemist must learn that certain absorption frequencies, chemical shifts, and so on, are indicative of specific composition or structural features and that these frequencies will vary depending on the molecular environment of the structural feature.

On the other hand, MS is really just chemistry and chemical reactions which the chemist can readily understand and interpret.[36] The mass spectrum is a representation of the abundances of various ions in the gas phase as a function of their mass/charge (m/z) ratios. Mass, charge, and ions are among the first ideas presented in the most basic chemistry courses and are second nature to the chemist. The mass spectrum is a spectrum of ions and all one needs to know to begin the interpretation of most spectra is the masses of the major isotopes of the elements, for example, H = 1, C = 12, N = 14, and O = 16. By summing these values for various combinations of atoms, the chemist can discover possible compositions of various ions. Multiple ions may be observed in mass spectra because of chemical reactions that convert an ion into one or more other ions with different m/z values. These chemical reactions in the gas phase in a mass spectrometer are often similar to thermal, catalytic, or photochemical reactions that are well known to the organic chemist. Like all of chemistry, experimental conditions can cause mass spectra to vary considerably, but this is an idea that chemists can readily understand and accept. Given that mass spectra are inherently about the masses of ions, the charges on ions, and chemical reactions, it is not surprising that MS has a special appeal to chemists and that mass spectra are not difficult for chemists to understand and interpret.[37]

Another major factor in the success of MS is its high sensitivity to very small quantities of material. Detection of nanogram (10^{-9}), picogram (10^{-12}), and even femtogram (10^{-15}) quantities of individual compounds is not unusual. High sensitivity allows the identification and measurement of substances at very low concentrations that are nevertheless significant in human health and ecological processes. Equally, if not more important, high sensitivity permits the use of small samples which are convenient, less costly to acquire and process, and sometimes all that is available.

The early and successful combination of GC with MS provided an on-line separation technique that is essential for many, if not almost all, types of environmental and other sample matrices. Many types of sample matrices are invariably complex, especially at the very low concentrations required for many investigations and regulations. At low concentrations, natural and anthropogenic background contamination is significant and generally must be at least partially separated from the analytes of interest to ensure accurate and precise identifications and measurements. The mass spectrometer is the most general of all chromatography detectors because ionization techniques are available for essentially every known type of substance. At the same time, the mass spectrometer is very flexible and can be operated in a mode that is far more selective than the most selective standard chromatography detector. The combinations and applications of on-line separation techniques with MS is the principal focus of this book.

Digital computer control, data acquisition, and data reduction systems were developed for MS in the late 1960s and these systems were improved and expanded much faster than for most other analytical techniques. This was likely caused by the need to process large volumes of MS data from combined GC/MS systems and made easy by the inherently digital nature of such data. Powerful software was developed to accomplish a number of data reduction tasks including spectrum background subtraction, several types of chromatographic displays, searches of reference spectra data bases, and quantitative analyses (several sections of Chapter 2). The early and rapid development of databases of reference mass spectra greatly assisted inexperienced users of MS. As early as 1972 a reference collection of 8124 spectra was available and interactive search software was developed to effectively utilize this database.[38] High quality collections of well over 100,000 reference mass spectra and efficient search software for personal computers are available today (Chapter 2, "Identification Criteria").

Validated standard analytical methods with integrated sample preparation and strong quality control were a major factor in the success of environmental MS.[28–35] The methods' descriptions provided step-by-step instructions for inexperienced analysts whose confidence increased rapidly with the successes of their work. These method descriptions also served as important models for applications in many other fields of investigation, for example, food and beverage analyses, forensic investigations, and pharmaceutical analyses. Some of the quality control provisions in these standard methods imposed discipline on some manufacturers who greatly improved the quality and reliability of their instrument systems (Chapter 2, "Standardization of Conventional GC/MS"). The early success of GC/MS systems also stimulated research to develop other combined separation/MS techniques. This led to the development of combined liquid chromatography (LC)/MS, supercritical fluid chromatography/MS, and capillary electrophoresis/MS instrument systems (Chapter 6). This research also produced powerful new sample introduction and ionization techniques that greatly expanded the scope of MS (Chapter 6).

One negative aspect of standard analytical methods is that regulatory offices of the USEPA are often slow to incorporate improved techniques into approved methods and slow to approve new analytical methods. This is also a problem with analytical methods published by private standard-setting organizations. The development of new and modified analytical techniques and methods was strongly stimulated by the environmental laws and regulations, and some new techniques and methods have important technical and economic advantages. However, there is always resistance to change especially when standard methods are used to build databases of environmental measurements which could be compromised by changes in analytical methods. These databases are used for regulatory and law enforcement purposes, and authorities in these areas are generally resistant to major or rapid changes in methods. Calls for less restrictive and more flexible regulatory analytical methods were published and some critics promoted the idea that any method that met defined performance criteria should be acceptable for regulatory use.[39,40] Very slowly and with considerable internal and external debate the USEPA has moved in the direction of less restrictive and more flexible regulatory analytical methods.

Abbreviations, Definitions, Notation, and Terms

Abbreviations, notation, and specialized terms are widely used in analytical chemistry and MS and these are frequent causes of confusion among specialists and nonspecialists alike. Mass spectrometry is a rapidly growing and diverse technology, and new techniques and devices are invented and named every year. Authors, inventors, and manufacturers usually give names, define terms, and create abbreviations that are logical and eventually widely accepted. Sometimes, however, two or more different names, terms, or abbreviations are given to the same or similar techniques or devices by different authors or manufacturers. When this occurs standard-setting organizations do not generally select and define the standard term and abbreviation in a timely way, and multiple names for the same or similar techniques are often found in the original research literature. A good example is the many terms that have been used to describe selected ion monitoring (Chapter 2, "Data Acquisition Strategies").

The abbreviations, definitions, notation, and terms used in this book are those endorsed and published by the Committee on Measurements and Standards of the American Society for Mass Spectrometry (ASMS). In particular, a report entitled *Standard Definitions of Terms Relating to Mass Spectrometry*, published in 1991, defines many terms, notations, and abbreviations that are used frequently.[41] Unfortunately, some manufacturers of MS and other equipment do not always adhere to standard abbreviations, notations, and terms when they are available. They sometimes seek a competitive advantage by choosing a different name for their implementation of a technique, device, or algorithm. Computer programmers who are often not familiar with standard abbreviations and terms exert a powerful influence when they invent or misuse terms in user interfaces of data system software. If an appropriate term is not available in the 1991 ASMS report, the most generally accepted and logical term, in the opinion of the author, is used in this book.

In order to minimize confusion, most abbreviations used in this book are defined at least once in every chapter and sometimes several times if they have not been used for several pages. However, some abbreviations, notation, and terms are used so frequently that they are not defined in the other chapters but are defined in this section.

- *Analysis*. The investigation of a sample of the physical world in order to learn about its components, composition, structure, or other physical or chemical characteristics. Only samples are analyzed. Elements, compounds, and ions are separated, identified, measured, or determined.
- *Analyte*. An element, compound, or ion that is the target of an identification or measurement in a sample matrix.

- *Base peak*. The most abundant ion in a mass spectrum which is usually displayed with 100% relative abundance in a bar graph or histogram representation of the spectrum.
- *Chemical noise*. Variable signals from the chemical background produced by instrument components, for example, GC septum bleed and GC or LC column bleed, and residues from previous samples or interferences that are sample components.
- *Congener*. A compound or ion that is a member of a series of related substances that differ only in the number of hydrogens that have been substituted by the same atom. For example, chlorobenzene, the dichlorobenzenes, and the trichlorobenzenes are congeners.
- *Dalton* (Da). The unit of atomic mass based on C = 12, which is not weighted by the natural abundances of isotopes; the unit of monoisotopic mass that is used in MS.
- *Detection limit*. The minimum quantity or concentration of an analyte that can be detected by an instrument or an analytical method. Criteria for detection must be specified with instrument and method detection limits, which are defined and discussed in more detail in Chapter 2.
- *Determine*. To identify and measure an element, compound, or ion in a sample.
- *Inside diameter* (ID). The inside diameter of a tube or enclosure.
- *Liquid–solid extraction*. The transfer of a substance from being dissolved in a liquid to being adsorbed on a solid in contact with the liquid.
- *Mass spectrum*. A bar graph (histogram) or plot of the abundances of ions in the gas phase as a function of mass/charge (m/z). In a bar graph the base peak is usually given an abundance of 100 and the abundances of all other ions are made relative to the base peak.
- m/z. The ratio of mass in daltons to charge, which is measured in a mass spectrometer
- *Outside diameter* (OD). The outside diameter of a tube or enclosure.
- *Sensitivity*. The electronic or other signal produced per unit amount of analyte. This term is best limited to specifications for instruments, instrument components, or other transducers that convert chemical information into electronic or other signals.
- *Signal/noise*. The ratio of the instrument or transducer signal to the background noise. Electronic noise was common in early instruments but is essentially not present in current digital data processing systems, but chemical noise is a factor.
- *Torr*. The basic unit of pressure in mass spectrometer vacuum systems. One Torr is the pressure required to support a column of mercury 1 mm in height and is equal to 133 kilopascals (kPa).

Concentrations of substances in a sample are expressed in mass of the substance per standard unit of sample volume for gases and liquids and standard unit of sample mass for solids. Since most concentrations used in analytical and environmental MS are very low, the masses of substances are usually expressed in milligrams (mg), micrograms (μg), nanograms (ng), picograms (pg), and femtograms (fg). The standard units of sample volume are cubic meters (m^3) for gases, liters (L) for liquids, and kilograms (kg) for solids. The terms parts per million (ppm), parts per billion (ppb), parts per trillion (ppt), and so on, are not used for several reasons. The words million, billion, and trillion may have different meanings in different countries and these terms are not clear when used to express concentrations in mass per standard unit of sample volume of a gas or liquid.

Literature Coverage

No attempt has been made to review the entire scientific literature for analytical and environmental applications of MS. The scientific journals that are most frequently referenced are the American Chemical Society publications *Analytical Chemistry* and *Environmental Science and Technology*, and the *Journal of the American Society for Mass Spectrometry* which began publication in 1990. The John Wiley and Sons journals *Biomedical and Environmental Mass Spectrometry* (1986–1990), *Biological Mass Spectrometry* (1991–1994), and the *Journal of Mass Spectrometry* (since 1995) are also referenced. The first two Wiley journals were only published for the years shown and the latter is a consolidation of several journals.

Original publications from many other journals are also cited in this book, but no attempt has been made to review in detail the many scientific journals that publish analytical and environmental MS related papers. It is noteworthy that for the years 1994–1995, *GC/MS Update, Part A—Environmental*, published by HD Science Limited, abstracted 1959 journal articles on environmental analysis by GC/MS alone. For complete information on the application of MS to specific substances or environmental matrices, a thorough search of the scientific literature is recommended. A particularly useful general review of MS is published in even years in the fundamental reviews issue of *Analytical Chemistry*. The 1998 review contains 1551 references to work published mostly since the previous review.[42] Reviews of pesticides, environmental analysis, air pollution, and water analysis are included in the application reviews issue of *Analytical Chemistry* in odd years.

Most chapters of this book contain one or more references to the *Federal Register* and the *Code of Federal Regulations* (CFR). The *Federal Register* is the official daily publication of rules, proposed rules, and notices of federal agencies and organizations of the United States Government. The *Federal Register* also contains executive orders and other presidential documents. The CFR is a codification of the general and permanent rules issued by the executive departments and agencies of the United States Government. The CFR is divided into 50 titles which represent the broad areas subject to federal regulation. Title 40 is concerned with the environment and environmental protection. Both the *Federal*

Register and the CFR are available in most large public and university libraries, government libraries, and law school libraries, and on the internet at URL http://www.access.gpo.gov.

REFERENCES

1. Beynon, J. H. *Mass Spectrometry and Its Applications to Organic Chemistry*; Elsevier: New York, 1960.
2. Biemann, K. *Mass Spectrometry: Organic Chemical Applications*; McGraw-Hill: New York, 1962.
3. McLafferty, F. W. *Interpretation of Mass Spectra*; W. A. Benjamin: New York, 1966.
4. Budzikiewicz, H.; Djerassi, C.; Williams, D. H. *Mass Spectrometry of Organic Compounds*; Holden-Day: San Francisco, CA, 1967.
5. Carson, R. L. *Silent Spring*; Houghton Mifflin: New York, 1962.
6. Weisburger, J. H.; Weisburger, E. K. *Chem. Eng. News*, 7 February 1966.
7. Lotspeich, F. B. *Science* **1969**, *166*, 1239–1245.
8. Abelson, P. H. *Science* **1970**, *169*, July 17 editorial page.
9. Abelson, P. H. *Science* **1970**, *170*, October 30 editorial page.
10. Ettre, L. S. *J. Chromatogr.* **1975**, *112*, 1–26.
11. Gohlke, R. S.; McLafferty, F. W. *J. Am. Soc. Mass Spectrom.* **1993**, *4*, 367–371.
12. Finnigan, R. E. *Anal. Chem.* **1994**, *66*, 969A–975A
13. Hites, R. A.; Biemann, K. *Anal. Chem.* **1967**, *39*, 965–970.
14. Hites, R. A.; Biemann, K. *Anal. Chem.* **1968**, *40*, 1217–1221.
15. Hites, R. A.; Biemann, K. *Anal. Chem.* **1970**, *42*, 855–860.
16. Reynolds, W. E.; Bacon, V. A.; Bridges, J. C.; Coburn, T. C.; Halpern, B.; Lederberg, J.; Levinthal, E. C.; Steed, E.; Tucker, R. B. *Anal. Chem.* **1970**, *42*, 1122–1129.
17. Ballinger, D. G. *Environ. Sci. Technol.* **1967**, *1*, 612–616.
18. Neher, M. B. *Summary Report on Evaluation of Gas Chromatograph / Mass Spectrometer / Computer Systems to Water Quality Office, Environmental Protection Agency*, Battelle Columbus Laboratories, 1 June 1971.
19. Alford, A. *Biomed. Mass Spectrom.* **1978**, *5*, 259–286.
20. Keith, L. H.; Telliard, W. A. *Environ. Sci. Technol.* **1979**, *13*, 416–423.
21. *Guidelines Establishing Test Procedures for the Analysis of Pollutants; Proposed Regulations, Federal Register* 3 December, **1979**, *44*, 69464–69575.
22. Jennings, W. *J & W Separation Times* **1991**, *5*(3), 12–15.
23. Novotny, M. *Anal. Chem.* **1978**, *50*, 16A–32A.
24. Dandeneau, R.; Zerenner, E. H. *J. High Resolut. Chromatogr. Commun.* **1979**, *2*, 351.
25. Alexander, T. *Fortune*, April 21, **1980**, 52–58.
26. Deegan, J. *Environ. Sci. Technol*, **1987**, *21*, 328–331.
27. Krieger, J. *Chem. Eng. News*, July 6, **1981**, 28–29.
28. *Guidelines Establishing Test Procedures for the Analysis of Pollutants under the Clean Water Act; Final Rule and Interim Final Rule and Proposed Rule, Federal Register* 26 October, **1984**, *49*, 43234–43439; Title 40, Code of Federal Regulations, Part 136.
29. Method 524.2, Revision 3.0 and Method 525.1 Revision 2.2 in *Methods for the Determination of Organic Compounds in Drinking Water*, USEPA Report EPA/600/4-88/039, December **1988**, Revised July **1991**; URL http://www.epa.gov/nerlcwww/methmans.html.

30. Method 524.2, Revision 4.0 in *Methods for the Determination of Organic Compounds in Drinking Water*, Supplement II, USEPA Report EPA/600/R-92/129, August **1992**; URL http://www.epa.gov/nerlcwww/methmans.html.
31. Method 524.2, Revision 4.1 and Method 525.2, Revision 2.0 in *Methods for the Determination of Organic Compounds in Drinking Water*, Supplement III, USEPA Report EPA/600/R-95/131, August **1995**; URL http://www.epa.gov/nerlcwww/methmans.html.
32. Method 200.8, Revision 4.4 in *Methods for the Determination of Metals in Environmental Samples*, USEPA Report EPA/600/4-91/010, June **1991**; URL http://www.epa.gov/nerlcwww/methmans.html.
33. Method 200.8, Revision 5.3 in *Methods for the Determination of Metals in Environmental Samples*, Suppl. I, USEPA Report EPA/600/R-94/111, May **1994**; URL http://www.epa.gov/nerlcwww/methmans.html.
34. *Compendium of Methods for the Determination of Toxic Organic Compounds in Ambient Air*, USEPA Report EPA/600/4-89-018, **1989**.
35. *Compendium of Methods for the Determination of Toxic Organic Compounds in Ambient Air*, 2nd ed., USEPA Report EPA/625/R-96/010b, January **1999**.
36. Meyerson, S. *Anal. Chem.* **1994**, *66*, 960A–964A.
37. Budzikiewicz, H.; Djerassi, C.; Williams, D. H. *Mass Spectrometry of Organic Compounds*; Holden-Day: San Francisco, CA, 1967, preface.
38. Heller, S. R. *Anal. Chem.* **1972**, *44*, 1951–1961.
39. Poppiti, J. *Environ. Sci. Technol.* **1994**, *28*, 151A–152A.
40. Newman, A. *Anal. Chem.* **1996**, *68*, 733A–737A.
41. Price, P. *J. Am. Soc. Mass Spectrom.* **1991**, *2*, 336–348.
42. Burlingame, A. L.; Boyd, R. K.; Gaskell, S. J. *Anal. Chem.* **1998**, *70*, 647R–716R.

2

Basic Strategies of Analytical and Environmental Mass Spectrometry

Contemporary mass spectrometry (MS) is a highly diverse technology encompassing a variety of mass spectrometer designs and many different experimental techniques that are applicable to the chemical, physical, and biological sciences. The purpose of this chapter is to define and describe some of the basic strategies used in analytical and environmental MS. Some of these strategies are closely associated with conventional gas chromatography/mass spectrometry (GC/MS) where some of these ideas were developed and put into practice first. Most of them, however, are also applicable to the broad range of other MS techniques described in this book. In addition to the basic strategies, some alternatives to conventional GC/MS are discussed and some important limitations of MS are described. The analysis of environmental and other samples inevitably requires many decisions which can and do significantly impact the quality and usefulness of the results. This chapter is about strategic thinking in analytical and environmental MS.

General Analytical Strategies and Samples

The goals of any project that requires chemical analyses will usually define the types of samples to be analyzed, for example, ambient air, drinking water, agricultural soil, blood, urine, and fish tissue. The project goals should also determine the general strategy for the chemical analyses. There are two general analytical strategies and these are designated the target analyte and the broad spectrum approaches.[1] These strategies are defined and discussed in detail in this section. Depending on the type or types of samples to be analyzed, and the analytical strategy, decisions can be made about the need for on-line separation techniques and other sample introduction systems. The analytical strategy will also determine the specific MS technique or techniques that should be used. A decision could be to use conventional GC/MS, which is a basic strategy defined in this chapter. Alternatively, one or more of the techniques described in Chapters 5–8 may be required, or some combination of GC/MS and other techniques. Inevitably many other issues must be considered early in the project planning process. These include sampling strategy, sampling techniques, sample preservation during shipment and storage, and sample processing before analysis. Sample handling issues are briefly described in this section.

Target Analyte and Broad Spectrum Strategies

Target analytes are specific substances which are either known or thought to be in the sample and which must be identified and measured (determined) to meet the project goals. An alternative strategy, which has existed since the beginning of chemical analysis, is called the broad spectrum (BS) approach.[1] The BS strategy is the extreme opposite of the target analyte (TA) approach, that is, the goal is to identify the sample components without a predetermined list of target substances. The question asked with the BS strategy is: what substances of interest are in the sample?

A major issue for any project is whether to use the TA or BS strategy or some combination of both. This strategic decision will, to a considerable degree, determine how much time and other resources will be required, what specific equipment and analytical capabilities will be needed, and how much the study will cost. Early in the environmental movement, during the 1960s and 1970s, the BS strategy was employed to discover the specific substances that contribute to environmental contamination.[2,3] During the regulatory era of the 1980s and 1990s in the United States, the TA strategy dominated environmental and many other types of analyses.[1] Both strategies are needed and should be carefully considered and understood by designers of environmental monitoring and other studies.

Target Analyte Strategy. A list of analytes, often with similar physical and chemical properties, drives this approach. If the target analytes are a diverse group with very different properties, they are usually subdivided into groups according to their similar properties. The target analytes can dictate the type of sample that is required. The water insoluble 2,3,7,8-tetrachlorodibenzo-*p*-dioxin is much more likely to be found in the creek sediment than dissolved in the water. Similarly, the target analytes can help determine the sampling strategy, for example, the location of sampling stations just downstream from an industrial discharge into a river. The TA approach has many advantages that favor its widespread application. Sample preparation procedures can be designed to extract the target analytes from the sample matrix with maximum efficiency, isolate them, and concentrate them in a solvent most suited to their chromatographic separation. Some interferences may be eliminated during the sampling or sample processing by pH adjustments, chemical derivatization, or by evaporation of unwanted components.

With a TA strategy, the chromatographic conditions and the mass spectrometric technique can be selected to give the best possible separation of target analytes, the shortest analysis time, optimum detection limits, and maximum information about the analytes. Precise chromatographic retention times of the target analytes can be measured using pure samples of the compounds. The mass spectrometer can be calibrated for quantitative analysis using a series of calibration solutions with different concentrations of the target compounds. Special reference compounds known as internal calibration standards can be used to enhance the precision of the measurements. The total analytical method can be tested and perfected using test sample matrices fortified with known quantities of pure target analytes. The result is an analytical method optimized for maximum selectivity, recovery, precision, and detection limits for the target analytes.

The TA strategy has been widely adopted in environmental and other fields and is often mandated by various regulatory and monitoring requirements. Compared to a BS strategy, the TA approach is far simpler to define and implement, easier to describe and understand, much easier to develop and estimate costs, and much more straightforward to test and evaluate the results. The vast majority of chemical analyses reported in the scientific literature use the TA strategy.

Broad Spectrum Strategy. The basic idea of this strategy is to analyze the sample, not just to look for and determine specific analytes. The approach is to minimize, as much as possible, sample processing to allow a broad variety of generally similar substances to reach the chromatographic separation system and the mass spectrometer. Preliminary separations and other measures to remove interferences are minimized, or not used, to avoid discarding potentially valuable and interesting components. Because of minimum sample processing, the BS strategy may be limited to less complex samples. If a solvent is used it should be methylene chloride or some other solvent in which a broad variety of organic compounds are soluble, rather than a solvent like *n*-hexane in which fewer compounds are soluble. The chromatographic separation should be of a general type to allow the widest variety of compounds to elute during a relatively long analysis time. A universal or broadly applicable MS ionization technique should be used to acquire mass spectra over a wide mass range and with appropriate sensitivity to maximize detection of eluting substances.

In a BS analysis, several or more different types of chromatographic separations may be required depending on the components of the sample and the breadth of information desired. Similarly, several or more kinds of mass spectrometric data may be needed. The identification of all or most substances in one or more chromatograms, even with extensive mass spectrometric and other spectroscopic data, can be a challenging and difficult process. These data may not be sufficient to even tentatively identify all the recognized components, especially if pure authentic samples of suspected substances are not available in the laboratory. For these reasons, and the general preoccupation with target analytes, the BS strategy is much less common, but is extremely important for the discovery of unknown environmental contamination or other problems.

Combination TA and BS Strategy. A strategy sometimes used in environmental and other studies is to develop a target analyte list that fits the objectives of the study and that gives some attention to other chromatographic peaks to identify potentially new or unexpected substances. This approach is well suited to conventional GC/MS and to some of the other mass spectrometric techniques described in subsequent chapters.

Samples

The issue of obtaining and processing samples can be far more complex and time consuming than it would appear and a well-planned strategy is required. Depending on the general analytical strategy, several or more samples in different containers with different preservation techniques may be required from exactly the same sampling station at the same time. This would occur, for example, if the analytical strategy included both elemental analyses (Chapter 7) and the determination of several groups of target analytes with different properties, for example, volatile and semivolatile compounds (Chapters 3 and 4). This section is intended to provide a brief overview of these issues which can have a significant impact on the quality of the mass spectrometric analyses.

Sampling Strategy. The sampling strategy is the general plan for the acquisition of the samples needed for a study. It is concerned with the number of sampling stations, the locations of sampling stations, the time of day of sampling, the frequency of sampling, the depth of sampling for environmental water or sediment samples, the meteorological conditions for air sampling, the need for replicate samples from the same station at the same time, the desirability of batch or composite samples, and a myriad of important related issues. A well-designed sampling strategy along with good analytical chemistry is essential to produce the most usable information about the state of the environment or any other system under study. The American Chemical Society's Committee on Environmental Improvement has published *Principles of Environmental Sampling*, which considers many issues of sampling strategy and provides references to other publications.[4] A summary of the first edition of this book is also available.[5] Many of these principles are general and are applicable to a broad range of studies other than environmental monitoring.

Sampling Techniques. Sampling techniques are the actual physical procedures used to acquire environmental or other types of samples. Important aspects of these procedures include sampling equipment, the containers used for various types of samples, techniques for cleaning sample containers and sampling equipment, the calibration of sampling equipment, compositing techniques, and other variables that impact the validity of the sample and the analytical results. Inadequate sampling procedures can easily invalidate analytical results obtained with excellent laboratory and MS methods. Sampling techniques, especially for ambient air, are described in Chapters 3, 4, 7, and 8, and details of sampling procedures are included in reference books and standard analytical method descriptions.[4–7]

Sample Preservation. Preservation is concerned with the measures taken before, during, and after sample shipment to the laboratory to protect the integrity and validity of the sample. Preservation must ensure that the analytical results reflect the actual condition of the sample at the time and place of sampling and not the result of some physical or chemical changes caused by the conditions of shipment or storage before analysis. Many types of samples are shipped or stored at low temperatures to retard microbiological activity or are treated with acids for the same or some other reason. Details of sample preservation are specialized and are contained in reference books and standard analytical methods' descriptions that are cited subsequently, especially in Chapters 3, 4, 7, and 8.[4–7]

Field and Continuous Analyses. Mass spectrometric measurements in the field at the sampling station or in a processing plant are alternatives to sampling and shipment of the sample to a remote or field laboratory for analysis. Measurements in the field are made by continuous sampling of a process stream or environmental medium and by obtaining and analyzing discrete samples. These mass spectrometric techniques, which are sometimes advantageous, have been developed and used and are described in Chapter 8.

Sample Preparation. Sample preparation is the physical and chemical processing of the sample, or an aliquot of the sample, in the laboratory before the final chromatographic separation and mass spectrometric measurements. Sample preparation is a major and complex area with hundreds or thousands of different options and approaches. Many of the issues of sample preparation are inexorably linked to the success or failure of the final separation and measurement. For example, interfering contamination may be introduced into the sample in the preparation laboratory or a solvent used that is incompatible with the chromatographic separation or mass spectrometric measurement. Complete descriptions of sample preparation techniques are beyond the scope of this book, but some important aspects are described in Chapters 3, 4, and 7 in connection with standard analytical methods described in those chapters. Trade-offs and compromises among sample preparation procedures, chromatographic separations, and mass spectrometric measurements are common issues in the development of analytical methods.

Conventional Gas Chromatography / Mass Spectrometry

A basic strategy of analytical and environmental MS is the application of conventional GC/MS to the problem. This is a basic strategy because conventional GC/MS has been for many years, is now, and will likely continue to be the single most important mass spectrometric technique used for organic chemical analyses in many fields. Conventional GC/MS is the *killer application* of MS in environmental chemical analysis. In the 1996 Public Broadcasting System's production *The Triumph of the Nerds*, a killer application was defined as a personal computer (PC) application program so compelling and useful that it motivated hundreds of thousands of individuals and businesses to buy a PC just for that single application. The first spreadsheet program *Visicalc* was the killer application for the Apple II PC, *Lotus 123* was the killer application for the original IBM PC, and desktop publishing was the killer application of the Macintosh. Tens of thousands of conventional GC/MS systems have been purchased worldwide just to implement a single environmental analytical method. The objective of an analysis with conventional GC/MS is the simultaneous identification and measurement of target analytes or the identification of nontarget or unknown substances in a sample.

Conventional GC/MS is defined as open tubular (capillary or wider bore) GC with a directly connected (on-line) mass spectrometer. The GC carrier gas is helium, automated sample injection is often employed, and the GC oven is usually temperature programmed. The mass spectrometer uses nominal 70-eV electron ionization (EI) and either wide mass-range data acquisition or selected ion monitoring (SIM) at unit mass/charge (m/z) resolving power (low resolution) up to about m/z 600. One or more microprocessors, associated electronics, and applications software are used to control the operations of the instrument system, acquire data, and reduce raw data to usable information. Figure 2.1 is a diagram showing the basic components of a conventional GC/MS system. Most contemporary conventional GC/MS instruments are readily accommodated on a standard laboratory bench top or table.

It is not possible within the scope of this book to describe in detail the broad range of important GC topics such as injector designs, column construction, column selection, elution behavior, and the optimization of separations.

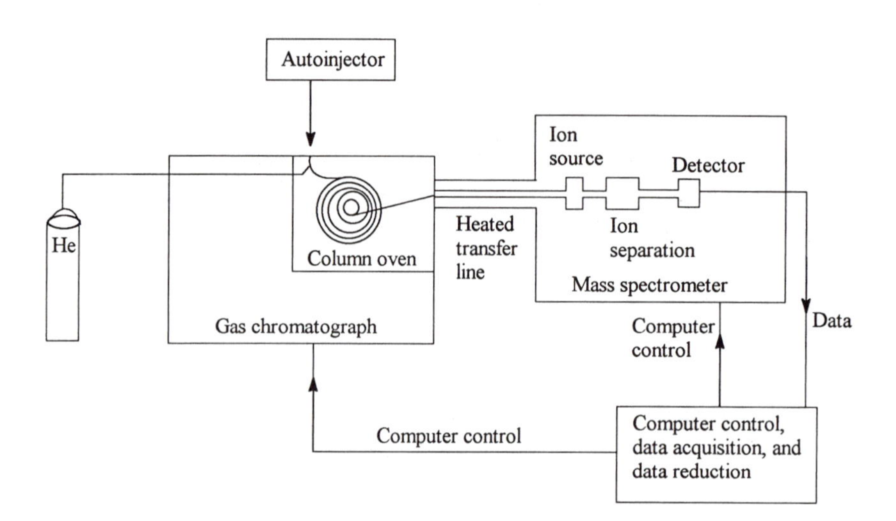

Figure 2.1 Diagram of the basic components of a conventional GC/MS system.

The reader is referred to other publications and manufacturers' current literature for detailed information on these subjects.[8–9] However, GC performance and techniques are very important in GC/MS and they are discussed as needed and appropriate. Similarly, it is not possible to include detailed descriptions of the design, construction, and principles of operation of various types of mass spectrometer. The strategies and techniques presented in this chapter, and many other parts of the book, are generally applicable to all types of mass spectrometers. Conventional GC/MS can be implemented with magnetic deflection,[10–14] linear quadrupole,[11–14] time-of-flight,[12,14–15] quadrupole ion trap (Paul ion trap),[12,16–18] Fourier transform (Penning ion trap),[12,14,19] or any other type of mass spectrometer one may choose to employ. The reader is referred to the cited publications for basic information on mass spectrometers. However, the performance characteristics of the mass spectrometer are very important and they are discussed as needed and appropriate.

Conventional GC/MS, as defined here, is the result of a number of strategic decisions that were made over the last 25 years by instrument manufacturers, instrument users, and analytical method developers. The laboratory instrumentation industry currently produces and markets a variety of instrument systems which can be broadly categorized as basic, conventional GC/MS systems. Often these basic systems have surprisingly different capabilities, specifications, and limitations which can be crucial to their successful application in environmental and other analyses. The GC/MS instrument features that are essential to the successful application of conventional GC/MS to environmental and many other analyses are described in the following sections.

Gas Chromatography

The basic components of a gas chromatograph used with a mass spectrometer are a carrier gas handling system, a sample injector, a column, and a column oven (Figure 2.1). While hydrogen has some clear advantages as a GC carrier gas, and was recommended and used with open tubular GC for many years,[20] helium, largely for reasons of safety, is the near universal choice as the carrier gas for environmental and other analyses. The most common and useful injection mode is splitless/split which means that during injection of an aliquot of sample, and for a brief period after injection, the split valve is closed (splitless mode) and sample vapors are forced into the GC column by the carrier gas flow. Subsequently, the split valve is opened and residual sample vapors are purged from the injection port (split mode) before the onset of temperature programming. The injection port liner is usually inert quartz and the instrument is designed so that sample vapors never come into contact with cold spots or with hot metal, such as stainless steel, which can cause catalytic decomposition of many compounds. Open tubular GC columns are constructed from fused silica with low-bleed stationary phases of varying polarity chemically bonded to the silica surface. Columns are typically 30–75 m in length and have inside diameters (IDs) no greater than about 0.75 mm. The column oven must be capable of precise temperature control and temperature programming at variable rates for variable times.

Interface to the Mass Spectrometer

When narrow-bore GC columns, for example, 0.32 mm ID or less, are used, many mass spectrometers have sufficient vacuum pumping capacity to allow insertion of the delivery end of the GC column directly into the mass spectrometer ionizer. If pumping capacity is insufficient, which usually occurs with higher carrier gas flows from wider bore GC columns, some other interface is required to reduce the gas flow into the vacuum of the mass spectrometer. This is usually the open split interface (Figure 2.2) or the glass jet separator (Figure 2.3). In both devices the operational principle is that the light carrier gas helium will tend to diffuse out of the main stream of gas whereas the heavier molecules will not diffuse as readily and will be carried on a more or less straight line into the mass spectrometer. The quadrupole ion-trap mass spectrometer requires a pressure of about 1×10^{-3} Torr of helium for proper operation. If this pressure cannot be maintained by the carrier gas flow, a supplemental supply of helium is required.

Mass Spectrometry

The basic components of a mass spectrometer are a vacuum system, sample inlet systems, a method of ionizing the sample, an ion separating device, and an ion detector (Figure 2.1). All mass spectrometers operate in a vacuum and vacuum requirements vary depending on the type of mass spectrometer. The gas chromatograph is the only sample inlet system considered in this section, but other systems are described in Chapters 6–8. Sample vapors emerging from the chromatograph must be ionized and in conventional GC/MS the ionization technique is electron ionization (EI). In EI, energetic electrons are emitted from an electrically heated filament in a vacuum and focused into the sample stream where ionization occurs. The enclosure in which ionization occurs is called the ion source. In conventional GC/MS, the nominal energy of the ionizing electrons is 70 eV, which is substantially greater than the ionization potential of all

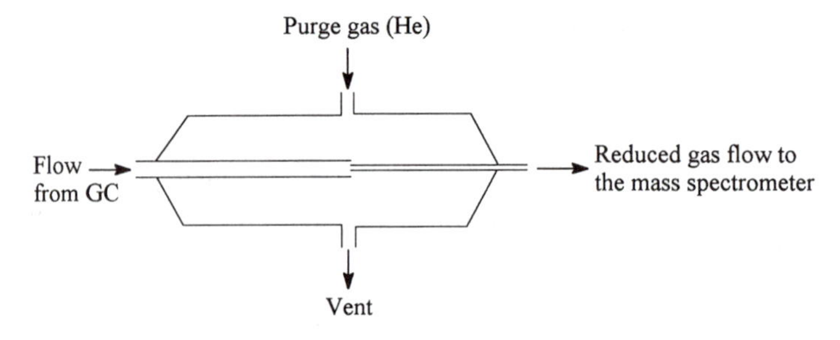

Figure 2.2 Diagram of an open split GC/MS interface.

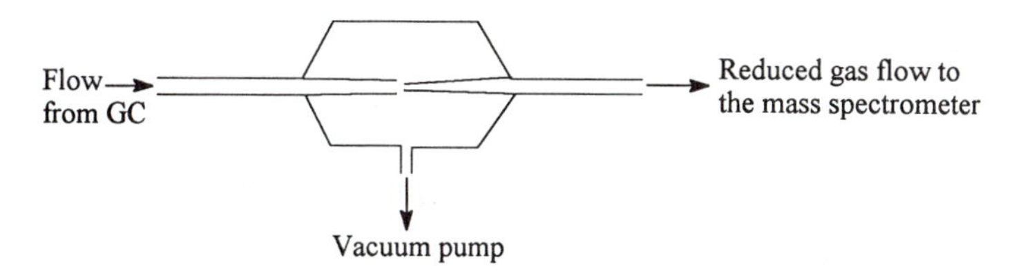

Figure 2.3 Diagram of a jet separator GC/MS interface.

compounds and elements. Because of various instrument design issues, the actual energy of the ionizing electrons in any given instrument may be anywhere from 50 to 200 eV or more. The exact value of the electron energy is not important because in the general range 50–200 eV most compounds give about the same total number of ions with about the same relative abundances. Therefore, the nominal value of 70 eV is usually specified.

Electron ionization of almost all molecules at 70 eV produces mainly positive ions and the positive-ion mass spectrum is measured in conventional GC/MS (Reaction 2.1). Electron ionization probably does not consist of direct hits by energetic electrons on molecules (M). Much more likely, an electron in a higher energy atomic or molecular orbital is excited by absorption of energy from a passing ionizing electron, and the electron is ejected from the molecule to form a positive molecular radical ion ($M^{+\cdot}$). The electron lost from the molecule comes from, if present, an unshared electron pair of an oxygen, nitrogen, sulfur, or other atom, or the π-electron system of an unsaturated molecule. By convention, the positive charge on an ion is shown localized on one of these atoms, bonds, or rings. If there are no unshared pairs or π-electrons in the molecule, the electron is ejected from a σ bond which has more devastating effects on the $M^{+\cdot}$. Depending on the stability of the $M^{+\cdot}$, it may persist in high abundance or undergo rapid unimolecular fragmentation reactions (Reaction 2.2) to give other ions. The fragmentations of the $M^{+\cdot}$ are rationalized using mechanisms similar to those used to describe liquid-phase reactions of organic compounds.[21,22]

$$M + e^-(70\ \text{eV}) \rightarrow M^{+\cdot} + 2e^- \qquad \text{(Rx 2.1)}$$

$$M^{+\cdot} \rightarrow \text{fragment ions and neutral atoms, radicals, etc.} \qquad \text{(Rx 2.2)}$$

Almost all mass spectrometers physically separate gas phase ions according to their mass-to-charge ratios (m/z), where m is the mass in daltons (Da) and z is the charge. The Fourier transform mass spectrometer detects ions with different m/z values by another process. The exact mechanisms for the extraction of ions from the ion source and the physical or logical separation by m/z is dependent on the type of spectrometer used.[10–19] Conventional GC/MS operates at about unit m/z resolving power or slightly better, which means the spectrometer is capable of separating ions that differ by 1 m/z unit. This is commonly called low resolution. The physically separated ions are detected by an electron multiplier whose output is connected to the computer-controlled data acquisition system.

Computer Control and Data Handling

Conventional GC/MS, and essentially all other MS systems, requires a directly connected computer control, data acquisition, and data reduction system. In practice this usually involves multiple central processing units (CPUs), large random access memory (RAM), large capacity disk drives (gigabytes), back-up storage media (magnetic tape, optical disks, etc.), video display equipment, and sophisticated systems and applications software. Analog voltages from the electron multiplier are converted into digital information and stored on magnetic media. The data reduction software must have many capabilities including multitasking data acquisition and data reduction, storing GC/MS data and processing stored data, recognizing a GC peak within any given target analyte retention time window, comparing mean mass spectra from GC peaks with spectral data in user-created and standard reference databases, and generating lists of tentatively identified target compounds, nontarget compounds, and supporting data. The software must be capable of defining GC peaks, subtracting background data, integrating ion abundances of any specific ion between specified time or scan number limits, calculating response factors and calibration lines or curves, calculating calibration statistics, and calculating analyte concentrations using the calibration data.

It is beyond the scope of this book to describe conventional GC/MS hardware and most software in greater detail as a complete description would be a large book in itself and ample reference material has been published.[11–14, 18] However, a number of essential software-driven data acquisition, data reduction, and data display capabilities are described in other sections of this chapter. A thorough discussion of the fragmentations of $M^{+\cdot}$ ions produced by EI is also beyond the scope of this book. Several excellent books on the interpretation of EI spectra are available.[12, 21–22] The interpretation of conventional EI mass spectra is also presented in a number of short courses on MS offered by the American Chemical Society, the American Society for Mass Spectrometry, and the Pittsburgh Conference on Analytical Chemistry and Applied Spectroscopy. In this and subsequent chapters the positive-ion EI mass spectra of many molecules are referenced or discussed.

Alternatives to Conventional GC/MS

When considering a strategy for an environmental or some other chemical analysis, a frequent issue is whether to use conventional GC/MS or select GC coupled with a standard detector. This issue is usually raised because of the expectation, or at least the hope, that chromatography with a standard detector will provide the analytical information needed at a cost lower than that of conventional GC/MS. In addition, other forms of spectrometry and other types of ionization for MS occasionally have been used as alternatives to conventional GC/MS for environmental and other analyses. In this section the strategy of using an alternative to conventional GC/MS is considered with emphasis on data quality and cost.

Standard Detectors

The scientific literature of the last 30 years or so is replete with reports of environmental and other analyses conducted using packed, or more recently, open tubular column GC and several standard GC detectors.[23] These standard GC detectors include flame ionization, electron capture, photoionization, electrolytic conductivity, flame photometry, and several element-selective detectors. Similarly, high-performance liquid chromatography (HPLC) with ultraviolet absorption or fluorescence detection has been used to separate and detect some compounds that may also be determined by GC/MS, for example, some polycyclic aromatic hydrocarbons.[24] All of these standard GC or HPLC detectors have at least some, or sometimes considerable, compound or element selectivity which reduces or eliminates responses from nontarget analytes, background, interferences, and so on. Some additional selectivity may be provided in these analyses by selective sample preparation procedures including solvent extraction and preliminary chromatography or other separation techniques. However, all analyses based on standard GC or HPLC detectors have the common feature of relying heavily on a chromatographic retention time or a relative retention time index to identify the target substances that are separated and cause a signal in the detector. Therefore, it is important to consider the value of the retention time as an identification tool either alone or in combination with other information.

Chromatographic Retention Data

Figure 2.4 shows a hypothetical but not unrealistic 30-min chromatogram containing a single 3-s wide (at half height) GC peak measured with a standard selective GC detector. Other peaks in the chromatogram have been deleted for this example. Assume that this chromatogram was obtained from the analysis of an environmental sample from a fairly contaminated source, for example, an industrial wastewater discharge, a river sediment downstream from a petroleum refinery, or a high organic soil near a waste disposal site. Such samples have a high probability of containing many compounds including some of environmental interest. Even a sample of water from a clean lake or an air sample from a small city has a good probability of containing many compounds, including some of environmental interest, if the detection limits of the analytical method are sufficiently low.

The question is, what is the probability that one can correctly identify as a target analyte the substance represented in Figure 2.4 using the measured retention time? In target compound analyses with standard detectors, retention data for all the target analytes are carefully measured using authentic samples of each of the target compounds. Furthermore, the variability of retention data is also

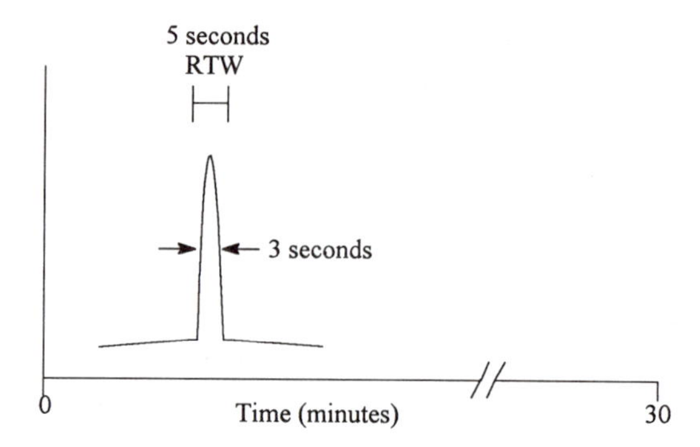

Figure 2.4 Hypothetical 30-min chromatogram containing a single 3-s wide (at half height) GC peak.

measured and is usually very low with open tubular GC columns when referenced to an internal standard. Finally, there is a reasonable expectation that the target analytes might be in the particular sample. The retention time of one of the target analytes exactly matches the measured retention time in Figure 2.4.

In considering the question, it is useful to estimate the number of possible substances that might have exactly the same retention time as the substance that caused the peak in Figure 2.4. By mid-1998 the Chemical Abstracts Service of the American Chemical Society had registered 18 million chemical substances.[25] If it is assumed that of the 18 million, 15 million are singular molecular species and that 10% of the 15 million are amenable to separation by GC, which are both reasonable estimates, then 1.5 million substances need to be considered possibilities. If one further assumes that 80% of the 1.5 million are either extremely unlikely to ever be in any sample, or can be eliminated from consideration because they are discarded in a selective sample preparation procedure, or do not cause a response in the selective detector, then only 300,000 possible compounds need be considered. If one assumes that the retention times of all 300,000 compounds are perfectly evenly distributed over the 30 min time of the chromatogram, 167 compounds would elute every second or 835 compounds in each 5-s retention time window (RTW). Use of a 5-s or wider RTW is standard practice and allows for some variability in the retention time measurement. In this hypothetical example, the retention time of only one target analyte matched the retention time of the unknown substance, but it is not comforting to know that, on average, as many as 834 other compounds also could elute in that same RTW. Of course, it is highly unlikely that the retention times of all 300,000 compounds would be perfectly evenly distributed over the 30 min. Therefore, in any given RTW there could be considerably more or less than 835 compounds expected.

Perhaps it is absurd to assume that all 300,000 compounds are real possibilities, but it may be reasonable to assume that 1% of these, or 3000, could really be present in the processed sample and cause a response in the selective detector. Even 3000 compounds perfectly evenly distributed over the 30 min of the chromatogram corresponds to an average of 8.3 compounds eluting during each 5-s RTW. In practice, there could be many more than 8.3 compounds in each RTW. Even with a selective sample preparation procedure, high resolution GC, and a standard selective GC detector, if the sample is not pristine and detection limits are sufficiently low, the probability is fairly high that at least two compounds will coelute in any narrow RTW. Therefore, the probability of false identifications, or false positives, is fairly high and retention time data alone are not adequate for the identification of target analytes in a chromatogram.

The reasonableness of the assumptions used in the foregoing analysis can be tested by considering several common environmental contaminants of great interest. The chlorinated biphenyls, chlorinated dibenzodioxins, chlorinated dibenzofurans, and the constituents of the pesticide mixture toxaphene all have similar physical and chemical properties. There are 209 possible chlorinated biphenyl congeners, 75 possible chlorinated dioxin congeners, 135 possible chlorinated dibenzofuran congeners, and an estimated 670 chlorinated toxaphene components[26] for a total of 1089 individual compounds. These many compounds have similar solubilities in organic solvents, similar polarities, similar chromatographic behavior, and most cause a signal in the selective electron-capture GC detector. Thus, if present in a sample, they will likely be in the same solvent extract and elute from a typical GC column over a roughly 45 min period with temperature programming. Chlorinated hydrocarbon pesticides, their isomers, congeners, and some degradation products also have similar properties and, if present in a sample, would be expected in the same solvent extract and chromatogram.

These chlorinated hydrocarbon materials are very common contaminants in many kinds of environmental samples including sediments from lakes and harbors, soil from various waste sites, and fatty tissue from mammals and fish. Given this type of sample, it is not unreasonable to expect that several hundred compounds could appear in the same chromatogram and cause signals with the same selective detector. Even with careful preliminary separations and fraction collecting, the chances of coelution of one or more compounds in a high-resolution GC are very high. In the late 1970s a Cincinnati, Ohio, drinking water sample was concentrated by reverse osmosis and analyzed by GC/MS, and 460 compounds were identified or tentatively identified.[27] In a national survey of pesticides in drinking-water wells, the false positive rate was 60–80% for three analytical methods which used standard GC detectors.[28] In general, the use of retention time data to identify compounds in environmental samples, even with a selective GC detector, is risky and is especially prone to false positives.

Coelution, Resolution, and Analysis Time

Coelution of several or more compounds in high-resolution GC is not uncommon, even with far fewer analytes than were discussed in the previous section. The probability of coelution is significantly enhanced by the analytical chemist's desire to reduce the analysis time as much as possible. This latter tendency is exacerbated in the commercial laboratory testing industry where *time is money*.

In this industry there is pressure to squeeze analyses into as little time as possible to maximize sample throughput, minimize capital equipment costs, and maximize profits. An example of the trade-off between resolution and analysis time is shown in Figure 2.5. In this study a mixture of only 61 volatile organic compounds, many of them solvents and other low molecular weight molecules, was investigated to find the best possible gas chromatographic separation in about 30 min.[29] The chromatogram in Figure 2.5 was obtained using multiramp linear temperature programming and a 30 m × 0.53 mm (ID) fused-silica GC column. The wide-bore GC column was used to accommodate the gas flow from the inert gas purge and trap procedure. Most of the compounds were separated from one another within about 33 min, but 12 pairs of compounds coeluted with none of the pairs separated by more than 12 s (Table 2.1).

The coeluting pairs in Table 2.1 cannot be recognized as such by any single standard GC detector. Given this problem, the standard analytical approach is to alter the chromatographic conditions in order to separate all the analytes. Much of the chromatography literature is devoted to techniques for the complete resolution of all substances in a test mixture, or actual sample, because complete resolution is essential for the commonly used standard GC detectors. Some of the conditions that could be changed to improve resolution in Figure 2.5 are the selection of a higher resolution narrower bore column, a longer column, a slower temperature programming rate, a smaller sample size, or a different polarity stationary phase. Some of these changes would lengthen the analysis time, raise detection limits, or cause other undesirable effects. In this example a more polar stationary phase was not available and a less polar stationary phase caused even more coelutions. With a 60-m column, a slower temperature programming rate, and three fewer analytes, more analytes were separated in a 53-min chromatogram, but five of the pairs in Table 2.1 were not separated by more than 12 s.[30] Regardless of how well separated a standard mixture is, an analyst can never be sure that a real sample will not contain

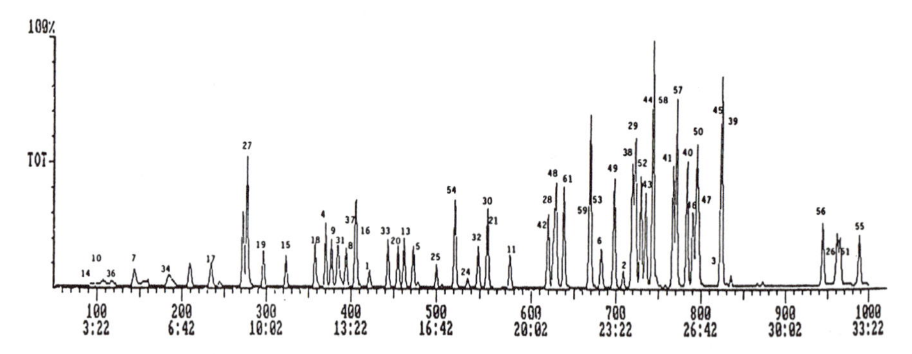

Figure 2.5 Normalized total ion chromatogram from a volatile compound mixture containing 25 ng of most compounds. (Reproduced from Ref. 29, *Journal of Chromatographic Science*, by permission of Preston Publications. A Division of Preston Industries, Inc. Copyright 1990 Preston Publications.)

Table 2.1 Coeluting compound pairs in the GC separation of 61 volatile compounds

Compounds	Retention times (min:s)	Identification numbers in Figure 2.5[a]
Tetrachloromethane and 1,1-dichloropropene	13:10 and 13:06	8 and 23
1,2-Dichloroethane and benzene	13:38 and 13:30	16 and 37
1,3-Dichloropropane and tetrachloroethene	18:42 and 18:36	21 and 30
Dibromochloromethane and 1,2-dibromoethane	19:12 and 19:24	11 and 12
1,1,1,2-Tetrachloroethane and ethylbenzene	20:52 and 21:00	28 and 48
m-Xylene and *p*-xylene	21:22 and 21:18	60 and 61
Styrene and *o*-xylene	22:24 and 22:16	53 and 59
1,1,2,2-Tetrachloroethane and 1,2,3-trichloropropane	24:04 and 24:08	29 and 35
4-Chlorotoluene and 1,3,5-trimethylbenzene	24:46 and 24:50	44 and 58
1,4-Dichlorobenzene and 4-isopropyltoluene	26:36 and 26:30	47 and 50
n-Butylbenzene and 1,2-dichlorobenzene	27:32 and 27:26	39 and 45
Hexachlorobutadiene and naphthalane	32:04 and 32:12	26 and 51

[a]Not all identification numbers are shown in Figure 2.5, but a complete list of identification numbers and retention times is given in the original publication (Ref. 29).

additional compounds that coelute with one or more of the 61 target compounds. Hundreds or thousands of additional compounds would need to be tested with the same chromatographic procedure and undoubtedly many more coelutions would be discovered.

Given this dilemma, the customary next step is to revert to a reasonably short and economical analysis time, about 20–30 min, but add more capability to the analysis to provide more information about the substances eluting from the column. Two routine strategies for target compound analyses with standard GC detectors are:

- A second GC column for confirmation
- Two standard GC detectors to provide additional information

Both approaches add to the cost and time of the analysis and usually fall well short of providing a fully satisfactory solution. A major problem with second column confirmation, besides the additional equipment, supplies, and time needed, is that often a second column merely shifts the retention times of most or all of the compounds to shorter or longer times without altering in a major way the order of elution of most of the compounds or resolving all coeluting pairs. In a national survey of pesticides in drinking-water wells, only about 70–80% of pesticides identified with second column confirmation could be verified with GC/MS.[28] The overall confidence in identifications is not usually improved greatly with the second column confirmation technique.

Two standard detectors can allow separate measurements of coeluting target compounds (A and B) if compound A causes a response in detector A but not in detector B and compound B causes a response in detector B but not in detector A. One detector arrangement used is a photoionization detector (PID) in series with an electrolytic conductivity detector (ELCD).[30] A critical limitation of this dual detector technique is that the first detector in series must be nondestructive to one of the analytes of each pair and allow it to pass undisturbed to a second detector. The PID responds to aromatic and some unsaturated compounds with ionization potentials less than about 10 eV and the ELCD responds to organohalides. The PID is placed first after the GC column and detects nonhalogenated aromatic compounds, but allows the saturated organohalides to pass undisturbed into the ELCD. Of the common standard detectors, only the PID provides this selective nondestructive capability. Unfortunately, this detector is not 100% efficient in ionizing aromatic and unsaturated compounds and removing them from the effluent. Therefore, halogenated aromatic and unsaturated compounds cause responses in both detectors.

If the dual PID/ELCD system was used with the separation shown in Figure 2.5 and Table 2.1, nine of the 12 coeluting pairs would cause a nearly simultaneous response in both the PID and ELCD, indicating that both compounds may be present. Three pairs, dibromochloromethane and 1,2-dibromoethane, *m*-xylene and *p*-xylene, and styrene and *o*-xylene, cause responses in one or the other detector but not both and therefore the members of each pair cannot be individually recognized as being present by this dual detector technique. Of the nine coeluting pairs that cause near simultaneous responses in both detectors, both compounds of three pairs cause a response in the ELCD but only one member of each pair causes a response in the PID. Therefore, if the PID signal is missing for any of these three pairs, and the ELCD signal is present, it can be concluded that only one member (the saturated organohalide) of the pair is present; but if both the ELCD and PID signals are present, one or both members of the pair could be present. Similarly, of the nine coeluting pairs that cause near simultaneous responses in both detectors, both compounds of four pairs cause a response in the PID, but only one member of each pair causes a response in the ELCD. Therefore, if the ELCD signal is missing for any of these four pairs, and the PID signal is present, it can be concluded that only one member (the nonchlorinated aromatic hydrocarbon) of the pair is present. Again if both signals are present, one or both compounds could be present. Only two pairs of coeluting compounds, 1,2-dichloroethane and benzene, and 1,1,1,2-tetrachloroethane and ethylbenzene, can be independently recognized and measured with this dual detector technique.

To avoid the need for a nondestructive detector, the GC effluent can be split between two different detectors to provide similar information. This would allow simultaneous use of, for example, a flame photometric detector and an ELCD, but would introduce further complications from the spliter in the analytical system. The exact time of arrival of the analytes at the two detectors in the split effluent may be uncertain because of slight differences in the path length to the two detectors. This can cause confusion in the identification of corresponding peaks at the two detectors, particularly in a complex mixture containing coeluting compounds, as in Figure 2.5. Also, all analytes may not be split between the two detectors in the same proportion, and detection limits may be raised by splitting the GC effluent between the two detectors. Splitting the effluent stream may also adversely effect the quantitative analysis of the sample.

From this analysis it can be concluded that two standard GC detectors in series or parallel can provide some additional information to help identify target analytes, but the additional information is usually not definitive and the issue of unexpected interferences by nontarget analytes is unresolved. Under these circumstances the value of standard detectors in environmental and other analyses is limited and should be carefully assessed in any research, monitoring, or other study, with particular attention to the risk of false positives.

Risk of False Negatives

A false negative occurs when the analyte is not detected and that result is false. While the variability of retention data on open tubular GC columns, particularly when referenced to an internal standard, is usually very low, unexpected problems can and do occur. Sudden or gradual drift in operating parameters such as carrier gas flow rate and column temperature can easily cause a GC peak to shift out of a narrow RTW. Component failures such as injection port liner contamination, stationary phase degradation, gas leaks at connectors, and loss of detector sensitivity can easily cause a GC peak to disappear. While quality control (QC) procedures should detect these problems, inadequate QC may not identify the failure in time, and economic pressures may prevent reanalysis of samples analyzed during the out-of-control periods. Use of standard GC detectors demands very intensive QC, which is often considered uneconomic. The risk of false negatives is significant when relying on retention time measurements for peak identifications.

Role of Standard Detectors in Environmental and Other Analyses

The results of chemical analyses can influence major ecological, economic, health, regulatory, and social decisions. Correct identifications of analytes in environmental and other analyses are absolutely essential. These identifications can have a major impact on, for example, a regulatory decision to ban or approve a product, a decision to construct a multimillion dollar treatment or manufacturing plant, or a decision to conduct further chemical analyses and long-term testing. Even environmental and other studies conducted for some apparently unrelated purpose can influence major decisions when the results are published and later used to support a decision on some issue totally unknown to the original investigator.

As illustrated in the foregoing sections, the risk of misidentification or false positives and false negatives is significant with standard detectors in many types of samples. On the other hand, standard chromatographic detectors are sometimes convenient, comparatively low in cost, and can provide reliable results when used in an appropriate way. Standard chromatographic detectors should be limited to use with samples taken from closed systems that are very well characterized and not subject to external inputs and unexpected interferences. For example, the products of a manufacturing operation are usually well characterized with regard to their purity, by-products produced, and manufacturing problems that may cause the production of alternative products or different mixtures of products and by-products. Under these circumstances, a standard detector can be used with reliable results since each analyte is known and the analysis can be optimized for the group of analytes with essentially no chance of interferences or unexpected complications.

Sometimes environmental and other studies are conducted using standard chromatographic detectors as screening tools with the provision that any tentative identifications will be confirmed by MS. This may be a reasonable approach, especially if large numbers of samples are to be analyzed and only one or a few GC/MS instruments are available. A good use of the near-universal flame ionization detector (FID) is to determine approximate concentration levels in samples or extracts without attempting to use the retention time data to identify any detector response peaks (concentration ranging). By determining approximate concentration levels with an FID, high concentration samples and extracts can be diluted to an appropriate concentration before the GC/MS analysis.

Relative Costs of Mass Spectrometry and Standard Detectors

Except for analyses of samples from closed systems and concentration ranging, the use of standard GC detectors for environmental and other analyses may significantly decline for economic reasons. Figure 2.6 is a rough estimate made in 1979 of the relative costs of using a mass spectrometer or conventional GC detectors for target compound analyses.[1] At that time, capital plus operating costs for a conventional GC/MS system were estimated at about $150,000

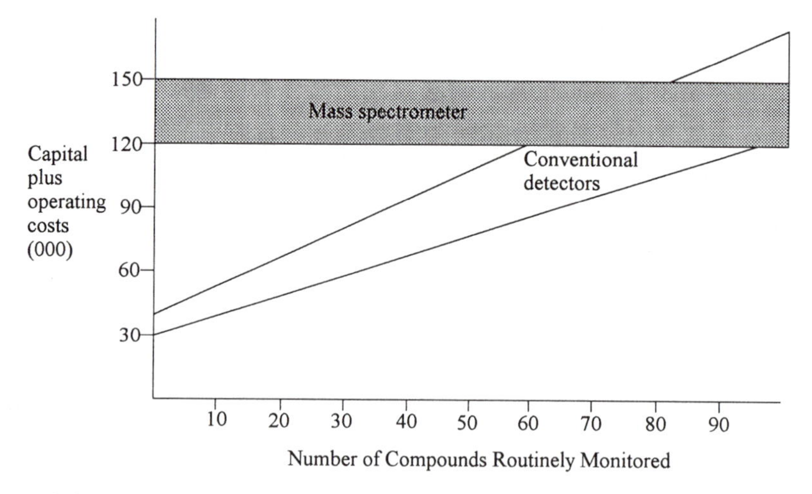

Figure 2.6 A 1979 estimate of capital plus operating costs for mass spectrometer and conventional GC detectors as a function of the number of target analytes. (Redrawn from Ref. 1, published in 1979 by the American Chemical Society.)

annually. This included roughly $50,000 per year for depreciation on a capital cost of $250,000 for a typical floor-standing instrument of that period. A typical system included a packed column GC, a single quadrupole mass spectrometer, the latest 16-bit minicomputer, several disk drives with removable 5-Mbyte disk packs, 32 kbytes of random access memory, an industry standard nine-track magnetic tape drive, and a reasonably fast impact line printer. Operating costs of $100,000 per year included the system operator, a technician for sample preparation, equipment maintenance, training, QC, supplies, and management. As seen in Figure 2.6, it was expected that this cost would remain approximately flat, as indicated by the broad band, regardless of the number of compounds monitored up to a reasonable limit because conventional MS is a universal detector. Additional compounds added to the target list cause only modest increases in cost as long as the sample preparation procedure is the same.

On the other hand, the capital equipment cost in 1979 for a gas chromatograph with a standard detector was much less than a GC/MS system. If the number of target analytes was small, the sample preparation procedure was the same for all analytes, and a single selective detector in a single gas chromatograph could be used for all the analytes, annual operating costs were estimated to be as low as $30,000. However, as the number of target analytes increased, it was likely that additional gas chromatographs and selective detectors would be required to accommodate the diversifying characteristics of the additional analytes. Only highly electronegative compounds and elements cause signals in electron capture GC detectors; only aromatic and unsaturated compounds with ionization potentials less than about 10 eV cause signals in photoionization GC detectors; only halogenated organics cause signals in electrolytic conductivity GC detectors; and only nitrogen- and phosphorus-containing compounds cause signals in NP (nitrogen–phosphorus) GC detectors. Therefore, as the number of target analytes increased, it was likely that the diversity of compounds would increase, and different selective detectors and gas chromatographs would be required to achieve adequate detectability for all compounds. It is impractical to easily and quickly change detectors in a single gas chromatograph, and all detectors would need to be in operation continuously to cover all the analytes.

Also, because of the selective nature of these GC detectors, several or more somewhat different sample preparation procedures would probably need to be developed to optimize the recoveries of the compound groups determined with different detectors and to eliminate potential interferences. Often more rigorous sample preparation procedures are required for standard GC detectors than with MS detection to reduce the number of potential interferences. Therefore, the expectation in 1979 for standard GC detectors was steadily increasing capital plus operating costs as the number of target analytes increased (Figure 2.6).

During the last 20 years there has been a significant reduction in the cost and size of conventional GC/MS equipment. New types of mass spectrometers have been developed, components miniaturized, and many instrument functions are controlled by inexpensive microprocessors and highly reliable integrated semiconductor circuits. In some commercial GC/MS instruments, single quadrupole mass spectrometers are integrated into the instrument package to the extent that the mass spectrometer occupies only an estimated 20% of the total volume. Bench-top mass spectrometers that can be easily moved from bench to bench by a single person are fairly common and cost as little as $50,000. This price includes the gas chromatograph, the computer hardware, and the software. Instead of kilobytes and megabytes of random access and disk memory, megabytes and gigabytes, respectively, are standard and at a much lower cost than in 1979. Microprocessors used in GC/MS computer systems are more powerful than mainframe computers of 20 years ago and most commercial GC/MS systems have two or more such processors optimized to control and operate different parts of the system. Because of advances in automation, computer hardware, and software, today's systems are far more powerful than those of even 10 years ago. Systems are typically so powerful that one operator can easily and simultaneously oversee the operations of six or more GC/MS systems doing routine target compound analyses.

Today, the estimated cost band in Figure 2.6 for GC/MS should be less than $100,000 annually, including amortization of the capital equipment. The intersection of the cost bands for GC/MS and conventional GC detectors now occurs at a much smaller number of target analytes. While the cost of GC instruments with standard detectors has also benefited from the reductions in the cost of microelectronics, the change has been less dramatic. The GC detectors and ovens, for example, have not changed that much and the instruments are simpler than GC/MS equipment and benefit less from miniaturized electronics. Significantly, the need for more rigorous sample preparation with standard GC detectors has not changed. Given the current cost structure, new laboratories may decide to dispense with most conventional GC equipment and standard detectors and purchase a larger number of low-cost GC/MS systems for routine analytical work. This would save the cost of all reinjections and reprocessing costs for confirmation with, for example, second columns.

Fourier Transform Infrared and Other Mass Spectrometry Ionization Techniques

Fourier transform infrared (FTIR) spectrometry and other MS ionization techniques are not alternatives to conventional 70-eV electron ionization (EI) MS, but they are used to provide valuable supplemental information for environmental and other analyses. Combined GC/FTIR spectrometry was developed during the 1970s and combined GC/FTIR/MS systems were demonstrated during the early 1980s.[31] A GC/FTIR/MS system was shown to provide more useful information about the components of environmental samples than either conventional GC/MS or GC/FTIR alone.[32] However GC/FTIR has several significant limitations. The sensitivity of FTIR is several orders of magnitude less than that of conventional GC/MS, the database of reference gas-phase FTIR spectra is not very large, and FTIR spectra are generally more difficult to interpret than mass spectra. The development of compact GC/FTIR instrument systems lagged the development of commercial bench-top GC/MS systems by a number of years, and quantitative FTIR techniques were also slower to develop.

Nevertheless, GC/FTIR is a valuable technique and may be indispensable for the determination of the structure of substances whose mass spectra are not contained in reference databases. The issue and complexity of the identification of unknowns is discussed in more detail in this chapter in the section "Identification Criteria".

Electron ionization with nominal 70-eV electrons is the basis of conventional GC/MS, but other techniques are used to generate gas phase ions after GC separation and before mass spectrometric analysis. Among the ionization techniques available, the most frequently used for environmental and other analyses are positive or negative ion chemical ionization and electron capture ionization. These techniques are not alternatives to conventional 70-eV EI because they selectively ionize molecules that react with reagent ions or capture thermal electrons. In addition, these soft ionization techniques produce few, if any, fragment ions that are needed to support identifications of analytes. However, the techniques do provide valuable supplemental information and, in some cases, may be the technique of choice for target compound analysis. Chemical ionization and electron capture ionization MS are described in Chapter 5.

Retention Data and Chromatographic Resolution with Mass Spectrometry

In the foregoing sections, considerable emphasis was placed on the unreliability of chromatographic retention time measurements for the identification of target compounds. However, when used in conjunction with other information, particularly mass spectrometric data, retention time measurements are extremely valuable and are routinely used to exclude compounds from further consideration in target compound analyses. This important role of retention time measurements when used with MS is discussed in more detail in the sections of this chapter on "Limitations of Mass Spectrometry" and "Identification Criteria". Chromatographic separations are also very important in conventional GC/MS and other on-line separation/MS techniques, but complete resolution of all substances in a sample or a sample extract is not nearly as critical with MS as it is with standard detectors. The capabilities of MS in recognizing and measuring unresolved sample components are discussed in the next section. More specialized techniques to increase analyte selectivity and resolve mixtures with MS are described in Chapter 5.

Data Acquisition Strategies

Before a sample prepared for mass spectrometric analysis can be analyzed, a data acquisition strategy must be selected. There are two general data acquisition strategies which are defined as *continuous measurement of spectra* (CMS) and *selected ion monitoring* (SIM). The CMS strategy is applicable to all sample introduction and ionization techniques and all types of mass spectrometers. The

SIM strategy is also applicable to all sample introduction and ionization techniques, but is only useful with mass spectrometers that operate on a continuous beam of ions, that is, magnetic deflection and linear quadruple mass spectrometers. The SIM data acquisition strategy is not meaningful with an ion storage spectrometer (ion trap or Fourier transform) in which ions are stored in batches or with a time-of-flight instrument in which ions are pulsed to the detector in batches. Both CMS and SIM are commonly used with GC/MS and with the other mass spectrometric techniques described in Chapters 5–8. The general analytical strategy, that is, target analyte, broad spectrum, or a combination of the two, significantly influences the selection of the data acquisition strategy. The two general data acquisition strategies for MS are described and their advantages and disadvantages discussed in this section.

Continuous Measurement of Spectra

Figure 2.7 is a diagram of the CMS mode of data acquisition. The curved, peak-shaped line represents the actual concentration of a substance eluting from a separation device, for example, a GC column. The series of diagonal lines below the curve represents repetitive scans of the mass spectrometer as the substance elutes from the column. Each scan begins at the same *m/z* value, usually about 30, and ends at the same highest *m/z* of interest, for example, 450. The quantitation mass shown in Figure 2.7 is described in the section of this chapter entitled "Quantitative Analysis". In the repetitive scans, ions are detected by *m/z*, and the *m/z* values and abundances are stored by the data system as complete mass spectra.

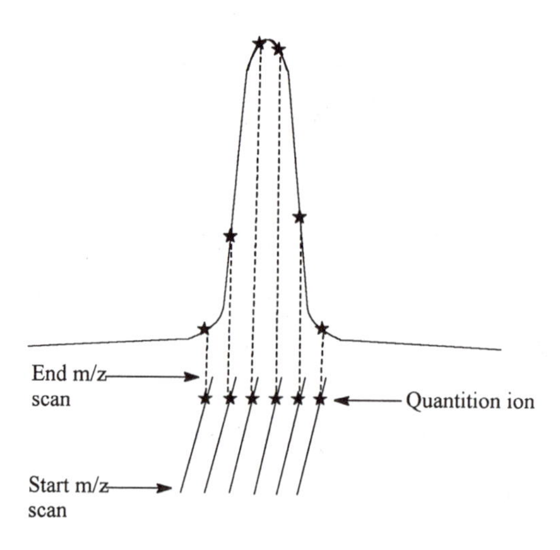

Figure 2.7 Diagram of the continuous measurement of spectra (CMS) mode of data acquisition. (Redrawn from Ref. 64, published in 1983 by the American Chemical Society.)

Repetitive scanning and data storage are usually started shortly after elution of the solvent in GC/MS and continue until all components of interest have eluted from the column. Analogous data acquisition starting points apply to other separation techniques. The total number of complete spectra acquired and stored during a chromatographic analysis with CMS depends on the starting and ending times of mass spectral data acquisition, the mass range scanned, the scan speed, and any delays between scans. A 30-min chromatogram with one spectrum acquired and stored per second will produce a data file of 1800 complete mass spectra. The CMS mode of data acquisition can be described as the continuous acquisition of complete mass spectra as analytes enter the mass spectrometer from a chromatograph or some other sample introduction system.

Two very important and widely used graphical displays can be constructed from CMS mass spectral data:

- *Total ion chromatogram (TIC)*: A TIC is a plot of the sum of the ion abundance measurements in each member of a series of mass spectra as a function of time and spectrum number. The ion abundances in each spectrum of the data file are summed, and the sums, frequently normalized to the largest, are plotted as a function of time and spectrum number. This plot is the mass spectral equivalent of an ordinary chromatogram, which is a plot of the standard detector signal as a function of time. The TIC is sometimes called a *reconstructed chromatogram* because the plot is often displayed after data acquisition is complete, but this term is less informative than TIC, which is the recommended term used in this book. The main difference between a TIC and a chromatogram from a standard detector is that all the individual m/z values and abundances in each spectrum of a TIC are stored in the data file. Any or all of these mass spectra can be retrieved from storage, displayed, and processed using a variety of mathematical algorithms and visual display formats. Figure 2.5 is a TIC of 61 volatile compounds. If a chromatographic separation is not used, this plot is sometimes called a *total ion profile*.
- *Extracted ion chromatogram (EIC)*: An EIC is a plot of the change in ion abundance of one or several ions as a function of time and spectrum number, using abundance data *extracted* from the total ion chromatogram. This plot is also known as a *mass chromatogram*, which is a less informative term that is not used in this book. The EIC may be formed by extracting the abundance of the same ion from each spectrum of the data file and plotting those abundances, usually normalized, as a function of time and spectrum number. Alternatively, abundances from several ions in each spectrum can be summed and plotted.[33] The number of possible EIC plots is enormous and depends on the imagination of the analyst, the capabilities of the software to extract and sum individual abundances, and the richness of the abundance data in the file. If a chromatographic

separation is not used, this plot is sometimes called an *extracted ion profile*.

The continuous measurement of mass spectra as substances eluted from a GC column was first used by Ron Hites and Klaus Biemann in the late 1960s.[34–36] Interestingly, their original purpose in using this data acquisition mode was *not* to take advantage of the visual and data interpretation possibilities of the TIC and EIC plots. In the late 1960s, when computer data processing was in its infancy, the first home-made GC/MS systems relied on an observer who monitored the output of a total ionization detector (TID) and initiated a mass spectrometer scan manually whenever a peak appeared on the TID chart recorder. The TID was usually a flame ionization detector which received part of the split GC effluent. This was a tedious process because the observer never knew exactly when the GC peak apex would occur and would often miss the maximum concentration of the eluting substance. Clearly this technique required the presence of an ever-alert and keen observer to initiate acquisition of a mass spectrum whenever something interesting eluted.

During the 1960s the single-focusing magnetic deflection mass spectrometer was the prevalent design in service. The spectrum was scanned by increasing or decreasing the electric current in the electromagnet, which changed the magnetic field strength and caused ions of different m/z to be sequentially focused on the detector slits. The spectrometer used by Hites and Biemann did not have a temperature-controlled magnet, such as one that circulated cold water through coils in the iron core, and the irregularly timed magnetic field scans gradually heated the magnet, which caused hysteresis and an irreproducible magnetic field at the beginning of many scans. This caused serious problems with m/z calibration and mass assignments. To overcome these two problems Hites and Biemann invented the CMS technique and saved all the mass spectra on magnetic tape. The ever-alert and keen observer was no longer needed to push the data acquisition button at the correct time, the magnetic field stabilized after several minutes at a slightly elevated temperature, and calibration of the m/z scale was greatly simplified. More importantly, they also invented and recognized the value of the TIC and EIC plots.

The TIC and EIC plots have tremendous practical value. The TIC can be examined like any other chromatogram to locate analyte peaks and determine retention times. By retrieving various mass spectra from the data system, target analytes may be identified and other peaks tentatively identified or labeled as unknowns. Coeluting substances may be recognized by careful examination of the spectra measured across a peak. By careful subtraction of one spectrum from another the individual spectra of the coeluting compounds may be obtained in reasonably pure form. Therefore, complete chromatographic resolution of all target analytes is not necessary with CMS. The coeluting pairs in Table 2.1 were located in Figure 2.5 by displaying the EICs for specific ions and examining the mass spectra measured across the TIC peaks. Using these techniques, an analysis time can be shortened by reducing chromatographic resolution, yet analytes can still be identified and measured using their underlying mass spectra. Target analytes may be located under a mountain of background and interferences by plotting EICs of ions specific to the target analytes. Using the EIC, the signal/noise of the target analyte is increased substantially by eliminating most, if not all, the chemical noise. Software has been developed, and implemented on some data systems, to deconvolute TIC peaks containing coeluting substances, display TICs with enhanced resolution, and extract reasonably pure mass spectra from clusters of coeluting compounds and background.[37–41]

The number of spectra acquired for any given chromatography peak depends on several operating parameters and other factors. As illustrated in Figure 2.7, roughly four spectra are taken during elution of the chromatographic peak. This may be an adequate number for some purposes but too few for many purposes. The operating parameters and other factors that must be considered include:

- The mean or typical chromatographic peak widths
- The scan speed of the mass spectrometer
- The mass range that must be scanned
- The probability of coeluting compounds
- The precision and accuracy needed for quantitative analysis
- The detection limits required
- The existence of programmed delays between scans

With a very narrow 1-s wide (at half height) GC peak and an average scan-speed mass spectrometer, only one or two 500 Da mass range spectra might be measurable before the compound is depleted from the spectrometer. The user of this data file would be seriously handicapped in locating coeluting compounds and would experience poor precision and accuracy in quantitative analysis. Changing the chromatographic conditions to give wider peaks would allow more spectra to be measured and produce better quantitative precision, but wider peaks would also increase the probability of coeluting compounds. A shorter mass range scan or a faster scan speed would also provide more spectra per peak and better peak area precision. A very fast time-of-flight mass spectrometer[14] could acquire thousands of mass spectra over a 1-s wide GC peak, and this technique can resolve coeluting compounds, but any individual spectrum would have a relatively low signal/noise. Some data systems have programmed delays between scans (to write data to memory, stabilize fields, etc.) and these dead times can be a significant factor in instrument performance. The analyst must decide what compromises will be made between chromatographic resolution, mass range, scan speed, signal/noise, accuracy, and precision. These issues are discussed later in this chapter under "Quantitative Analysis".

Selected Ion Monitoring

The other major data acquisition strategy is called *selected ion monitoring* (SIM), which is defined as the measurement in real time of ion abundances at only selected m/z values as substances enter the mass spectrometer. With SIM, complete mass spectra are not acquired. The SIM strategy is to integrate the signals of the selected ions over a longer time than is possible with CMS, and therefore increase the signal/noise of the selected analytes. In Figure 2.7, SIM could be represented by a few dots at selected *m/z* in place of the diagonal lines which represent complete spectra.

The SIM data acquisition strategy is only meaningful with mass spectrometers that operate on a continuous beam of ions. With this type of instrument, as any given ion is focused on the electron multiplier detector, all the other ions are discarded and literally wasted. In a magnetic deflection instrument all ions not being detected are discarded on the partially closed slits or elsewhere in the instrument. In a linear quadrupole instrument, all ions not being detected are discharged on the rods. During CMS with these spectrometers, analyte ions are discarded during those parts of the *m/z* scan in which they are not being measured. These losses are greatly reduced with SIM. With ion-storage mass spectrometers, all analyte ions can be stored and later measured and SIM is not meaningful in the same sense. However, selected analyte ions can be stored and their concentrations increased while other ions are discarded and this can provide an increase in signal/noise analogous to that of SIM. With a time-of-flight spectrometer essentially all ions are measured and SIM is not a viable concept.

Selected ion monitoring has been given many names over the years by various instrument developers and manufacturers. Some that appear in the scientific and commercial literature are:

- Accelerating voltage alternation (magnetic deflection)
- Ion specific detection
- Mass fragmentography
- Multiple ion analysis
- Multiple ion detection
- Multiple ion monitoring
- Multiple peak monitoring
- Multiple specific ion detection
- Selective ion detection
- Selected ion peak recording
- Single ion monitoring
- Tuned ion analysis

The only term recommended by the American Society for Mass Spectrometry[42] is *selected ion monitoring*, which does not imply a limit to the number of ions (single or multiple) or their type (molecular or fragment), but does infer a real-time process (monitoring). The corresponding chromatographic display should be called a *selected ion chromatogram* (SIC) to distinguish it from the TIC and the EIC. If a chromatographic separation is not used, this plot should be called a *selected ion profile*.

Not all implementations of SIM on different types of magnetic or quadrupole mass spectrometers give the same degree of signal/noise enhancement. Equipment manufacturers, instrument designers, and software engineers have used a variety of hardware devices and software algorithms to implement SIM. The only way a user can assess the actual enhancement is to make a direct comparison between CMS and SIM. Enhancement factors between 2 and 1000 or more may be observed with any given instrument system.

Clearly SIM is suitable only for the target analyte strategy. Specific ions to be monitored must be designated by the analyst before sample injection. If too few ions are monitored or too little related analytical information is available, the reliability of the identification of the analyte could be compromised. This issue is discussed later in the section of this chapter entitled "Identification Criteria". Mass spectrometer computer-control systems allow multiple changes in real time of the m/z values and total number of ions being monitored. With this capability a number of analytes eluting at different times can be monitored. The advantages of SIM may be summarized:

- High selectivity for specific target analytes and selectivity that can be changed during the chromatographic separation (real time).
- Improved signal/noise through signal accumulation and averaging over longer data acquisition times compared to CMS, but the degree of enhancement is hardware and software dependent.
- Elimination of most chemical noise from coeluting interferences and background.
- Increased accuracy and precision for quantitative analysis compared to CMS because of many more data points per chromatographic peak.
- Possible reduction in sample preparation before analysis because of high selectivity and chemical noise elimination.

Simultaneous Identification and Measurement

If the goal of an environmental or other chemical analysis is target compound determination, then a strategic decision is whether to use CMS or SIM data acquisition. A great strength of MS is simultaneous qualitative and quantitative analysis, but compromises are required with both data acquisition strategies. The CMS approach provides a strong qualitative analytical capability because complete mass spectra are measured and stored in the data system. With CMS, the analyst is not required to select target analytes or the corresponding ions to be monitored, a large number of substances can be determined in a single analysis, and nontarget substances may be identified and measured. However, CMS is not generally a high accuracy and high precision quantitative technique because too few data points may be available for each analyte. The SIM approach provides a strong quantitative analytical capability because the focus is on just the analytes of interest, but the price is the loss of potentially valuable information that confirms or solidifies the identification of the analytes. In some cases SIM may be the only option because of the complexity of the sample matrix, the detection limits required, or the quantitative accuracy and precision needed. The high selectivity and signal/noise afforded by SIM is needed when coeluting interferences or a high background from the sample matrix and low analyte concentrations cause a low signal/noise for the target

Table 2.2 Summary of strategies for analytical and environmental mass spectrometry

General analytical strategy	Data acquisition strategy	Strengths and weaknesses
Target analytes only	CMS or SIM	CMS—strong qualitative and relatively weak quantitative analysis SIM—strong quantitative and relatively weak qualitative analysis
Broad spectrum sample analysis	CMS	Strong qualitive and relatively weak quantitative analysis
Combination of TA and BS	CMS	Strong qualitative and relatively weak quantitative analysis

analytes. The analytical methods used to determine 2,3,7,8-tetrachlorodibenzo-*p*-dioxin and related compounds, which are described in Chapter 4, are good examples of the application of SIM to increase signal/noise and lower detection limits. Table 2.2 contains a summary of strategies for analytical and environmental MS.

Limitations of Mass Spectrometry

Much has been written, including this book, about the power and sensitivity of MS for chemical analysis.[10–19, 21–23, 43–46] However, MS has some important limitations and some of these are reviewed in this section. Some of these limitations are particularly evident in conventional GC/MS, but are partially overcome by some of the techniques discussed in subsequent sections and chapters. Some of these limitations are obvious when measured spectra are compared with database spectra or with spectra measured on other instruments under different or even similar conditions. These limitations may be particularly bothersome when mass spectra and related data are interpreted by inexperienced or untrained analysts or introduced as evidence into legal proceedings.

Volatility and Thermal Stability

For successful GC, analytes must have sufficient vapor pressures at the GC injection port and at column oven temperatures and must be thermally stable. For general purpose use, GC injection port temperatures are typically 200–300 °C and column ovens are often heated to this same range and maintained there for 20–30 min or longer to promote elution of some analytes. Many compounds are thermally stable at these temperatures, but some have low vapor pressures which makes them difficult or impossible to separate by GC. These compounds tend to stay in the stationary phase of the GC column and if they elute they appear as broad and sometimes tailing peaks with long retention times. Polycyclic aromatic hydrocarbons (PAHs), particularly those with five or six fused rings such as benzo[*a*]pyrene and benzo[*g,h,i*]perylene (Chart 2.1), tend to behave this way. Larger PAHs often have too low a vapor pressure even at more elevated temperatures and must be separated by other techniques. These vapor pressure and thermal stability requirements also apply to other thermal vaporization sample-inlet systems.

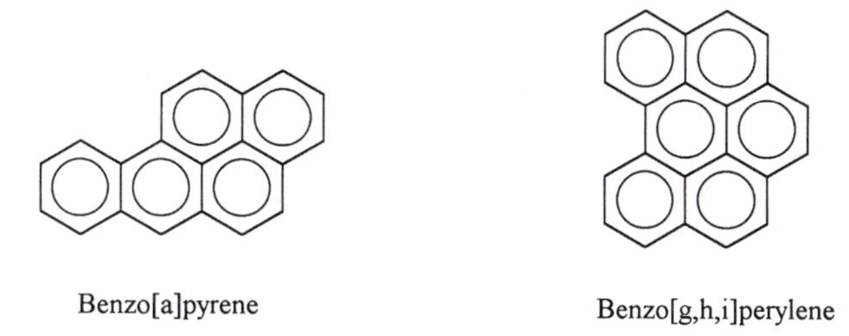

Chart 2.1

For successful conventional GC/MS, compounds must also be thermally stable in the gas phase in the mass spectrometer. Electron ionization filaments operate at temperatures > 1500 °C and radiate heat into the ion source. Ion source enclosures, which are typically constructed of stainless steel, are often maintained at 200–300 °C to prevent condensation of analytes and the accumulation of deposits on ion source walls. Under these conditions, collisions of marginally stable vapor phase molecules with hot metal walls can cause thermal or metal catalyzed decomposition of analytes and other substances in the sample. Carbonaceous deposits and other nonvolatile reaction products gradually accumulate on the ion source walls and these can produce background ions in mass spectra. These background ions are often called artifacts or interferences and they contribute to the variability sometimes observed in mass spectra. Ion source contamination can also cause degraded performance (resolution and sensitivity) by modifying or insulating voltages used for extraction of ions from the ion source.

Thermal decomposition in the GC inlet system and ion source is a major factor which greatly limits the utility of GC/MS. Many types of compounds undergo thermal reactions at typical injection port, column, and ion source temperatures to produce new compounds that are stable. These reaction products are subsequently ionized to give molecular ions at a lower m/z than that of the original compound. For example, alcohols lose the elements of water and produce olefins, chlorides lose HCl, and thiols lose hydrogen sulfide. The identification of these thermal decomposition products can give an incorrect assessment of the nature of the compounds in the original sample. A mass spectrum may be interpreted in terms of an olefin, but the substance in the sample is really an alcohol that has undergone thermal dehydration before ionization. Oxidation or reduction reactions and alkyl group transfers can also occur and produce ions with masses larger than those of the original analytes.

Acids and bases, which may be reasonably thermally stable, often adhere to basic and acidic sites, respectively, in injection port liners and columns and this can promote chemical reactions or decomposition. Important compounds of environmental interest that are susceptible to thermal reactions and decomposition include the phenoxyacetic acid herbicides, the pesticide precursor hexachlorocyclopentadiene, *N*-nitrosamines, many *N*-methyl and other carbamate pesticides, and the chlorinated hydrocarbon pesticide endrin. Various types of innovative on-column and other GC injectors have been developed, in part, to overcome some of these difficulties and these may provide some relief with some compounds. However, most compounds of this type must be analyzed using other separation and mass spectrometric techniques. The determination of compounds not amenable to GC is discussed in detail in Chapter 6.

Samples and sample extracts may contain many thermally unstable compounds or nonvolatile substances which decompose on or adhere to the injection port liner walls. These carbonaceous and other deposits may catalyze the decomposition of marginally stable analytes or adsorb compounds that are normally well behaved. Injection port liners or the first part of a GC column should be suspected any time a normally well-behaved analyte in a calibration mixture either suddenly disappears or the GC/MS system sensitivity to that analyte slowly degrades. Liners should be changed regularly and the first few meters of a GC column should be removed to restore the efficiency of the column.

In spite of these vapor pressure and thermal stability requirements, an impressive variety of biologically active or otherwise interesting compounds are amenable to GC/MS. These compounds are often less polar, or contain nonpolar regions, but do not contain thermally unstable or water solubilizing functional groups. Compounds amenable to GC/MS are often lipophilic, that is, fat soluble, and accumulate in fatty tissue. These are just the compounds that are either volatile air pollutants or are readily extracted from water, particulate matter, sediments, and soils by organic solvents such as methylene chloride. Many examples of these GC/MS analytes are given in Chapters 3 and 4.

Temperature Dependent Fragmentation of Molecular Ions

It has been known for a long time that the relative abundances of molecular and fragment ions from thermally stable compounds may vary widely depending on the temperatures of the mass spectrometer inlet system and ion source.[21] This variability can be confusing when comparing measured EI spectra to spectra of the same compounds published in the scientific literature or in historic databases.

The earliest mass spectra, before the early 1960s, were nearly always measured by introduction of the sample into the mass spectrometer through a very small orifice from a heated glass or metal reservoir. These heated reservoir inlet systems provided a constant pressure of hopefully pure compound in the ion source. The temperature of the reservoir was maintained at 150–300 °C to ensure that the compound was completely vaporized. The net effect of this heating was the excitation of sample molecules to higher vibrational, rotational, and translational energy levels. It is reasonable to expect that electron ionization of molecules in excited vibrational energy states will lead to more rapid fragmentation of the molecular ion than would electron ionization of vibrationally cool molecules. Molecular ion abundances of many thermally stable compounds, such as heptadecane, are very low ($< 5\%$ relative abundance) when measured with a heated reservoir inlet system.[21]

During the 1960s a sample introduction technique was introduced to allow the measurement of mass spectra of compounds with low vapor pressures ($\sim 10^{-7}$ Torr) or were solids at ambient temperature and could not be readily introduced into a heated reservoir inlet system.[47–48] The direct insertion probe (DIP) or solids probe expanded greatly the range of compounds that could be easily introduced into the mass spectrometer. With a DIP, a few microliters of a solution of a compound is placed on the tip of the probe (usually in a short glass capillary sealed at one end). The device is inserted into the vacuum of the instrument, and the solvent is allowed to evaporate. The probe tip is then inserted further and heated until the sample residue evaporates directly into the ionizer of the mass spectrometer. Often no heating is needed because the heat radiated from the filament and the ion source walls is sufficient to vaporize enough of the sample to measure the mass spectrum. However, in contrast to the heated reservoir inlet system, spectra obtained with a direct probe are often of vibrationally cooler molecules and often these spectra show significantly more molecular ion abundances and less fragmentation than do spectra of the same compounds obtained with a heated reservoir inlet system.

Figure 2.8 shows the effect of temperature on the partial EI spectra of the thermally very stable molecule bis(perfluorophenyl)phenylphosphine which has a molecular ion at m/z 442. In the DIP spectrum, the molecular ion $M^{+\cdot}$ is the base peak at 100% relative abundance (RA) and a major fragment ion at m/z 198 is the second most abundant ion at about 84% RA.[49] In the spectrum obtained by GC/MS with a heated injection port, heated GC column, and heated GC/MS interface, the $M^{+\cdot}$ abundance is less than 50% and the fragment ion at m/z 198 is now the base peak.[50] The RAs in mass spectra measured by GC/MS are often similar to spectra obtained with a heated reservoir inlet system and somewhat different from DIP spectra. Cholesterol, which can undergo dehydration either before or after ionization, is often used to evaluate the thermal characteristics of DIP inlet systems.

Elevated temperatures can cause complete fragmentation of some molecular ions, eliminating this valuable information from their mass spectra. This may lead to interpretation difficulties especially when using literature and database spectra that do not include information about the nature of the inlet system used to measure the spectra or the temperatures of the inlet system or ion source. A great many literature and database mass spectra have been measured with a DIP because of the convenience and ease of use of this sample introduction system. However, these reference DIP spectra may not compare well with conventional GC/MS spectra. The various effects on spectra caused by the sample introduction system and ion source temperatures are

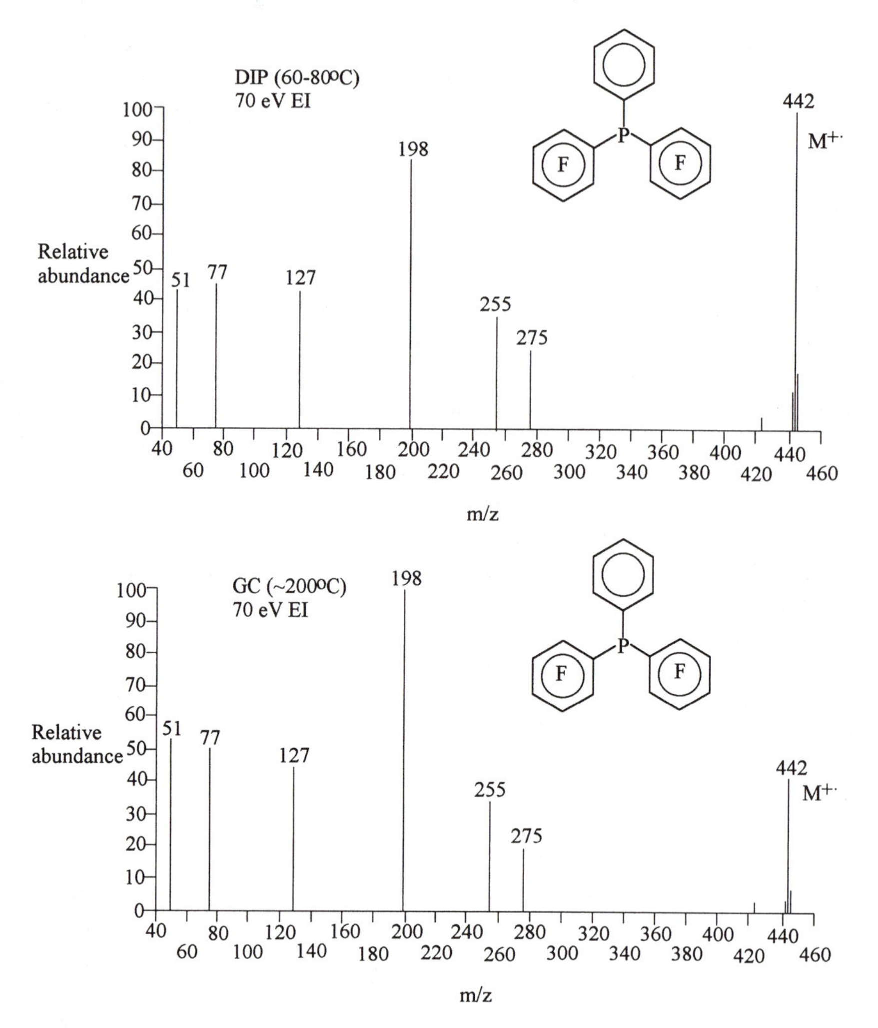

Figure 2.8 Effect of inlet system temperature on the partial EI spectra of the thermally very stable molecule bis(perfluorophenyl)phenylphosphine.

important limitations to conventional GC/MS and the interpretation of mass spectra.

Sample Pressure Dynamics

In chromatographic separations, particularly with high resolution GC, rapid changes in the concentrations (pressures) of compounds occur as the analyte elutes from the column and enters the mass spectrometer. Relative abundances (RAs) of ions can vary widely depending on the rate of change of sample pressure during data acquisition. In Figure 2.7, the third scan from the left is taken as the pressure of the analyte is rapidly increasing. The RAs in this scan will be abnormally low in the first part of the scan and abnormally high in the last part, giving a distorted mass spectrum. Similarly, the second scan from the right is taken on the back side of the peak where the analyte pressure is rapidly decreasing and the spectrum will be distorted with abnormally high RAs for low-mass ions and abnormally low RAs for high-mass ions. The narrower the chromatographic peak and the slower the mass scan the more the distortion. Mass spectrometrists have developed several strategies for dealing with this problem:

- Use only the spectrum at the top of the peak, for example, the third from the right in Figure 2.7. This provides some relief but works best for the broader peaks produced with packed-column GC or reverse-phase liquid chromatography.
- Average spectra across the peak using the three or five closest to the centroid of the peak. This is usually a satisfactory solution and is supported by most data systems.
- Increase the scan speed to acquire multiple spectra while the analyte pressure is not changing as rapidly. This is effective but could reduce signal/noise and increase variability because fewer ions are measured in any one scan.
- Reduce chromatographic resolution to give wider, slower eluting peaks. However, this increases the chances of coelution and may also reduce signal/noise.

Analysts should be aware of this limitation of all forms of chromatography combined with MS and the strategies available to overcome it. The impact of rapidly changing analyte pressures on quantitative analysis is presented later in this chapter.

Mass Spectrometer Design and Tuning

The basic design of the mass spectrometer can have a significant impact on the ions observed and their relative abundances. Before the mid-1970s the vast majority of mass spectra of organic compounds were measured with single- or double-focusing magnetic deflection mass spectrometers. Although there are several different basic designs for these instruments,[10–14] operators can usually increase or decrease RAs somewhat by adjusting m/z resolving power, ion source potentials, and lens focusing voltages at various m/z values. The tuning of these operating parameters affects the extraction of ions from the source and their transmission through the magnetic and, if present, electric fields to the electron multiplier detector. By optimizing these parameters at low, mid-range, or high m/z, the corresponding areas of the spectrum may be enhanced or repressed somewhat. Magnetic fields are also used to focus ionizing electron beams in ion sources, and these fields can variably effect the extraction of ions from the source and their measured abundances. The early MS literature contains much discussion of mass discrimination with different instrument designs and under various operating conditions.

The quadrupole mass spectrometer was introduced as a GC detector by the Finnigan Corporation in the late 1960s.[11–14] This instrument was considerably less costly than traditional magnetic deflection instruments and had some clear advantages as a GC detector, but its performance characteristics were largely unknown. A 1972 multilaboratory study demonstrated that the quadrupole design had variably different ion transmission characteristics and more tuning flexibility than the traditional magnetic deflection instruments.[51] The source of the tuning flexibility was various ion source potentials and the quadrupole rod radio frequency/direct current (RF/DC) voltage ratios that were adjusted by the operator during the instrument tuning process. In addition, the Finnigan Corporation pioneered the use of early minicomputers for data acquisition and control of quadrupole mass spectrometer systems. The interfacing hardware and software algorithms used in these data systems were also a factor in the instrument tuning process. Users of some quadrupole GC/MS systems literally had tuning flexibility to produce just about any RA at virtually any mass. The implications for the reproducibility and comparability of quadrupole mass spectra were very serious:

- Chromatographers and others inexperienced in MS were pressed into service as GC/MS system operators during the 1970s. Expecting sensitivity similar to that of an electron-capture GC detector, these operators tended to use insufficient m/z resolving power to gain sensitivity and paid little or no attention to traditional spectral quality criteria such as natural isotope abundance ratios. Others tended to use too much resolving power and sacrifice ion abundance unnecessarily. The more complete the separation of ions of adjacent m/z values (mass resolution), the fewer the ions that are available for detection because ions must be discarded to increase m/z resolution. This tuning flexibility led to production of spectra that often had considerably different RAs from those of magnetic deflection instrument spectra in databases and the scientific literature.
- With quadrupole instruments, transmission or detection of higher mass ions, that is, those greater than about 200 Da, was sometimes inadequate for environmental and other analyses. Many important analytes are polychlorinated and/or polybrominated and have molecular weights in the 200–900 Da range. Higher mass transmission was poor on some early quadrupole systems or low mass detection was so superior that higher mass detection seemed poor. This led to production of spectra that were missing some ions or had very low abundances of higher mass ions.
- Quadrupoles have active separating elements (rods) that were directly contaminated during the separation process by the discharge of filtered ions. This contamination affected performance, particularly mass resolution and the transmission of high m/z ions, and

again led to the production of spectra with RAs different from those of spectra from magnetic deflection instruments in databases and the scientific literature.

Generally, 70-eV electron ionization (EI) fragmentation patterns from quadrupole GC/MS systems were somewhat or significantly different from those produced by the established magnetic deflection mass spectrometer systems. This threatened the usability of the existing database of EI spectra and the information in the MS literature. It also cast doubt on the credibility and value of the quadrupole mass spectrometer. Much discussion of this issue occurred and many speakers at scientific meetings in the 1970s included qualifying phrases like *quadrupole spectra* and *magnetic spectra*. During the 1980s instrument manufacturers gradually introduced automated tuning programs to adjust instrument operating parameters for optimum performance. These software-controlled programs provided a major improvement in consistency compared to the highly variable individual operator's manual adjustments. However, the automated systems were based on proprietary and generally unknown criteria and sometimes accentuated the differences in spectra produced by different models of quadrupole instruments.

With the introduction of the ion-trap mass spectrometer as a GC detector by the Finnigan Corporation in 1985, another performance issue surfaced. Space charging and self-chemical ionization in the ion trap affected the ions observed and their RAs in 70-eV EI mass spectra.[18, 52–55] Many investigators were concerned with the differences observed in quadrupole and ion trap spectra. Soon the phrase *ion trap spectra* came into use. As other types of mass spectrometers, such as time-of-flight or Fourier transform instruments, are more widely used in environmental and other analyses, other differences in spectra may appear. In response to these design and tuning issues, initiatives were made to standardize conventional GC/MS spectra within defined limits, and these efforts are described later in this chapter and in the analytical methods descriptions in Chapters 3 and 4.

Structural Isomers and Very Stable or Highly Fragmented Molecular Ions

A well-known limitation of MS is that many structural isomers give exactly, or almost exactly, the same mass spectra. Figures 2.9–2.11 show the structures and EI mass spectra of three volatile dichloroethene isomers.[50] The three spectra are, for all practical purposes, identical and these isomers cannot be distinguished by their mass spectra alone. If, however, the three isomers are separated chromatographically, the combination of mass spectra and GC retention times provides the correct identifications. The three dichloroethenes shown are readily separated by 1 min or more on several GC columns and are routinely identified and measured in environmental and other samples.[7, 29]

Cycloheptatriene and toluene (Chart 2.2) are structural isomers that give essentially identical EI mass spectra.[50] Upon electron ionization each $M^{+\cdot}$ ion

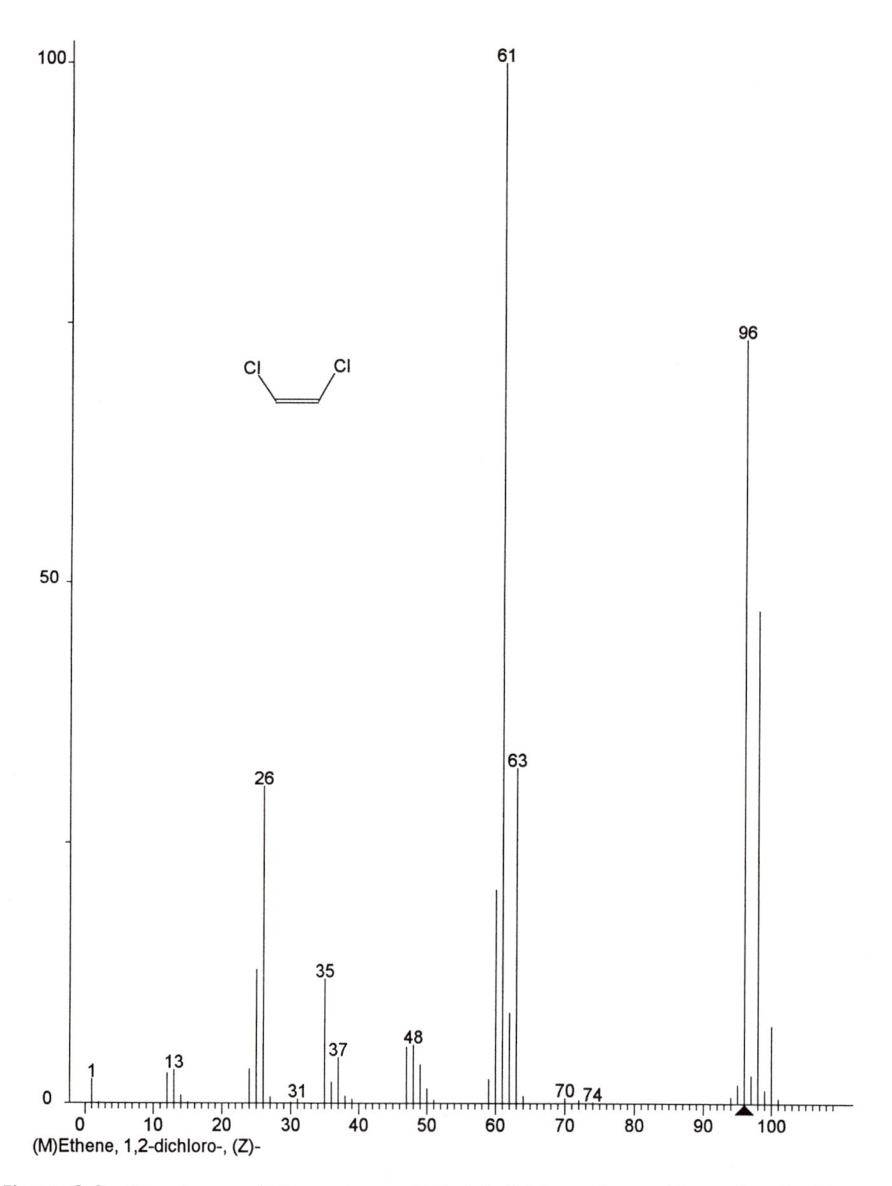

Figure 2.9 Structure and EI spectrum of *cis*-1,2-dichloroethene. (Reproduced with permission from Ref. 50. Copyright 1992 U.S. Department of Commerce on behalf of the United States.)

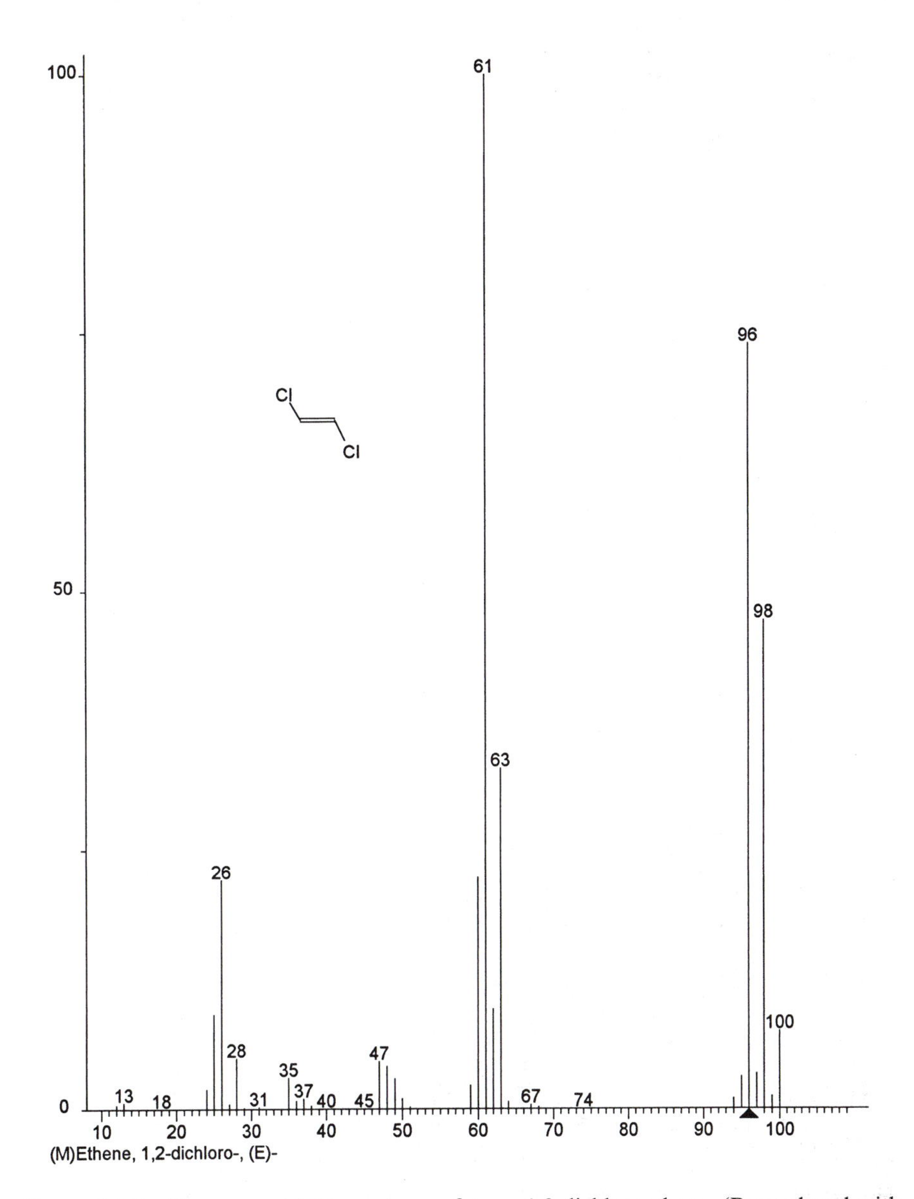

Figure 2.10 Structure and EI spectrum of *trans*-1,2-dichloroethene. (Reproduced with permission from Ref. 50. Copyright 1992 U.S. Department of Commerce on behalf of the United States.)

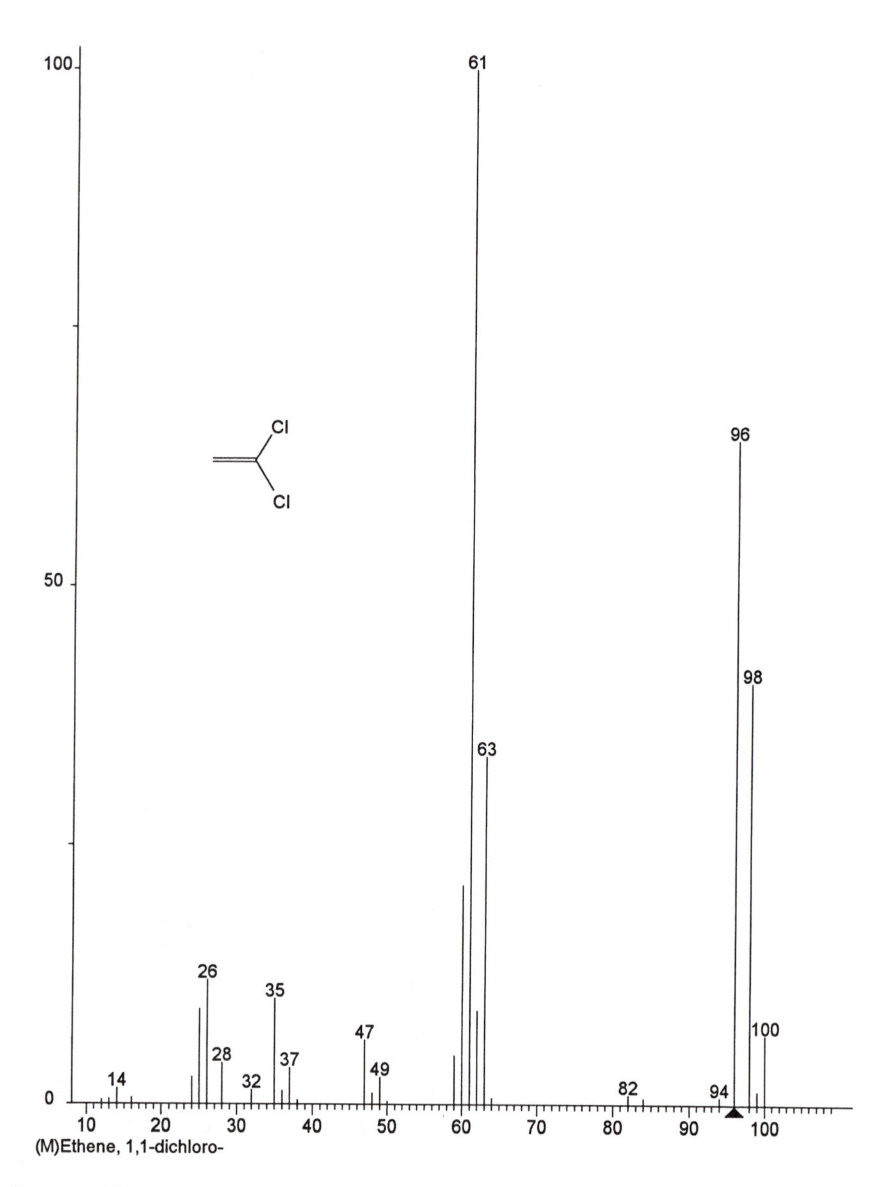

Figure 2.11 Structure and EI spectrum of 1,1-dichloroethene. (Reproduced with permission from Ref. 50. Copyright 1992 U.S. Department of Commerce on behalf of the United States.)

loses a hydrogen atom and forms the symmetrical tropylium ion which is the base peak in each spectrum. This common intermediate undergoes fragmentation to produce the same lower *m/z* ions in both spectra. The three xylenes (1,2-, 1,3-, and 1,4-dimethylbenzene—lower part of Chart 2.2) also give essentially identical EI mass spectra.[50] After initial electron ionization, each $M^{+\cdot}$ ion rearranges to a methylcycloheptatriene ion which loses a methyl radical to give the same base peak tropylium ion at *m/z* 91. Fragmentation of *m/z* 91 gives the same lower *m/z* fragment ions in the three mass spectra. Although *o*-xylene is readily separated by GC, the *meta* and *para* isomers invariably coelute and cannot be distinguished by GC/MS.

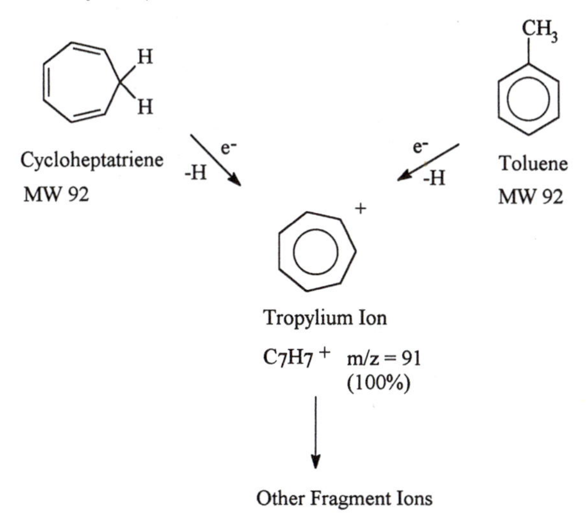

CH3
CH3
e-
CH3
e-
CH3
CH3
e-
CH3
MW 106
H H
+
CH3
m/z = 106
- CH3
+
Tropylium Ion
C7H7 + m/z = 91
(100%)

Chart 2.2

Databases of mass spectra are replete with other examples of structural isomers that cannot be distinguished by MS.[50] Often identical mass spectra of structural isomers is one reason that identifications on the basis of mass spectra alone can be incorrect. This limitation is a significant factor in identification criteria, which are discussed in a later section of this chapter. By contrast, some structural isomers give sufficiently different mass spectra that allow them to be distinguished. The isomeric phenyl-*n*-pentanes in Chart 2.3 can be distinguished by the significantly different and generally reproducible RAs of several fragment ions in their EI spectra.[50] Sometimes a stereochemical feature facilitates a rearrangement or fragmentation in one isomer and not the other and this gives different EI spectra of isomers.

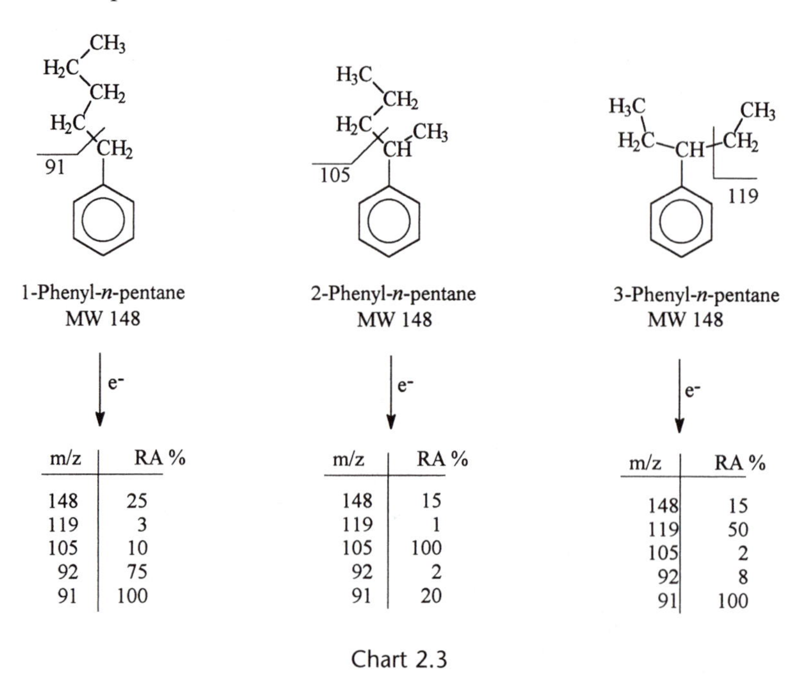

m/z	RA %
148	25
119	3
105	10
92	75
91	100

m/z	RA %
148	15
119	1
105	100
92	2
91	20

m/z	RA %
148	15
119	50
105	2
92	8
91	100

Chart 2.3

Another important limitation of MS is that some $M^{+\cdot}$ ions are too stable, and give few or no fragment ions, and others are very unstable and give a large number of different fragment ions of low *m*/*z*. Much of the power of MS for compound identification and structure determination comes from the fragmentation of some, but not all, of the $M^{+\cdot}$ ions. Primary fragment ions usually decompose to give smaller ions which are also very important. The identification of the compositions of the fragment ions and the compositions of the substructures lost during various fragmentations provides valuable clues to the composition and structure of the molecule itself. This process is analogous to the classical method of structure determination by chemical degradation to known smaller substances and the eventual piecing together of the full structure of the original molecule. A difference is that classical chemical degradation often required many years of research, but a mass spectrum can be measured and displayed in a few seconds. The process of interpreting mass spectra can be compared to the work of an archeologist who finds pieces of pottery, broken masonry, and other remains and constructs a model of an ancient village. Another analogy is the assembly of a jigsaw puzzle with a reasonable number of pieces with different characteristics. In MS, the more diagnostically useful ions, that is, ions that clearly suggest a specific composition or structure, the better, but too many ions, especially many small pieces, adds confusion and complications. Little or no fragmentation of the $M^{+\cdot}$ ion is equally undesirable.

Polycyclic aromatic hydrocarbons and many heterocyclic aromatic compounds containing N, O, or S atoms are examples of compounds that typically give very stable $M^{+\cdot}$ ions and few fragment ions.[50] Their EI spectra are, therefore, not very helpful in determining their structures, and various isomers cannot be distinguished. By contrast, other compounds appear to simply go to pieces on EI and produce a great variety of ions at mostly lower m/z values. The highly chlorinated pesticide endosulfan is a good example of a compound whose EI spectrum is difficult to interpret because of excessive fragmentation of the $M^{+\cdot}$ ion and the production of many small and medium size ions.[50,56]

Except for the inability to distinguish isomers, the EI spectra of the dichloroethene isomers in Figures 2.9–2.11 are examples of nearly ideal mass spectra. The $M^{+\cdot}$ ions, which are the two-chlorine natural abundance isotope patterns at m/z 96, 98, and 100, are present in all three spectra. Major fragment ions are also present which can be readily interpreted as caused by consecutive losses of Cl atoms from the $M^{+\cdot}$ and the $(M–Cl)^+$ ions. The m/z 26 ion, which has no chlorines, is likely a $C_2H_2^+$ and, therefore, the molecule is a dichloroethene. Finally, all the spectra are simple and not cluttered with a large number of ions.

The identification of analytes can also be made without interpreting the fragmentation process by comparing the measured spectra with spectra in a database. As long as there are sufficient ions in the spectra to compare, there can be reasonable confidence in the result. Database spectrum matching is discussed in the section of this chapter entitled “Identification Criteria”.

Standardization of Conventional GC/MS

Some of the limitations of MS described in the previous section, for example, the general weakness in differentiating structural isomers with conventional GC/MS, are beyond control. Other mass spectrometric or other analytical techniques are required to overcome these limitations. However, because of the great importance of conventional GC/MS in environmental analysis, the issue of spectrum variability due to mass spectrometer design and tuning has been studied carefully and considerable efforts made to minimize the effects of this limitation. This section is concerned with strategies for the minimization of design and tuning variability and the standardization of EI spectra from conventional GC/MS. With standardization, spectra of the same compounds measured with different instruments under similar conditions should at least be similar.

A multilaboratory study was conducted in 1972 to determine the variability of mass spectra obtained with contemporary magnetic deflection and quadrupole GC/MS systems.[51] A test compound, bis(perfluorophenyl)phenylphosphine, was introduced into four magnetic deflection instruments and 11 Finnigan model 1015 quadrupole instruments through their packed-column GC inlet systems. This test compound, which is also known as decafluorotriphenylphosphine (DFTPP), was selected because it had a number of favorable properties including an appropriate molecular weight and a useful, but not too complicated, EI spectrum. In addition to an abundant molecular ion at 442 Da, the spectrum contained about 15 major fragment ions fairly evenly distributed over the mass range 50–442 Da. The compound was available as a pure crystalline solid to facilitate the preparation of standard solutions, it was thermally stable, soluble in many common organic solvents, and easy to chromatograph. It was known to oxidize slowly on contact with air to form the phosphine oxide, but this was considered a positive attribute. The detection of the phosphine oxide in the chromatogram would indicate that the standard solution had been exposed to air, solvent evaporation may have occurred, and a fresh standard solution was needed.[57] The laboratories participating in the study were asked to adjust or tune their instruments according to whatever procedures were commonly used in their laboratories and to introduce the test compound through the heated GC inlet system.

The results of the study clearly indicated the need for standardization of the tuning of these GC/MS systems and for a systematic program of spectral quality control. Differences in relative abundances (RAs) in spectra from all systems were significant, and even among the quadrupole systems of the same design, differences in RAs were substantial. The variability in the spectra reported by the laboratories was intolerable for instrument users, especially inexperienced analysts who would eventually need to depend heavily on standard databases of reference spectra for compound identification. Without better controls and some standardization of spectra there was little hope that GC/MS could be employed for large-scale environmental monitoring, especially for regulatory purposes where mass spectra could be introduced as evidence in legal proceedings. Because the 11 quadrupoles were the same model from the same manufacturer, it was clear that instrument design was not the only factor causing the differences in spectra. The availability of tuning flexibility and the absence of consistent tuning guidelines were very significant causes of the variability.

There was some precedent for standardization of mass spectra. Magnetic sector mass spectrometers were used for many years, and still are, for quantitative analyses of petroleum hydrocarbon mixtures introduced with heated reservoir inlet systems. Standard methods for these analyses published by the American Society for Testing and Materials (ASTM) describe some procedures that appear to be designed to address this issue.[58] ASTM standards D2424, D2786, and D2498 called for either maximizing the abundance of the molecular ion at m/z 226 in the spectrum of hexadecane or adjusting the ratios of the abundances of several fragment ions and the molecular ion. These tuning adjustments were made using the ion-source voltage controls and other operational variables of the instrument. These standards were developed by analytical professionals who understood that some standardization would be beneficial in comparing multilaboratory data. However, hexadecane and other aliphatic hydrocarbons are unsuitable as reference compounds for tuning and standardization with GC/MS.[51]

Decafluorotriphenylphosphine (DFTPP) was proposed as a GC/MS system performance test or reference compound to implement 70-eV EI GC/MS spectra standardization.[51] The proposal was that, after all calibrations and tuning adjustments were completed, the user would introduce DFTPP into the GC/MS system via the same GC inlet that would be used for sample analysis, and verify that the instrument was performing according to proposed specifications (Table 2.3). The specifications were based on a set of key DFTPP ions and corresponding RA ranges and included checks on resolving power, relative abundances, natural abundance isotope ratios, and the threshold for data acquisition. The RA criteria were developed from a statistical analysis of the data obtained in the 1972 multilaboratory study. The compound DFTPP was never intended as an actual calibration compound although it has frequently been mistakenly referred to as such. These proposals were made with several objectives in mind:

- The regular overall performance check would provide the needed GC/MS system quality control (QC), a concept not popular at the time but one that would later have its day.
- The RA criteria would provide guidance to instrument users who would tune their equipment to give RAs in the recommended ranges. This would in turn provide some minimum standardization for 70-eV EI spectra from GC/MS systems and provide more confidence in identifications by comparisons of measured spectra with historical database spectra.

Table 2.3 Original DFTPP key ions and relative abundance criteria

m/z	Relative abundance criteria	Purpose of checkpoint
51	30–60% of *m/z* 198	Low *m/z* relative abundance
68	Less than 2% of *m/z* 69	Low *m/z* resolving power
70	Less than 2% of *m/z* 69	Low *m/z* resolving power
127	40–60% of *m/z* 198	Low–mid *m/z* relative abundance
197	Less than 1% of *m/z* 198	Mid *m/z* resolving power
198	Base peak, 100% relative abundance	Mid *m/z* resolving power and relative abundance
199	5–9% of *m/z* 198	Mid *m/z* resolving power and isotope ratio
275	10–30% of *m/z* 198	Mid–high *m/z* relative abundance
365	1% of *m/z* 198	Baseline threshold adjustment
441	Less than *m/z* 443	High *m/z* resolving power
442	Greater than 40% of *m/z* 198	High *m/z* resolving power and relative abundance
443	17–23% of *m/z* 442	High *m/z* resolving power and isotope ratio

Source: Data from Ref. 51.

- The RA criteria would serve to guide instrument designers and manufacturers and encourage the production of commercial GC/MS systems that gave at least similar EI spectra.

After publication in 1975, these standardization and quality control proposals were evaluated in a few laboratories, but they were generally ignored until they were incorporated into United States government regulatory proposals in 1979.[59] In 1984 a second reference test compound, 4-bromofluorobenzene (BFB), was introduced for use in analyses where it was chromatographically inconvenient to use DFTPP.[60] Ion abundance criteria for BFB and more information on the government regulatory use of mass spectral standardization and QC are given in Chapter 3. One major difficulty with the DFTPP proposal was that during the late 1970s and early 1980s only about half of the quadrupole GC/MS systems produced by the Finnigan Corporation could achieve the DFTPP specification of 40% RA for the molecular ion at *m/z* 442 (Table 2.3).[61] Since Finnigan was the leading quadrupole GC/MS supplier at the time, this was a serious problem.

The conversion dynode electron multiplier was an important technical development for detection of negative ions, but this device also greatly improved the detector response to higher *m/z* positive ions.[62] This development and other design changes in the 1980s increased the efficiency of extraction, transmission, and detection of higher mass ions in quadrupole GC/MS systems. However, some manufacturers of quadrupole GC/MS instruments introduced in the 1980s became a little too enthusiastic for higher *m/z* ion abundances. These instruments produced DFTPP spectra consisting of a base peak at *m/z* 442 and few other fragments above 10% RA. There is evidence that the RAs of higher *m/z* ions were enhanced in one instrument by an ion-source extraction system biased against low-mass ions.[63] This extreme opposite of previous experience reinforced the idea that standardization of tuning and spectra was not only a good idea, but was also essential if the quadrupole GC/MS system was to have any major impact on environmental analyses.

In recognition of the improved designs, several adjustments to the DFTPP ion abundance criteria were proposed and some of these and other unpublished proposals were incorporated into several government regulatory programs as described in Chapter 4.[57] Although some of these regulatory programs adopted somewhat different ion abundance and performance criteria, these mostly minor deviations did not negate the basic objectives of 70-eV EI GC/MS mass spectral standardization and QC. Currently, many commercial GC/MS data systems incorporate software to check test spectra against performance criteria and produce performance reports.

Managers of environmental and other analytical programs that use conventional GC/MS should adopt a mass spectral standardization and QC technique. One option is the techniques and criteria developed for regulatory environmental analyses (this section and Chapters 3 and 4). Alternatively, criteria should be developed using some appropriate compound. Managers should not rely on the generally undocumented criteria used by GC/MS manufacturers in automated tuning and calibration software. This decision could impact the value of the analytical results especially when spectra are compared quantitatively during database searches, used for quantitative analyses, interpreted by inexperienced or untrained analysts, or introduced into legal proceedings. Some analysts may argue that standardization is unnecessary when selected ion monitoring (SIM) is used for data acquisition. However, even with SIM, performance measures such as resolving power, relative abundances at specific m/z values, data acquisition threshold adjustment, and accurate natural abundance isotope ratios are important. Verification of, for example, appropriate tuning, m/z, and abundance scale calibration is very useful and provides documentation of system performance with SIM.

Identification Criteria

Identifications of compounds in environmental and other samples are widely reported in scientific journals, government documents, and private laboratory reports. These identifications are based on a great variety of analytical data and other evidence which ranges from weak to very strong. Chemists, peer reviewers, journal editors, regulatory authorities, and others must deal with the issue of how much and what kind of evidence is needed to provide a high degree of confidence, for example, 99% or better, in an identification. The purpose of this section is to describe strategies and criteria for the identification of target and nontarget compounds in environmental and other samples.

The strategies and criteria in this section are applicable to target compounds amenable to separation by GC (Chapters 3 and 4) and to target compounds not amenable to GC (Chapter 6). In addition, a strategy for the identification of nontarget or unknown compounds in broad spectrum analyses is described. Nontarget or unknown compounds could range from well-known compounds to new compositions of matter whose structures are not reported in the chemical literature. This strategy depends on the measurement of conventional 70-eV EI mass spectra, which may not be possible with some analytes not amenable to GC. If it is not possible to obtain a conventional EI spectrum, alternative approaches must be used including combinations of the techniques described in Chapters 5 and 6.

In a civil court trial, cases are decided on the basis of the preponderance of evidence, that is, 51% of the evidence could be sufficient for a judgment. In a criminal trial, certainty beyond reasonable doubt is required to convict an accused and this might require a great deal of evidence that supports the decision. In the identification of elements, compounds, ions, or other chemical substances there are no generally accepted rules or universal guidelines for how much and what kind of evidence is needed to make unequivocal identifications. Within each community of scientists or analytical chemists some standard practices or recommendations may exist, but most identifications rely heavily on the judgment and technical expertise of individual analysts and the concurrence of peer reviewers, journal editors, and scientific peers. Sometimes the editors of scientific journals, in conjunction with editorial advisory boards or peer reviewers, may set general guidelines for publication of identifications in the journal, but often the author's judgment, particularly a respected author, is not questioned. If an identification is sufficiently important in science, commerce, or the law, a convincing amount of evidence, one way or another, will eventually be produced. Some regulatory agencies have developed guidelines and identification criteria and these are included in the analytical methods described in Chapters 3 and 4. In analyses with MS, identification criteria can and should be formulated as a function of the analytical and data acquisition strategies.

Target Analytes with CMS Data Acquisition

In the target analyte strategy, authentic samples or standards of the target analytes are available in the laboratory, and the performance of the analytical methods with regard to these target analytes is either known in detail or can be readily determined.[64] Accordingly, the sample media, sampling strategy, sampling technique, sample preservation, sample preparation, chromatography, and MS can be optimized with respect to the target analytes. The precise mean retention times of the target analytes can be measured on the same chromatography column that will be used for the sample analyses. Complete mass spectra can be measured under exactly the same conditions that will be used for the sample analyses. Mean method recoveries can be determined using the established selective analytical method. The variabilities of these measurements can also be determined and confidence levels calculated. This information allows the analyst to predict, with, for example, 99% confidence, that a particular compound will elute from a column within a retention time window, for example, between 15.2 and 16 min. Finally, some information about the nature and history of the sample should be available and this can sometimes add a measure of credibility to the identification. Given this background information, it is not difficult to establish identification criteria for the target analytes. Meeting all these criteria provides reasonable and convincing evidence for the presence of the target compound in the sample. For example, target analyte identification criteria can be stated as:

- The specific sample preparation procedure used is known to separate the target analyte from the sample matrix and concentrate it in the extract which is analyzed. If the target analyte is present in the sample matrix, it is practically certain, except for some unusual and unknown matrix effect, that it is concentrated in the specific solution that is injected into the chromatograph.
- The chromatographic retention time is within a reasonably narrow and precisely defined range (window) which is known to be the retention time range of the target analyte with 95–99% statistical confidence.
- All of the ions in the reference mass spectrum, that is the average spectrum measured in the calibration process, are contained in the

background-corrected spectrum from the sample. Some allowance is made for missing ions of very low abundance, perhaps less than 1%, because of some variability in the data acquisition threshold or a concentration near the detection limit, but all other ions are present with RAs in reasonable agreement with the RAs measured in the calibration process. Reasonable agreement could be 20% or more given the variabilities of, for example, dynamic sample pressure and quantities near the detection limit. Some other ions may be present in the sample spectrum, but these should be accounted for by coeluting substances, variable column bleed, or variable general instrument background that was not eliminated by background subtraction. Reference and sample spectra which have only one or a few ions for comparison provide weaker evidence for an identification than spectra with a reasonable number of ions.

- Due consideration is made for the potential for coeluting structural isomers which may have identical mass spectra, for example, *m*-xylene and *p*-xylene.
- The location of the sampling station, knowledge of the general environment in that area, the upstream or upwind conditions, or the nature of the sample are compatible with the identified target compound. It would be reasonable to find benzene in the air downwind from a petroleum refinery, chloroform in a chlorinated drinking water supply, or atrazine in a stream near an agricultural area. It would be unreasonable to identify benzoyl chloride or trimethyl borate in an aqueous sample because they rapidly react with water to form other compounds.

Most contemporary MS data systems offer automated search and identification software for target compounds. Most, if not all, of these programs operate using some variation of the reverse search principle.[65] In a reverse search, the software compares all the ions in a reference mass spectrum with the background-corrected spectra from the sample data file to determine whether all the ions in the reference spectrum are contained in the measured spectra. The search of the sample data file is usually limited by considering only sample spectra measured during the retention time window (RTW) of the target analyte. These identifications must not be accepted on blind faith. Analysts must visually compare the reference spectrum to the software identified spectrum from the sample to ensure that the mass spectral criteria have been met. Some software may not implement the full spectrum reverse search and may use only one or a few ions to make the identification. This can lead to erroneous identifications.

Target Analytes with SIM Data Acquisition

If SIM is the method of data acquisition, all the criteria described in the previous section are applicable except that full mass spectra will not be available for comparison. The key question with SIM is: how many ions must be monitored to ensure the correct target compound identification? Monitoring a single ion for each target analyte is occasionally adequate when the sample is very well characterized and consistent in composition. With most complex samples, monitoring a single ion for each analyte at unit *m/z* resolution is insufficient evidence for a specific analyte. Even with selective sample preparation procedures, multiple substances can elute in the same RTW (see the section of this chapter entitled "Chromatographic Retention Data"), and several of these substances could have the same monitored ion in their mass spectra. Some additional information is almost always required with most samples. A number of approaches have been suggested and used and several of these are described:

- A strategy that has been used in some environmental studies is monitoring two ions for each analyte and selecting ions which have some predictable abundance relationship, such as ions with different amounts of naturally occurring isotopes. Many compounds of environmental interest contain chlorine and/or bromine atoms and their naturally occurring isotopes provide highly characteristic ion groups with predictable abundance ratios (see Table 2.4). This approach provides four measured quantities, including the retention time, plus the selective sample preparation and sample history on which to base a reasonable identification of a target analyte.
- A three-ion criterion, plus retention time, has been proposed and widely used in the pharmaceutical industry regulatory area.[66] This criterion is based on an analysis of the probability that the same three *m/z* values will be present in the spectra of multiple compounds. In the 62,235 spectra 1992 NIST/EPA/NIH mass spectral database,[50] 19,097 spectra have the very common *m/z* 91 ion (5–100% RA) and 8451 spectra have the common *m/z* 149 ion (5–100% RA). However, only 295 spectra, or less than 0.5% of the total, have a *m/z* 91, a *m/z* 149, and a randomly selected *m/z* 238 ion present in their spectra in the 5–100% relative abundance range. With less common ions there is a substantial reduction in the number of spectra with the same three ions.
- Other mass spectrometric techniques are increasingly being used to provide more selectivity with SIM while monitoring less than three ions. Three of these, which are briefly mentioned here, are described in more detail in Chapter 5.
  - — Selective ionization techniques such as positive-ion chemical ionization, negative-ion chemical ionization, and electron-capture ionization can greatly reduce the number of ions formed from background and coeluting interferences and can themselves be somewhat optimized for the target analytes.
  - — High-resolution MS is used to measure the abundance of only those monitored ions that fall in a narrow exact mass range, for example, *m/z* 200.0693–200.0713.[67] This technique can substantially increase the confidence in the identification since only one

Table 2.4 Abundance ratios for bromine and/or chlorine containing ions at masses X, X + 2, etc., where X is the mass that contains only ^{79}Br and/or ^{35}Cl[a, b]

Atoms	X	X + 2	X + 4	X + 6	X + 8	X + 10	X + 12
Br	1.02	1					
Br_2	1.05	2.05	1				
Br_3	1.07	3.13	3.07	1			
Br_4	1.10	4.28	6.28	4.09	1		
Cl	3.07	1					
Cl_2	1.53	1	0.16				
Cl_3	3.13	3.06	1	0.1			
Cl_4	1	1.31	0.64	0.14	0.01		
Cl_5	1	1.63	1.06	0.35	0.06	0.037	
Cl_6	1.44	2.82	2.30	1	0.24	0.03	0.002
Cl_7	1	2.28	2.23	1.22	0.40	0.08	0.01
Cl_8	1.26	3.29	3.76	2.45	1	0.26	0.04
ClBr	3.14	4.09	1				
$ClBr_2$	1	2.28	1.59	0.31			
$ClBr_3$	1	3.26	3.82	1.87	0.30		
Cl_2Br	1.34	2.19	1	0.14			
Cl_2Br_2	1.20	3.14	2.81	1	0.12		
Cl_2Br_3	1	3.58	4.88	3.11	0.91	0.10	
Cl_3Br	2.88	5.64	3.68	1	0.10		
Cl_3Br_2	1	2.93	3.19	1.59	0.37	0.03	
Cl_3Br_3	0.52	2.02	3.13	2.44	1	0.21	0.02
Cl_4Br	1.31	2.99	2.51	1	0.19	0.01	
Cl_4Br_2	1	3.26	4.14	2.63	0.89	0.15	0.01
Cl_4Br_3	0.29	1.22	2.11	1.93	1	0.30	0.05

[a]Ratios are normalized to different ions in each group to produce small numbers that are readily compared to relative abundance ratios measured in groups of ions in mass spectra.
[b]Raw data used to compute ratios is from Beynon, J. H. *Mass Spectrometry and Its Applications to Organic Chemistry*, Elsevier: New York, 1960, p. 298.

or a few reasonable ion compositions fall into this narrow a mass range.

— Tandem MS can be used with collision induced dissociation to fragment a monitored ion and generate the equivalent of a complete mass spectrum with all the advantages of SIM data acquisition.

It is very difficult to rank these strategies and techniques, or some others, in any order of power or usefulness. While some of these techniques may be more selective than others, their selectivity may also depend on the nature of the sample, the sample preparation, and the nature of the analyte itself. Some measurements can be made with more precision and accuracy, which can add more weight to the value of the measurement. The number of potential interferences and the detection limits required are major factors in considering how much evidence is needed. Interferences invariably increase as detection limits are lowered. Clearly there are a large number of combinations of evidence which can be used to provide high-confidence identifications. Given a specific situation, one technique may provide more convincing evidence than others. Multiple pieces of evidence of the kind described in this section are usually required to firmly establish the identity of an analyte.

Assuming it is reasonable for the analyte to be in the sample, and a selective sample preparation procedure is used, the measurement of at least four chemical or physical properties that are related to molecular structure usually gives sufficient evidence to identify the target analyte. The properties measured could be, for example, the *m*/*z* values of the $M^{+\cdot}$ or fragment ions, a chromatographic retention time, the ratio of the abundances of two ions caused by naturally occurring isotopes of the same element, or an infrared absorption frequency. As more properties of the analyte are measured, more confidence can be given to the identification. There may also be situations where fewer than four measurements are available, but the identification can be considered reasonable and acceptable. Ultimately, the judgment of the analytical community will decide the validity of an identification if it is an important question. The identification criteria used for 2,3,7,8-tetrachlorodibenzo-*p*-dioxin and related compounds measured by SIM are described in Chapter 4.

Nontarget Compounds with CMS Data Acquisition

Nontarget compounds will often appear in total ion chromatograms among the target analytes, especially if the sample was taken from a source subject to contamination and the sample preparation was oriented to a broad spectrum analysis. These nontarget substances can range from well-known chemicals which are easily identified to completely new structures of matter that have not been reported in the chemical literature. Identifications of these substances can sometimes be accomplished by a simple extension of the target analyte approach. At the other extreme, identifications can require an array of mass spectrometric and other analytical techniques and full structure determinations. Final identifications will usually not be complete until authentic samples of the substances have been obtained, perhaps by a total synthesis, and tested under the analytical conditions used during their discovery.

The identification of nontarget compounds can be, and often is, far more complex and challenging than target analyte determinations. This simple fact explains why the target analyte strategy is used for most chemical analyses and why new environmental pollutant discoveries are unusual and infrequently reported. A full treatment of the uses of mass spectrometric and other spectroscopic techniques for the elucidation of the structure of an unknown compound is beyond the scope of this book. Some guidelines and suggestions are given in this section and some very helpful techniques are described in Chapter 5. The basic strategy of unknown identification is to narrow the possibilities to a few so that authentic samples of specific substances can be obtained and tested using the identification criteria described for target analytes. A practical approach to tentative identifications of unknowns is given in subsequent sections.

Computer Searching of Databases of Spectra. At present, this strategy is limited to nominal 70-eV EI spectra obtained with conventional GC/MS or particle beam liquid chromatography/MS (Chapter 6). Several major databases of nominal 70-eV EI spectra are available and various search algorithms have been described and compared.[68,69] Search software and EI mass spectra databases are available on all contemporary mass spectrometer data systems and they are also widely used on desktop computers and computer networks. Collections of other types of reference spectra may become available in future years. If an unknown is discovered in a total ion chromatogram which was not obtained with nominal 70-eV EI, computer searching will be limited to whatever appropriate databases are available, but the evaluation principles discussed in the next section will apply.

The reverse search principle for identifying target compounds in files of measured spectra was defined in the section on target analytes with CMS data acquisition. When a spectrum of an unexpected and unknown substance is obtained, a forward search to find a match in a file of reference spectra is the recommended first approach. In a forward search, the software attempts to find among the spectra in a reference database a spectrum which is the same as or very similar to the spectrum of the unknown substance (a match). Before starting the search, instrument background, from a baseline spectrum just before or after the unknown peak, should be subtracted from the unknown spectrum by using standard data system software. It is also very important to examine the spectra across the unknown chromatography peak and ascertain that the peak is from a single compound and does not represent the spectra of two or more coeluting compounds. If several coeluting compounds are discovered, an attempt should be made to create as pure a spectrum as possible by subtracting spectra from one another. Finally, the average spectrum across the peak should be used in the database search to minimize RA distortions from varying sample pressure.

The crucial part of all search systems is the algorithm that computes a number which allows the reference spectra to be ranked according to the best match and, hopefully, the most probable correct substance. Results from various search systems and databases can vary from excellent, where the highest ranked reference spectrum is the correct identification of the unknown, to dismal, where the correct identification of the unknown is ranked low or missed.[70,71] Problems with databases and computer searching systems which lead to incorrect ranking of spectra include:

- Condensed or abbreviated spectra are used in the database.
- The database contains incorrect reference spectra or misidentified reference spectra.
- The database spectra were measured using impure samples or the samples were introduced with nonchromatographic inlet systems.
- Database or sample spectra contain artifacts of thermal decomposition, temperature dependent fragmentation, sample pressure dynamics, spectrometer tuning, or spectrometer design.

- Search systems cannot distinguish structural isomers which have very similar spectra.
- Search systems typically do not use relative ion abundance data.
- Sample spectra contain ions from unrecognized coeluting compounds, uncorrected background, and so on.
- A reference spectrum of the unknown is not in the database.
- Poor quality sample spectra are produced by failures in maintenance, calibration, quality control, and so on.
- The ranking algorithm is not effective.

For these reasons, as in the somewhat more controllable reverse search, the highest ranked, or any other search result, must not be accepted on blind faith. Analysts must visually compare search software ranked database spectra with the spectrum of the unknown and critically evaluate the information in the spectra (see next section). The results of a forward search can provide important clues which can lead to the acquisition of an authentic sample of a substance for further testing using the target analyte criteria. If a critical evaluation of search results is not possible, and something must be reported, the search results should be labeled as search results and not as tentative identifications.

Evaluation of the Search Results. An evaluation strategy is to compare key ions in the spectra of the compounds ranked high in the database search with corresponding ions in the mass spectrum of the unknown. This evaluation strategy depends heavily on the availability of standardized high quality mass spectra of the compounds in the database and the unknown. The results of these comparisons are then used to select the compound or compounds found in the database search that are most consistent with the unknown spectrum. In the ideal case, all the compounds ranked high in the forward search except one will be rejected as inconsistent, and a tentative identification will be made. More often, however, several reasonable possibilities are found or none of the compounds found in the database search are consistent with the spectrum of the unknown. If this happens, more information is needed to narrow the possibilities or suggest alternative structures. Other mass spectrometric techniques can be applied to the sample to provide additional information to further eliminate candidates or arrive at a tentative identification. In the end, a tentative identification must be confirmed by acquiring an authentic sample of the substance and applying the target analyte identification criteria described previously.

In EI mass spectra the $M^{+\cdot}$ ion is the most important ion and it should be evaluated first. The $M^{+\cdot}$ is formed by removal of one electron from a molecule without fragmentation of the molecular structure. The $M^{+\cdot}$ gives not only an indication of the molecular weight of the compound but also provides clues to its composition. These clues come from the presence of naturally occurring isotopes of some important elements and from the m/z of the $M^{+\cdot}$ itself. Occasionally, multiply charged ions such as M^{2+} are also formed in EI spectra and they can be recognized and evaluated. Similar evaluations can be applied to other ions which contain molecular weight information, for example, $(M+1)^+$, $M^{-\cdot}$, or $(M-1)^-$

ions from other ionization techniques (Chapters 5 and 6), but appropriate adjustments are required to the subsequent statements about even and odd *m/z* values and even and odd numbers of electrons.

In order to evaluate the $M^{+\cdot}$ ion, the *m/z* values of the $M^{+\cdot}$ ions of the compounds ranked high in the database search must be retrieved from the database or calculated. However, because carbon and other elements commonly found in organic compounds have naturally occurring isotopes, there will always be two or more ions that correspond to the $M^{+\cdot}$ ion. Multiple $M^{+\cdot}$ ions are caused by the natural distribution of isotopes of carbon and other elements in the ions. For small molecules with molecular weights less than about 1000, multiple $M^{+\cdot}$ ions are often readily separated by mass spectrometers with low resolving power. In order to avoid confusion and simplify the specification of the mass of the $M^{+\cdot}$, it is defined, by convention, as the sum of the masses *of the most abundant naturally occurring isotopes* of the various elements that make up the molecule.[42] For the elements commonly found in organic compounds, that is, C, H, N, O, S, Si, Cl, and Br, the most abundant naturally occurring isotope is also the isotope with the lowest mass. Therefore, the lowest *m/z* ion of a group of $M^{+\cdot}$ ions is typically called the $M^{+\cdot}$ ion. The other common elements in organic compounds are F, I, and P, but these have no naturally occurring isotopes. Only a few compounds, for example, PF_3, are formed from atoms with no naturally occurring isotopes and only these give a single $M^{+\cdot}$ ion. Conventional atomic weights are natural abundance-weighted averages and are not used in MS.

The $M^{+\cdot}$ ions containing less abundant isotopes are, however, very important and provide valuable information used in the evaluation and interpretation of mass spectra. The relative abundances (RAs) of the ions in the $M^{+\cdot}$ group are determined by the natural abundances of the isotopes of the various elements and the related statistical probabilities of various combinations of isotopes in the ions of the group. These statistical probabilities are readily calculated and can be compared with actual RAs of the ions in the $M^{+\cdot}$ group of the unknown substance and compounds ranked high in the database search. This comparison is absolutely necessary because database search systems typically do not consider RAs or isotope distribution patterns. Spectra can be ranked high in the database search because they have an ion with an *m/z* that is the same as an ion in the unknown spectrum. However, the high ranked spectrum may not be correct because the $M^{+\cdot}$ group of ions does not have the same RA pattern as the pattern in the unknown spectrum.

The following is a suggested step-by-step procedure for evaluation of the $M^{+\cdot}$ group of ions and other key ions in the spectra of the unknown substance and the compounds ranked high in the database search. This procedure may also produce information about the unknown even when there are no high ranked spectra from the database search.

1. For the compounds ranked high in the database search, obtain from the database or calculate the integer monoisotopic masses of the $M^{+\cdot}$ ions by using the integer masses of the naturally occurring isotopes with the highest abundance and lowest mass. For this C = 12, H = 1, N = 14, O= 16, S = 32, Si =

28, P = 31, F = 19, Cl = 35, Br = 79, and I = 127. The rules of chemical bonding dictate that the mass of the $M^{+\cdot}$ ion will always be an even number unless the compound contains an odd number of nitrogens. This is sometimes known as the nitrogen rule. Attempt to find these ions in the mass spectrum of the unknown and in the spectra of the compounds ranked high in the database search. If none of the candidate $M^{+\cdot}$ ions are present in the unknown spectrum or in the reference spectra, the $M^{+\cdot}$ may be too unstable for observation or the correct reference spectrum has not been located. If possible, an excellent strategy is to reanalyze the sample using a softer, lower-energy ionization technique such as positive- or negative-ion chemical ionization or electron capture ionization (Chapter 5). Frequently one of these techniques will be successful in identifying an ion closely related to the $M^{+\cdot}$ ion. If this is not possible, some of the lower ranked spectra in the database search should be checked for possible $M^{+\cdot}$ ions that are in the unknown spectrum. If this fails, skip to Paragraph 4 and look for isotope patterns among the fragment ions.

2. If a possible $M^{+\cdot}$ ion is present in the unknown spectrum and not in the high ranked reference spectrum, the reference spectrum may have been measured under conditions that did not allow the observation of the $M^{+\cdot}$ Alternatively, one of the lower ranked reference spectra may contain the $M^{+\cdot}$ Another possibility is that the ion in the unknown is a fragment, impurity, or background ion and not the $M^{+\cdot}$ ion. If the $M^{+\cdot}$ ion is present in a reference spectrum and not in the spectrum of the unknown, the search identified compound may be a candidate for elimination from consideration.

3. If a tentative $M^{+\cdot}$ is found, it should be checked for reasonableness by using the abundance of the $(M+1)^{+\cdot}$ ion, that is, the ion whose mass is 1 m/z unit higher than the mass of the $M^{+\cdot}$ ion. The $(M+1)^{+\cdot}$ ion is largely caused by the 1.1% natural abundance of ^{13}C and its mass will be odd unless there is an odd number of nitrogens in the compound. The abundance of the $(M+1)^{+\cdot}$ ion as a percentage of the abundance of the $M^{+\cdot}$ should be approximately equal to 1.1 times the number of carbon atoms in the molecule plus 0.36 times the number of nitrogen atoms (M + 1 can also be caused by ^{15}N). The expected $(M+1)^{+\cdot}$ RAs of the compounds identified in the database search should be calculated and the values compared with the RAs in both the ranked reference spectra and the unknown spectrum. If inconsistencies are discovered in any of the reference spectra, some candidates may be eliminated. If the $(M+1)^{+\cdot}$ abundance in the unknown spectrum is more or less than that in all the reference spectra, this same relationship can be used to estimate the number of carbons/nitrogens in the unknown. For example, if the $(M+1)^{+\cdot}$ abundance is 10% that of the $M^{+\cdot}$ ion at mass 170, a reasonable estimate of the number of carbon atoms in the unknown would be nine. While this same ^{13}C + ^{15}N relationship can hold for fragment ions, extreme caution must be used because fragment ion abundances can be distorted by other fragment ions with different compositions but the same integer mass.

4. Many compounds of great importance contain chlorine and/or bromine atoms and their naturally occurring isotopes produce highly characteristic groups of ions with predictable RAs. These easily recognized groups can be used to determine the number of chlorine and/or bromine atoms in a $M^{+\cdot}$ ion or a fragment ion. The RA ratios of some combinations of ^{35}Cl (75.5% natural abundance), ^{37}Cl (24.5%), ^{79}Br (50.5%), and ^{81}Br (49.5%) are shown in Table 2.4. These ratios have been normalized using different ions of the groups to produce small numbers that are readily compared to the RA ratios within groups of ions in the unknown spectrum and the spectra of the high ranked compounds found in the database search. If any of the compounds found in the search contain bromine and/or chlorine, the $M^{+\cdot}$ ion group in their reference spectra should have RA ratios that match the ratios in Table 2.4 reasonably well ($M = X$ in Table 2.4). In fragment ions, the ratios could be distorted by unresolved ions with different compositions but the same integer masses. The number of bromines and/or chlorines in the $M^{+\cdot}$ ion of the unknown can be predicted and compared with the compounds found in the database search, and inconsistencies eliminated. For example, a cluster of ions suspected to be the $M^{+\cdot}$ group is observed in the unknown spectrum at even masses, with smaller ^{13}C ions interspersed at odd masses. The ratio of the abundances of the first two even mass ions (lower masses of cluster) is close to 1:3.26. Using Table 2.4, it could be concluded that the $M^{+\cdot}$ cluster is due to either $ClBr_3$ or Cl_4Br_2. An examination of the higher mass abundances in the cluster of ions should differentiate these two possibilities. An examination of the RAs in the fragment ion groups should suggest the number of chlorines and/or bromines in the fragment ions and confirm the composition of the $M^{+\cdot}$ ion. Again caution must be exercised because some distortion in RAs of fragment ions is a potential problem. These same isotopic patterns should appear in the reference spectra of any potentially correct substances.

5. If any of the compounds found in the database search contain sulfur or silicon, the natural abundances of ^{33}S (0.78%), ^{34}S (4.4%), ^{29}Si (5.1%), and ^{30}Si (3.3%) provide opportunities for similar comparisons. Other elements such as B, Hg, and many other metals have multiple naturally occurring isotopes whose isotope abundance patterns may be recognized. Expected RAs within groups of ions containing combinations of various isotopes can be calculated using computer programs supplied with commercial mass spectrometer data systems or by scientific software developers.

6. The higher mass fragment ions, which are formed by low mass losses from $M^{+\cdot}$, should be examined in the spectra of the high ranked compounds from the database search and in the unknown spectrum. The $M^{+\cdot}$ ion is an odd-electron ion. If it fragments by loss of a neutral radical or atom, a frequently more abundant even-electron ion is formed. This even-electron fragment ion will have an odd mass unless it contains an odd number of nitrogens. Common and easily recognized neutrals lost from $M^{+\cdot}$ include F, Cl, Br, I, H, CH_3, phenyl, and other odd-mass hydrocarbon radicals. If, on the other hand, even-mass higher mass fragments are observed (in the absence of an odd number of nitrogens), this indicates losses from $M^{+\cdot}$ of small whole neutral molecules such as water, hydrogen sulfide, carbon monoxide, carbon dioxide, HCl, and nitrogen oxides. The even-mass fragments will frequently be less abundant because they are odd-electron ions, but the molecules lost are indicative of groups such as alcohols, thiols, ketones, anhydrides, chlorides, and nitro compounds, respectively. Any losses of these types observed in the unknown spectrum should be present in the spectra of any viable candidates from the database search. Reference spectra with inconsistencies may be eliminated.

After all these considerations, it may be possible to reduce the list of compounds found in the database search to one or a few. These may be considered candidates for the structure of the unknown or even a tentative identification, depending on the strength of the data. However, a firm identification will not be possible until an authentic sample of the candidate compound(s) is obtained and essentially treated as a target compound and shown to meet the criteria established for the identification of such compounds.

If insufficient information is available to make a reasonable candidate list, a decision must be made either to abandon the attempted identification or to seek more evidence from mass spectroscopic or other techniques. Perhaps the single most powerful mass spectrometric technique that can be applied is an exact m/z measurement with high-resolution MS. This technique, which is described in Chapter 5, may allow assignment of a unique composition or a small number of possible compositions to specific ions. This would add significantly to the information available about the unknown and aid in the assessment of the search results. Tandem mass spectrometry, which is also described in Chapter 5, can provide information about the pathways of fragmentation of the $M^{+\cdot}$ ion and fragment ions and provide clues to the structure of the unknown. The unknown may contain elements that are not recognized in the conventional GC/MS spectrum, for example, sulfur. An elemental analysis using the techniques described in Chapter 7 may be very useful.

Fourier transform infrared (FTIR) spectrometry can provide valuable information about certain functional groups in unknowns. The availability of FTIR spectra also allows the searching of databases of infrared spectra for candidate compounds. Combined GC/FTIR/MS systems have been developed which allow the on-line separation of components of samples and the acquisition of infrared spectra and conventional mass spectra from the same injection.[31,32] Identifications of new disinfection by-products in drinking water have been supported by GC/FTIR measurements.[72]

Quantitative Analysis

Quantitative analysis is extremely important in government regulatory work, industrial quality control, and in many areas of the chemical, physical, and biological sciences. Unfortunately, there are no generally accepted criteria for what constitutes a quantitative analysis, that is, criteria for how accurately and precisely a quantity must be measured for the measurement to be considered quantitative. Criteria for acceptable quantitative analysis must be established by the user of the information in conjunction with the analytical chemist. Inevitably there are trade-offs among accuracy, precision, selectivity, and detection limits on the one hand and the cost and time required for the quantitative analysis on the other. This section is primarily concerned with three very important strategic issues that impact the accuracy, precision, selectivity, and detection limits of quantitative analysis with MS. These issues are the data acquisition technique, the method of allocating ion abundance data to specific analytes, and the method of calibrating the ion abundance data in terms of concentrations of the analytes in the sample. The techniques and options discussed in this section are applicable to conventional GC/MS as well as the other mass spectrometric techniques described in subsequent chapters of this book.

Both continuous measurement of spectra (CMS) and selected ion monitoring (SIM) can be used for quantitative analysis. Undoubtedly the most popular of the two is SIM because it can provide quantitative measurements that are more accurate, precise, and generally more sensitive than those of CMS. However, one of the significant advantages of MS, compared to other techniques, is the ability to obtain complete mass spectra of eluting compounds and, simultaneously, reasonable measurements of the concentrations of target analytes. Therefore, a CMS data acquisition strategy can be advantageous especially when complete mass spectra are needed for qualitative assessment of the sample, the concentrations of analytes are sufficiently high, and high accuracy and high precision measurements are not required.

In the past there has been a body of opinion, and there still may be, that quantitative measurements, if not done by SIM, should be conducted using a standard chromatographic detector. The standard detector could be located after a second column in a second instrument (off-line) or parallel with the mass spectrometer after splitting the column effluent. A standard detector can sometimes provide more accurate and precise measurements of concentrations than a mass spectrometer provides in the CMS mode. However, the standard detector approach introduces significant complications including the increased cost of the additional equipment, supplies, and staff required to operate and maintain additional systems.

A major problem with either the off-line or split options is the need to correctly correlate the peaks in the total ion chromatogram with the peaks from the standard detector. Serious errors can result from timing differences between detectors. Coeluting substances and complex samples exacerbate the problem and can cause serious measurement errors even with selective standard detectors. For these reasons, the strategy of simultaneous qualitative and quantitative analysis using GC/MS and CMS data acquisition has been carefully studied.[64] The conclusion is that, although the accuracy, precision, and detection limits are not usually as good as with SIM, they are often quite adequate for many analytical programs. The CMS quantitative analysis strategy provides economy of operation, qualitative accuracy, and relative simplicity of operation. Compared to an off-line or parallel standard detector, CMS data management is less complex and quantitative accuracy is far better with complex samples.

Quantitation Ions and Peak Integration

Absolute ion abundances are usually measured and stored in MS data systems as analog-to-digital converter counts, direct ion counts, or some analogous parameter. These values range from a few hundred to millions and are normalized to generate the RAs displayed in mass spectra. Data system counts, in arbitrary units, are convenient measures of absolute ion abundances in MS. The allocation of ion abundance data to specific analytes is dependent on the data acquisition technique and the ions used for quantitation. Peak height measurements should not be used because of the greater variability associated with a few ion abundance measurements at the peak centroid compared to the lower variability associated with the integration of multiple ion abundance measurements over the peak area. Also, variable chromatographic peak widths and shapes often make peak height measurement a less accurate measure of amount than peak area measurement.

The curved peak-shaped line in Figure 2.7 represents the actual concentration of a substance eluting from a chromatographic column. The series of diagonal lines below the curve represent repetitive scans of the mass spectrometer as the substance elutes from the column (CMS data acquisition). Each scan begins at the same m/z value and ends at the same highest m/z. One option for peak integration is to sum the absolute ion abundances of all the ions in each spectrum under the chromatographic peak. This approach has the advantage of potentially great sensitivity because the absolute abundances of all the ions in all the spectra are summed. However, this integration technique has not been used very much because the total ion abundance can include background ions and ions from coeluting compounds. Therefore, this technique is susceptible to serious measurement errors. In addition, the technique is probably not supported on many commercial data systems. However, given the excellent resolving power of open tubular GC columns, the low column bleed of bonded stationary phases, and the ability to subtract background ion abundances, total ion integration probably deserves more attention. It would appear especially useful in situations where all analytes are separated and both complete mass spectra and high sensitivity are needed.

The standard technique for peak integration with CMS is the summation of the absolute ion abundances of a single quantitation ion for each analyte. This technique, which is illustrated in Figure 2.7, mimics a standard approach with SIM, except with SIM, data is acquired for just the single ion. The single-ion integration technique minimizes the chances of interferences from coeluting compounds or instrument background. The quantitation ion can be chosen for maximum selectivity from all the ions in the spectrum of a substance. Coeluting analytes can be simultaneously measured as long as their respective quantitation ions are not present in the spectra of the other substances eluting in the same chromatographic peak. Background and extraneous eluting material is not generally an interference as long as quantitation ions are selected from molecular ions or higher mass fragment ions.

Clearly, the selection of a quantitation ion can be critically important with either CMS or SIM. While generally a very abundant ion in the spectrum is chosen because it will give higher sensitivity compared to less abundant ions, this is not always the best choice. For example, if the target analyte was the compound whose structure and mass spectrum are shown in Figure 2.12, the m/z 149

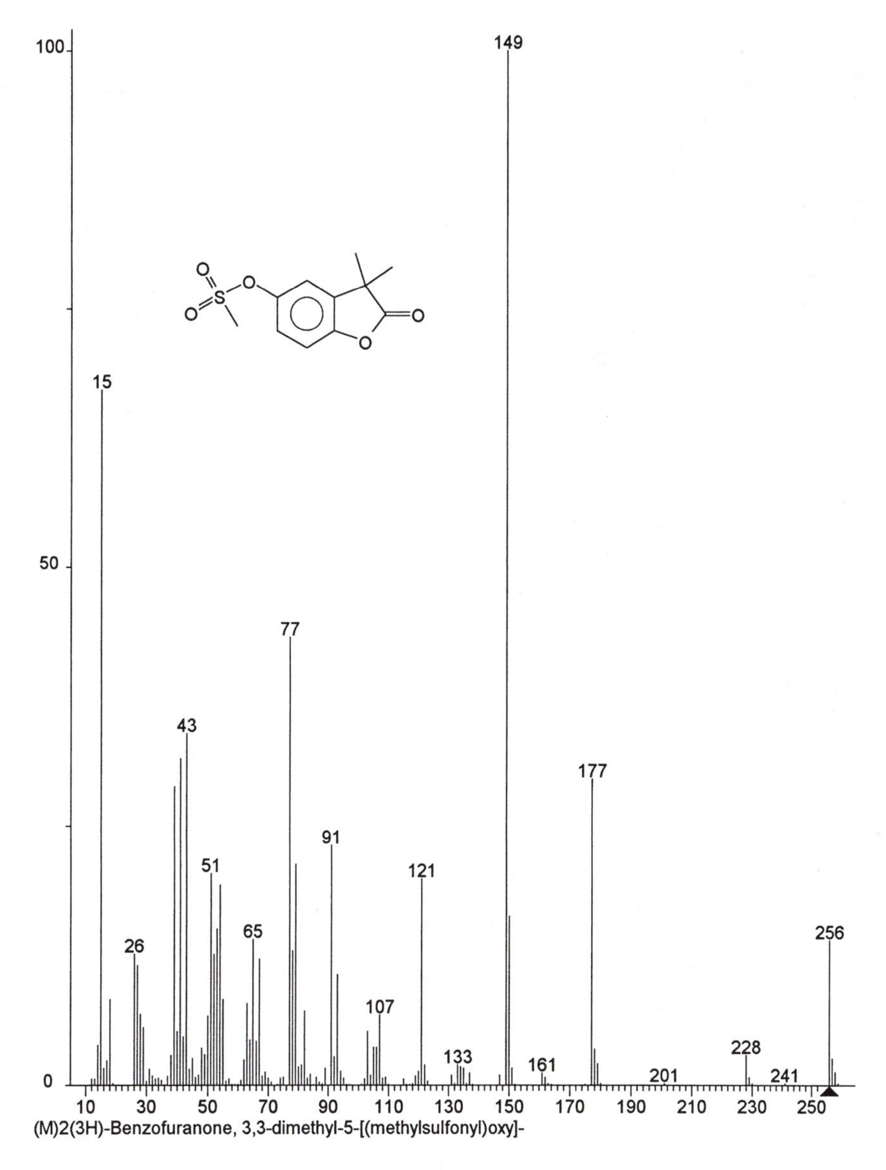

Figure 2.12 A target analyte whose most abundant ion is not an appropriate choice for the quantitation ion. (Reproduced with permission from Ref. 50. Copyright 1992 U.S. Department of Commerce on behalf of the United States.)

base peak ion could be selected as the quantitation ion for maximum sensitivity.[50] However, this would be an unwise choice because many common phthalic acid esters, which are important industrial plasticizers and very common laboratory background contaminants, have *m*/*z* 149 as the base peak in their spectra. Therefore, selection of *m*/*z* 149 as the quantitation ion would result in an increased probability of an erroneous measurement. Similarly, the *m*/*z* 15 and 77 ions (Figure 2.12) are relatively abundant, but both are common fragment ions that could exist in the spectrometer background. In this example, it would be more appropriate to select a less abundant but more selective quantitation ion, that is, either the $M^{+\cdot}$ ion at *m*/*z* 256 or the fragment ion caused by the loss of H_3C–SO_2 at *m*/*z* 177. In general, the higher the *m*/*z*, the less the probability of a background interference.

The single quantitation ion technique originated in the 1970s when packed-column GC was widely used, GC peaks were broad compared to those from capillary columns, coelution was quite common, and as many as 10–15 spectra could be acquired under each GC peak. The selectivity of a single quantitation ion was needed with packed columns and reasonably accurate peak areas were obtained with 10–15 absolute abundance values for each analyte in a GC peak. With the widespread use of high-resolution open tubular GC columns, and other high-resolution separations, peaks are often only 2–5 s wide and the probability of coelution is significantly reduced. Narrow peaks result in a smaller number of absolute abundance values per analyte and a less accurate peak area for the analyte. A faster scan speed can increase the number of data points per peak, but will result in lower signal/noise because fewer ions are measured in the shorter time allowed for each ion in the scan. Under these circumstances there would seem to be some potential benefit from using more than one quantitation ion (or monitored ion in SIM). For a target analyte with, for example, 20 ions in its spectrum, a large number of combinations of multiple quantitation ions are possible, and some of these could provide an excellent balance between signal/noise, peak integration accuracy, and low risk from interferences from coeluting substances. This strategy is used in some methods for 2,3,7,8-tetrachlorodibenzo-*p*-dioxin and related compounds, which are described in Chapter 4.

Quantitative Analysis Calibration Techniques

With the addition of quantitative analysis, there are three levels of calibration in MS. The first level is calibration of the mass scale, which is usually part of the instrument manufacturer's automated tuning program. Mass scale calibration programs for GC/MS systems frequently use the reference standard perfluorotributylamine (PFTBA) which gives many ions of known composition. Mass scale calibration files are generated and retained on the instrument's computer system and these are often stable for many months. Other mass scale calibration standards are used with other techniques, for example, perfluorokerosine (PFK) is frequently used with high-resolution MS (Chapter 5).

The second level of calibration, which is currently only applicable to conventional GC/MS, is tuning the instrument to produce RAs of ions that are in agreement with abundance ranges and other criteria established for the test compounds DFTPP or BFB (previous section of this chapter and Chapters 3 and 4). The first two levels of calibration have been combined in some manufacturer's automated tuning programs, but DFTPP, BFB, or some other test compound still must be injected into the GC/MS system to verify that the mass and abundance scales have been calibrated correctly. For quantitative analysis, a third level of calibration must be added to correlate integrated absolute ion abundances with amounts or concentrations of analytes injected into the mass spectrometer and ultimately with the concentrations of analytes in the original sample.

Once the data acquisition and quantitation ion strategies have been decided, several quantitative analysis calibration techniques are available. These are based on either external or internal standards and include the special case of isotope dilution. These terms are defined in the following sections along with a description of their uses, advantages, and disadvantages. While calibration for quantitative analysis of target analytes is generally accomplished before the samples are analyzed, all the techniques except isotope dilution are amenable to calibration after a preliminary assessment of the sample. With CMS data acquisition, this first allows the identification of the target analytes and nontarget compounds in the sample, then calibration for quantitative analysis of the target analytes and the nontarget compounds.

External Standardization. External standardization means that all calibration standards are external to the sample or sample extract. Nothing is added to the sample or sample extract to aid the calibration process. A series of calibration solutions containing the target analytes at several concentrations in an appropriate solvent are prepared and analyzed. The integrated quantitation ion abundance of each analyte is plotted as a function of amount or concentration of analyte. The calibration plot for each analyte is usually linear through the origin, but intercepts and calibration curves are sometimes observed. The concentrations of the same analytes in the sample or sample extract are determined by measuring the corresponding quantitation ion abundances and evaluating the amounts or concentrations in the sample either mathematically or graphically. This is the simplest and most straightforward of the calibration schemes.

External standardization has the advantages of not requiring the addition of any substance to the sample and of simple calculations. The major disadvantage of this approach, which is widely used with standard GC and liquid chromatography detectors, is that errors can be introduced by instrument instability or drift between the calibration measurements and the sample measurements. If, for example, the instrument loses some sensitivity during a series of analyses, there is no intrinsic mechanism to correct for this loss. The external calibration standards can either be made in an appropriate solvent and analyzed in that solvent or made in a simulated sample matrix and processed through the entire analytical method including the extraction and other sample preparation steps. If, as is more common, the calibration standards are not processed through the extraction and other method steps, the final measured concentrations in the sample will often be less than 100% of full recovery and will reflect any losses that occurred during the sample preparation process. If the calibration standards are processed through the entire analytical method, the final measured concentrations in the sample will be corrected or normalized for any losses that occur in the sample preparation.

Internal Standardization. Internal standardization is probably the most common calibration strategy used with conventional GC/MS and some other mass spectrometric techniques. A series of calibration solutions is prepared which contain the analytes in different concentrations in each solution and one or more nonanalytes at constant concentration in all the solutions. These nonanalytes are called internal standards (ISs) and they must not be present in the actual samples. The ISs are measured along with the target analytes in the calibration solutions, and IS response factors are defined as:

$$\mathrm{RF} = \frac{(A_{\mathrm{x}})(Q_{\mathrm{is}})}{(A_{\mathrm{is}})(Q_{\mathrm{x}})} \tag{2.1}$$

where RF is the response factor, X is the measured analyte, IS is the internal standard, A is the absolute abundance of the quantitation ion integrated over the chromatographic peak, and Q is the absolute amount or concentration injected in consistent units.

Rearrangement of this definition gives the form:

$$A_{\mathrm{x}}/A_{\mathrm{is}} = (Q_{\mathrm{x}})(\mathrm{RF})/Q_{\mathrm{is}} \tag{2.2}$$

A plot of $A_{\mathrm{x}}/A_{\mathrm{is}}$ as a function of Q_x should be linear through the origin with a slope equal to $\mathrm{RF}/Q_{\mathrm{is}}$, but intercepts or curves are sometimes observed. Most mass spectrometer data systems have options for handling linear and second-order calibrations. Measurement of $A_{\mathrm{x}}/A_{\mathrm{is}}$ in a sample or sample extract containing a known amount of IS allows the determination of the quantity of the analyte by using this calibration equation.

One major advantage of the IS technique is that any intermediate term (hours to days) instrument instability, drift, sensitivity change, and so on, that causes a change in A_{x} should produce a corresponding change in A_{is} and the area ratio is unaffected. Another major advantage is that injection and extract volumes need not be measured or known accurately because calculations are independent of these quantities. Multiple ISs are usually employed and each analyte is calibrated using the IS that elutes closest to the time of elution of the analyte. If the instrument system has good short to intermediate term stability, as is common with many contemporary GC/MS systems, a single IS is quite adequate for most analyses although it is always prudent to include a back-up IS for use if needed.

The IS does not need to be a stable isotope-labeled compound, but it commonly is to ensure that the IS is not present in the sample. The calibration solutions containing the analytes and the IS can either be made in an appropriate solvent for analysis or made in a sample matrix (e.g., water) and processed through the total analytical method including the extraction and other sample preparation steps before analysis. If the calibration measurements are made, for example, without extraction, the RFs will not include any bias that would result from differences in the extraction efficiencies of the IS and analytes. If the calibration measurements are made after processing the calibration solutions in a sample matrix through the entire analytical method, including the extraction and other sample preparation steps, the RFs will include any bias that results from differences in the IS and analyte extraction efficiencies, and so on. These different techniques are critical and will have a significant impact on the analytical results, which will also be influenced by the way the IS is used with the environmental samples. There are four important strategies in using ISs:

- *Internal standards are not extracted from samples.* The instrument system is calibrated using analytes and IS in an appropriate solvent without any prior extraction or other sample preparation steps. The ISs are added in known amounts (Q_{is}) to the final extracts of the samples after all sample preparation steps are complete. The A_x/A_{is} is measured for each analyte in the sample extract and Eq. 2.2 is evaluated graphically or mathematically for the quantities of analytes in the extract (Q_x). The concentrations of analytes in the original sample are calculated by relating Q_x to the size of the sample in appropriate units. The measured concentrations will usually be less than 100% of full recovery because of inefficiencies and losses that occur during the extraction and other sample processing. This is a valuable option that is commonly used during analytical method development to determine analyte recoveries with samples containing known quantities of analytes. This approach is widely used in environmental analyses and usually gives concentrations that are lower than the true values in environmental samples. An example of the application of this IS technique is Method 625 in Chapter 4.
- *Internal standards are extracted from samples.* The system is calibrated using analytes and ISs in an appropriate solvent without any prior extraction or other sample preparation steps. The ISs are added in known amounts (Q_{is}) to the samples before any sample processing. Normal sample processing is conducted and A_x/A_{is} is measured for each analyte in the final extract. Equation 2.2 is evaluated graphically or mathematically for the quantities of analytes in the extract (Q_x) assuming 100% recovery of the ISs. The concentrations of analytes in the original sample are calculated by relating Q_x to the size of the sample in appropriate units. The measured concentrations will be the true concentrations in the sample if the extraction efficiencies of the analytes and ISs are the same or very similar. This will be true even if the actual extraction efficiencies are low, for example, 50%. As long as the recoveries are the same, the concentrations of the analytes will be corrected to 100% recovery by the assumption of 100% recovery of the ISs. Of course, there will be the normal variability of the results around the measured concentra-

tions. To the extent that the extraction efficiencies of the analytes and ISs are different, the mean measured concentrations of the analytes will be biased high or low. This option is valuable when a number of analytes and ISs have similar extraction efficiencies because it provides concentrations corrected for sample processing losses using just a few ISs. An example of the application of this IS technique is Method 525.2 in Chapter 4.

- *Isotope dilution.* This is the same as option 2, but made perfect by the use of multiple ISs that include a stable isotope-labeled IS that is the same compound as each analyte. This is the classic *isotope dilution* analysis, that is, each analyte is diluted with itself, that is, labeled with stable isotopes. Stable isotopes most often used are ^{13}C, ^{2}H (D), and ^{15}N, although others are possible. If a sufficient number of atoms in the ISs are replaced with isotopes with a high percentage of enrichment, the m/z of the IS will be shifted far enough from the m/z of the analyte containing only natural amounts of each isotope, and no interferences will occur. The isotope dilution strategy ensures that the extraction and other sample processing steps of the method will have exactly the same effect on the analyte and the IS since they are the same compound. The calibration step is usually conducted to compensate for small impurities in the analyte and IS, but the RF should be close to unity. All measured concentrations in the sample are corrected to true concentrations regardless of the percentage recovery in the extraction or any other sample processing losses. Of course there will be normal variability of the results around the true concentration. Isotope dilution calibration is used in methods for 2,3,7,8-tetrachlorodibenzo-*p*-dioxin and related compounds (Chapter 4), but is not widely used because of the high cost of most isotope-labeled ISs.
- *Internal standards are extracted from standard solutions and samples.* This option corrects all measured concentrations for extraction inefficiencies and other sample processing losses without using the more costly isotope dilution technique. The system is calibrated using analytes and ISs in a sample matrix or simulated sample matrix, for example, distilled water, and the calibration standards are processed through the entire analytical method. The ISs are added in known amounts (Q_{is}) to the environmental samples before any sample processing. Normal sample processing is conducted and A_x/A_{is} is measured for each analyte in the final extract. Equation 2.2 is evaluated graphically or mathematically for the quantities of analytes in the extract (Q_x) assuming 100% recovery of the IS. The concentrations of analytes in the sample are calculated by relating Q_x to the size of the sample in appropriate units. The measured concentrations will be the true concentrations in the sample because both the RFs determined in the calibration, and the area ratios determined in the sample extract, include the sample processing bias which divides to unity in

Eq. 2.2. This will be true even if the actual extraction efficiencies are low, for example, 50%. Of course, there will be the normal variability of the results around measured concentrations. This option is valuable when it is reasonable and practical to process all the calibration standards through the entire analytical method. This approach is used in some environmental analyses and is sometimes referred to as calibration with procedural standards. An example of the application of this IS technique is Method 524.2 in Chapter 3.

Selection of Internal Standards

Except in the special case of isotope dilution, the structures of ISs are not required to be similar to the structures of the analytes. For example, it is not necessary to use a chlorinated hydrocarbon IS to calibrate the mass spectrometer signals of a group of chlorinated hydrocarbon pesticides. As far as the mass spectrometer is concerned, one ion appears to be as good as another. Internal standards must be thermally stable if used with GC/MS and chemically unreactive for all mass spectrometric techniques. They should elute from the chromatograph in the same general time frame as the analytes in order to compensate for short-term drift in instrument sensitivity. The absolute abundances of quantitation ions of the ISs should not be much larger or much smaller than the abundances the analytes to avoid excessively large or small RFs. If the ISs are extracted from the samples, but the RFs are determined without extraction, the ISs and the analytes should behave as similarly as possible during the extraction and other sample processing to minimize bias in the measurements of the concentrations of the analytes in the sample. Any time the ISs are extracted, they should be recovered in high yield to provide as strong and precisely measured a signal as possible.

Stable isotope-labeled analytes are often selected as ISs for a variety of other analytes because they have essentially zero probability of being in the environmental sample. However, the value of such a standard can be negated by a coeluting sample interference with an ion in its mass spectrum with the same m/z as the quantitation ion of the IS. For this reason it is always wise to include one or more back-up ISs which can be used in the event of an interference in the sample or sample extract. Other nonisotope-labeled compounds that have a low probability of being in the sample should not be neglected for consideration as ISs.

Number of Calibration Points in a Concentration Range

Textbooks on quantitative analyses may provide some guidance on the optimum number of calibration points in a given concentration range. This guidance would probably be based on the degree of linearity of the calibration, the precision of replicate determinations of calibrations points, and other statistical tests. In general, with MS, it is recommended that at least three appropriately spaced calibration points be used for each factor of 10 in concentration range that is calibrated. For concentration ranges greater than a factor of 10 with some well-behaved methods and analytes, fewer than the recommended number of calibration points may give acceptable results. For example, acceptable data may be obtained using three calibration points for a factor of 20 in concentration, four points for a factor of 50, and five points for a factor of 100. Judgment, experience, and the desired accuracy and precision of the analyses will usually indicate the optimum number of calibration points in a concentration range. In general, more points at different concentrations are more desirable than replicate measurements of a calibration solution at the same concentration.

Surrogate Analytes

The concept of a surrogate analyte was invented for quality control purposes and not as a calibration device.[59] A surrogate analyte is defined as a pure, well-behaved method analyte which is extremely unlikely to be found in a sample, and which is added to every sample in known amount before extraction or other sample processing. The surrogate is measured with the same procedures used to measure the other sample components and its purpose is to monitor method, instrument, and analyst performance with every sample. This definition sounds very much like that of an IS that is added to the sample before sample processing. The difference is that the surrogate is not used to calibrate the response of the mass spectrometer with respect to the other analytes. The surrogate is measured, using one of the standard calibration procedures, to monitor method performance in every analysis, and is especially valuable when a number of consecutive samples have no other analytes present. The surrogate recovery gives a continuing check on the proper conduct of the analytical method and the performance of the equipment. Of course, in the case of an interference problem with the IS, the surrogate analyte can serve as a back-up IS. Many examples of surrogate analytes are included in the USEPA regulatory analytical methods that are described in Chapters 3 and 4.

Detection Limits

One of the most important measures of analytical method performance is the detection limit. The detection limit is also very important in all kinds of environmental and other studies including exposure assessment, the determination of health effects, risk assessment, studies of air and water quality, and regulatory compliance monitoring. Detection limits are frequently reported in the scientific literature and claimed by equipment manufacturers, but the method used for their determination often is not described. The use of detection limits in environmental regulation in the United States has been somewhat controversial and a number of definitions have been proposed.[73] Generally, two fundamentally different detection limits are needed in analytical chemistry:

- *Instrument detection limit* (IDL). The IDL is generally accepted to mean the minimum quantity of a substance that can be identified and measured by the instrument using a pure sample of the substance in a pure solvent and some appropriate measurement criterion, for example, a signal/noise of 3. The IDL is valuable for evaluating and comparing instrument performance and is frequently reported in the scientific and commercial literature. The signal/noise or other criteria used to define a positive measurement must be documented along with the IDLs.

- *Method detection limit* (MDL). The MDL is much broader concept than the IDL and means the minimum concentration of a substance that can be identified and measured in a simulated or real sample matrix with the complete analytical method. The MDL includes any losses of analyte that occur during sample extraction, solution transfers, solvent evaporation, chromatographic injection, or any other method steps where losses can occur. The MDL should be higher than the IDL and is a realistic assessment of total analytical method performance with a specific sample matrix. For most chemical analyses the MDL is a much more useful concept than the IDL.

Often, reports of IDLs and/or MDLs are either based on a single measurement or the number of measurements is not given. Because variability is an inherent part of all physical measurements, detection limits which do not consider measurement variability are not very useful. As stated by John Taylor, "A measured value becomes believable when it is larger than the uncertainty associated with it."[74] Several investigators have developed statistical approaches to account for variability in the determination of detection limits and one of these is presented here and used in subsequent sections of this book.[73]

The MDL can be redefined as the minimum concentration of a substance that can be identified and measured in a specific sample matrix with the complete analytical method and reported with 99% confidence that the substance is present in the given matrix. The variability of the analytical method is determined by replicate analyses of the sample matrix which contains the substance at a concentration about three to five times the estimated MDL. If the substance cannot be detected in the matrix, the matrix is fortified before the MDL determination at a concentration three to five times the estimated MDL, but sufficiently high so that all of the identification criteria are met. Several techniques to estimate the MDL have been described.[73] The MDL is then defined mathematically and calculated as follows:

$$\mathrm{MDL} = t_{(N-1,1-\alpha=0.99)}S \tag{2.3}$$

where:

t = Student's t value for a one-tailed test at the 99% confidence level with $N-1$ degrees of freedom. The value of t is 3.143 for seven replicate measurements, which is a reasonable number for the determination of the MDL.
N = the number of replicate analyses.
S = the standard deviation of the replicate analyses.

With this calculation the MDL is always larger, by a statistically rational factor, than the measured variability of the method, and the MDL is consistent with Taylor's hypothesis.[74] A rationale for this approach to the MDL is illustrated in Figure 2.13. As the concentration of a substance decreases and approaches zero, replicate measurements at each concentration are expected to show increasing variability as indicated by the ranges of the normally distributed data points. The increasing variability is expected because of a progressively lower instrument response to the analyte and the variable signals from background chemical and other noise. However, in the region of Figure 2.13 at approximately three to five times the estimated MDL, the experience of many analysts is that as the concentration of the substance decreases, the standard deviation S either declines slowly or remains more or less constant over a range of concentrations. This is the target region for the concentration of the substance for the determination of the MDL. Since S is more or less constant or declining slowly in the region, the measured MDL will be about the same regardless of the concentration of the substance.

If the fortified concentration of the substance is too low, the identification criteria may not be met, which invalidates the measurement. If the criteria are met, a larger S will produce an MDL higher than the fortified concentration. If the fortified concentration is too high, that is, much higher than shown in Figure 2.13, the value of S will be high in absolute terms and the MDL will be meaningless. If the measured MDL is more than about a factor of three from the fortified concentration, a second determination is suggested using the first measured MDL as the fortification level for the second determination. Until adequate experience with the method is available, this iterative procedure may be needed to find a good estimate of the MDL.

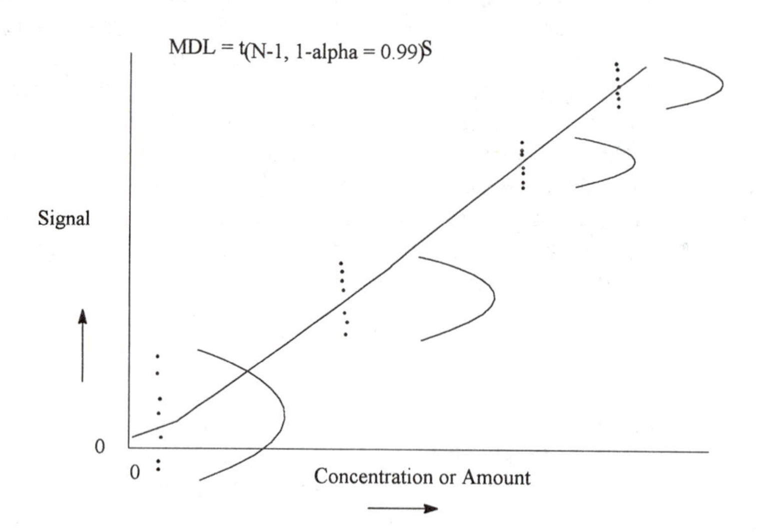

Figure 2.13 Representation of measurement variability as the concentration or amount of analyte approaches zero.

The relative standard deviation (RSD = S/concentration × 100) should be ~30–33% at the MDL. The MDL is an average detection limit and it will vary with repeated determinations because all the conditions of all the measurements cannot be duplicated exactly, but the variability should not be very large unless something in the method has changed significantly. About 50% of measurements of fortified concentrations at the MDL may not detect the analyte because a 50% false negative rate is an implicit part of the MDL definition.

The MDL should be determined using the analytical method in exactly the same way it will be used for routine sample analyses. No special precautions or sample handling should be used to obtain a more favorable (less variable) result unless those same precautions and sample handling are used in routine sample analyses. Similarly, it is good practice to make the replicate measurements used in the MDL determination over several days to introduce normal day-to-day variability into the MDL determination. When the MDL is determined in this way it provides a good measure of the capability of the analytical method during routine sample analyses. If special precautions or sample handling are used in the MDL determination, the result will not be a valid measure of analytical method capability during routine sample analyses.

This approach to measuring detection limits has a number of advantages including an operational definition that is easy to understand and implement. The procedure has a sound statistical basis and takes into account the variability inherent in all physical measurements. When comparing analytical methods or laboratory performance, a sample matrix readily available in many laboratories, for example, distilled water, is often used to measure MDLs. These distilled water MDLs should be lower than MDLs for the same substances in a matrix such as wastewater, and distilled water MDLs must not be used for other matrices. The MDLs for some specific substances are included in the regulatory analytical methods described in Chapters 3 and 4. This statistical approach could also be used to determine IDLs.

Analytical Method Validation and Analytical Quality Control

An analytical method is defined as the integration of the individual strategies and procedures used to analyze a sample. A complete analytical method defines the types of applicable samples, the general analytical strategy, and all the necessary steps required to produce identifications and measurements with realistic estimates of accuracy and precision. The analytical method description should contain all necessary details including:

- Definitions of new or unusual terms
- A statement of known and potential interferences
- Safety precautions
- Equipment, supplies, chemical reagents, and standard solutions
- Calibration and standardization procedures
- Analytical quality control procedures

- Sampling technique and sample preservation procedure
- Pollution prevention and waste management information
- Sample preparation procedure
- Separation and measurement procedures
- Identification criteria
- Quantitative analysis procedure
- Method performance data

Method Validation

Method validation is the process of demonstrating that an analytical method is capable of producing identifications and measurements of some degree of quality. The degree of quality is usually defined in terms of accuracy, detection limits, precision, recovery, selectivity, specificity, and other variables. Unfortunately, there is no generally accepted degree of quality that defines a validated analytical method. The quality of data needed for validation depends on the needs of the data users some of whom may be readers of the chemical literature years later. Criteria for method validation must be established by the original data users in conjunction with the analytical chemist. Invariably there are trade-offs among accuracy, precision, recovery, detection limits, and selectivity on the one hand and the cost and time required for the analysis on the other.

Method validation is accomplished through a series of analyses of simulated and real samples. Some of the types of samples used in method validation are summarized in a later section "Method Validation and AQC Materials and Tests." These analyses are usually completed over a short time, a few days to a few weeks, before any analyses of real samples. There are many possible levels of method validation ranging from a single analyst analyzing just one type of sample matrix with a single instrument to many analysts analyzing many types of sample matrices with different types of instruments. The exact level of method validation should be specified along with the presentation of the validation data. Method revalidation may be required after a significant change to the analytical method, after training a new analyst, or after acquisition of a new analytical instrument. The results of some multilaboratory method validation studies have been published and one study of Method 524.2 is summarized in Chapter 3.

Quality Assurance and Analytical Quality Control

Quality assurance (QA) and analytical quality control (AQC) encompass a wide variety of concepts and activities designed to ensure analytical results of the desired level of quality. Before the 1970s the concept of QA for analytical measurements was an obscure topic. While traditional teaching of analytical chemistry certainly included some statistical concepts and encouraged the evaluation of measurement accuracy and precision, these ideas did not penetrate the environmental community to any great extent until nearly the end of the1970s. About that time, many in government, industry, and the university communities began to realize that major environmental policy decisions, with potentially large economic and/or social impacts, depended on the results of chemical analyses. If the quality of these analyses was poor, inappropriate or bad decisions with major economic and social consequences could result. Considerable emphasis was placed on QA and the application of quality control techniques to analytical measurements. These programs became somewhat controversial and were much discussed and debated. The legislative and regulatory driving forces that led to this emphasis on quality and some related issues are described in Chapters 1 and 3.

A general treatment of QA and AQC is beyond the scope of this book. An excellent reference book is available and should be consulted for complete information.[74] However, QA and AQC are emphasized throughout this book and are particularly evident in the sections of this chapter entitled "Alternatives to Conventional GC/MS", "Limitations of Mass Spectrometry", "Standardization of Conventional GC/MS", "Identification Criteria", "Quantitative Analysis", and "Detection Limits". Many QA and AQC concepts are standard analytical chemistry practices such as control of interferences in glassware, reagents, solvents, and laboratory equipment. Standard analytical methods and methods manuals often contain detailed information about the purification of reagents and solvents, the required purity of calibration standards, and the decontamination of glassware and equipment.[6,7,59,60] This is the first line of quality control because many environmental and other analyses utilize small samples containing low concentrations of analytes. Serious errors can result from impure materials or contaminated supplies and equipment.

Beyond these standard practices and precautions, AQC is the process of ensuring on a continuous and real-time basis that an analytical method is in control and producing results that are consistent with the degree of quality established in the method validation process. The information needed to determine whether a method is in control is obtained from a series of tests of method performance at regular intervals during the course of the analyses of the real samples. Many of the types of analyses used to validate an analytical method are also used for AQC. The main difference is that method validation is usually accomplished first, before any real samples are analyzed, and often includes replication of analyses to establish meaningful method performance statistics and statistically significant control limits. The continuous and real-time performance tests are often single analyses to determine whether the analytical method is performing within established control limits. An AQC system should detect trends in analytical performance in real time and supply information needed to correct unfavorable trends before the analytical system goes out of control. Analytical quality control also provides a long-term record of the performance of methods, instruments, personnel, and the laboratory.

The decision to implement an AQC program is an important strategic decision because AQC invariably adds, often significantly, to the time and cost of the analyses of samples. Sometimes, the decision is to minimize cost and AQC, particularly when the study is not in direct support of some predictably important regulatory action. Unfortunately, it is not always possible to predict how environmental and other measurement data will be used. Data that are not supported by sufficient AQC may eventually be discarded as unreliable and the entire cost of the study will be lost. Published data that are supported by a sound AQC program have some lasting value because the user of those data, perhaps years later, can estimate their quality by reviewing the AQC results. When considering the potential economic impact or social, health, and legal uses of various identifications and measurements, it may not be unreasonable to devote as much as 50% of the individual measurements to method validation and AQC analyses. Several highly publicized criminal trials of the 1990s have clearly demonstrated that, often, the first line of defense is to discredit the analytical data, and this is made easy when the QA and AQC are weak or nonexistent.

Method Validation and AQC Materials and Tests

This section contains brief descriptions of some of the types of materials, solutions, and performance tests that are used during method validation and for AQC. Some of these are designed to detect laboratory or field contamination, determine recovery, check precision, or accomplish other goals. Standard analytical method descriptions should be consulted for examples of applications of these techniques in analytical methods.[6, 7, 59, 60]

Laboratory Reagent Blank (LRB). The LRB is the primary tool used to determine laboratory background contamination. It is defined as an aliquot of blank sample matrix, for example, reagent water, that is analyzed exactly as a real sample including exposure to all glassware, equipment, solvents, reagents, ISs, and surrogates that are used with other samples. The LRB is used to determine if method analytes or other interferences are present in the laboratory environment, solvents, reagents, supplies, or the equipment. If background contamination is sufficient to prevent measurements with the established accuracy and precision of the method, the condition must be corrected before proceeding to sample analyses. In general, background from target analytes should be below their MDLs. Common background includes methylene chloride from solvent extraction operations in nearby laboratories and phthalate esters, which are ubiquitous in many construction materials and supplies and are frequent laboratory contaminants. The LRB should be analyzed at regular intervals during the working day and at least several times during a full day of sample analyses. The results of these analyses should be maintained with the records of the sample analyses to document this essential background information. Normally the results of sample analyses are not corrected by subtracting the concentrations in the LRB because of the typically highly variable nature of the background substances in the LRB.

Field Reagent Blank (FRB). The FRB is similar to the LRB and serves a similar purpose. It is defined as an aliquot of blank sample matrix, for example, reagent water, that is placed in a clean sample container in the laboratory and treated as an environmental sample in all respects. It is shipped to the sampling site, exposed to field conditions, stored with the environmental samples, preserved, and analyzed with the other samples. The purpose of the FRB is to determine if method analytes or other interferences that are present in the field environment can contaminate the samples during sampling, preservation, shipment, and storage. The FRB is particularly important for measurements of gases and volatile liquids that can readily impact a sample during field operations, for example, by infusion through a plastic seal. The FRB can also detect contaminated sample containers, contaminated field sampling equipment, and contaminated field preservation reagents. One or more FRBs should accompany most shipments of sample containers to a field sampling site.

Performance Check Solution (PCS). The PCS is a solution of one or more method analytes, surrogates, ISs, or other test substances used to evaluate the performance of the instrument system with respect to a defined set of method criteria. If the analytical method uses conventional GC/MS, the PCS should contain the mass spectrometer performance test compounds decafluorotriphenylphosphine (DFTPP) or 4-bromofluorobenzene (BFB) or another reference standard. The PCS should also include method analytes that are known to cause problems, for example, the pesticides endrin and DDT are susceptible to thermal decomposition during GC. This solution should also contain several closely eluting, but resolved, method analytes that can be used to evaluate chromatographic performance. An aliquot of this solution should be injected into the instrument system at least every working day, and perhaps several times a day, to evaluate the performance of the system. The absolute values of the integrated areas of the quantitation ions of the ISs and surrogates should be monitored to evaluate system sensitivity and the ratios of these areas should remain reasonably constant. The surrogate data can also be monitored in every sample analyzed. Reduced system sensitivity, as indicated by steadily declining quantitation ion areas, is a signal that system maintenance is required. The results of these analyses should be maintained with the records of the sample analyses to document this information.

Laboratory Duplicate Sample Aliquots (LD1 and LD2). The LDs are two aliquots of the same sample taken in the laboratory and analyzed separately with identical procedures. The difference in the concentrations measured in the two aliquots is an approximation of the standard deviation S obtained from analyses of multiple aliquots. These analyses provide an indication of the precision associated with laboratory procedures, but not with sample collection, preservation, shipment, or storage procedures. The precision of the laboratory method should be established first by analyses of replicate aliquots during the method validation process. Periodic analyses of duplicate aliquots from a randomly selected sample provides an inexpensive test to determine whether the laboratory analytical process is in control and whether reasonable precision is being maintained in the laboratory.

Field Duplicate Samples (FD1 and FD2). The FDs are two separate samples collected at the same time and place under identical conditions and treated exactly the same throughout field and laboratory procedures. The difference in the concentrations measured in the two samples is an approximation of the standard deviation S obtained from analyses of multiple samples. These analyses can provide an indication of the precision associated with sample collection, preservation, shipment, storage, and laboratory procedures. However, the results must be interpreted carefully as differences could also be caused by an inhomogeneous sample. If it is certain that the samples are true FDs from a homogeneous source, and the precision of the combined field and laboratory procedures is established by analyses of replicate samples during method validation, periodic analyses of FDs provides an inexpensive test to determine whether the combined field/laboratory analytical process is in control and whether overall precision is being maintained. Clearly, FDs are more useful when used in conjunction with laboratory duplicates as that may allow the isolation of a specific problem, for example, inadequate sample preservation in the field.

Laboratory Fortified Blank (LFB). The LFB is an aliquot of blank sample matrix, for example, reagent water, to which known quantities of the method analytes are added in the laboratory. A blank sample matrix provides a controlled matrix that is readily available in all laboratories and one that is free of real environmental contamination. This allows tracking of laboratory performance over time and interlaboratory performance comparisons. It is strongly recommended that the method analytes added to the blank sample matrix be obtained from a source different from the source of the calibration standards. The LFB is analyzed exactly like a real sample and its purpose in a methods validation study is to establish method recovery, accuracy, and precision by analyses of multiple aliquots of the LFB. During the course of the analyses of real samples, periodic measurement of a LFB will determine whether the analytical method is in control and whether the laboratory is making acceptable identifications and measurements. As with every AQC test, the results of these analyses should be maintained with the records of the sample analyses to document this important information.

Laboratory Fortified Sample Matrix (LFM). The LFM is an aliquot of a real sample to which known quantities of the method analytes are added in the laboratory. The LFM is similar to the LFB except that the LFM may have real sample matrix effects which adversely affect method performance. The LFM is analyzed exactly like a sample to determine method performance with real samples and to determine whether the sample matrix contributes bias to the analytical results. The background concentrations of the analytes in the sample matrix must be determined in a separate aliquot and the measured values in the LFM corrected for background concentrations. A LFM should be analyzed with each group of samples that represent a significantly different matrix that could cause extraction or other analytical problems, for example, soil samples collected from a sandy area and from a high organic area.

Surrogate Analyte (SA). The SA was described at the end of the section "Quantitative Analysis" and was distinguished from the internal standard. The SA is a pure, well-behaved method analyte, which is extremely unlikely to be found in the sample, and which is added to every sample aliquot in known amount before extraction or other sample processing. It is measured with the same procedures used to measure other sample components, and its measured concentration, compared to the added concentration, provides an indication of method and laboratory performance with every sample. The SA recovery data are valuable because some samples may have all target analytes below the method detection limits, and the surrogate recovery provides some confidence that the method is in control with all samples.

Method and Instrument Detection Limits. The MDL and IDL are defined in a previous section of this chapter and these are often determined during the method validation process. During the course of the analyses of real samples, the maintenance of detection limits should be monitored by making concentrations of some analytes and surrogates near the detection limits in PCS, LFB, and LFM solutions.

Documentation

The results of the method validation and AQC performance tests should be considered an essential part of the documentation of the sample analyses. These data should be reported, at least in summary form, with the real sample analytical results and with the time relationships between the analyses of the samples and the performance tests. Control charts which plot the measured AQC parameters as a function of time are an excellent way of documenting the results of the AQC tests.[74] The validity of a sample analysis is enhanced and strongly supported when the documentation shows, for example, that the LRB, PCS, LFB, and LFM were all analyzed with satisfactory results within 24 h of the analysis of a sample.

REFERENCES

1. Budde, W. L.; Eichelberger, J. W. *Anal. Chem.* **1979**, *51*, 567A–574A.
2. Jungclaus, G. A.; Lopez-Avila, V.; Hites, R. A. *Environ. Sci. Technol.* **1978**, *12*, 88–96.
3. Hites, R. A.; Lopez-Avila, V. *Anal. Chem.* **1979**, *51*, 1452A–1456A.
4. *Principles of Environmental Sampling*, 2nd ed., Keith, L. H., Ed.; American Chemical Society: Washington, DC, 1996.
5. Keith, L. H. *Environ. Sci. Technol.* **1990**, *24*, 610–617.
6. *Compendium of Methods for the Determination of Toxic Organic Compounds in Ambient Air*; 2nd ed., USEPA Report EPA/625/R-96/010b, January 1999.
7. *Methods for the Determination of Organic Compounds in Drinking Water*, Suppl. III, USEPA Report EPA/600/R-95/131, August 1995; URL http://www.epa.gov/nerlcwww/methmans.html.
8. Grant, D. W. *Capillary Gas Chromatography*; John Wiley: New York, 1996.

9. Grob, R. L., Ed. *Modern Practice of Gas Chromatography,* 3rd ed.; John Wiley: New York, 1995.
10. Chapman, J. R. *Practical Organic Mass Spectrometry*, 2nd ed.; John Wiley: New York, 1994.
11. McFadden, W. *Techniques of Combined Gas Chromatography/Mass Spectrometry: Applications in Organic Analysis*; John Wiley: New York, 1973.
12. Watson, J. T. *Introduction to Mass Spectrometry*, 3rd ed.; Lippincott-Raven: Hagerstown, MD, 1997.
13. Message, G. M. *Practical Aspects of Gas Chromatography/Mass Spectrometry*; John Wiley: New York, 1984.
14. Holland, J. F.; Enke, C. G.; Allison, J.; Stults, J. T.; Pinkston, J. D.; Newcome, B.; Watson, J. T. *Anal. Chem.* **1983**, *55*, 997A–1012A.
15. Guilhaus, M. *J. Mass Spectrom.* **1995**, *30*, 1519–1532.
16. Todd, J. F. J. *Mass Spectrom. Rev.* **1991**, *10*, 3.
17. McLuckey, S. A.; Van Berkel, G. J.; Goeringer, D. E.; Glish, G. L. *Anal. Chem.* **1994**, *66*, 689A–696A.
18. March R. E.; Todd, J. F. J., Eds. *Practical Aspects of Ion Trap Mass Spectrometry*; CRC Press: New York, 1995, Vols 1–3.
19. Marshall, A. G.; Grosshans, P. B. *Anal. Chem.* **1991**, *63*, 215A–229A.
20. Grob, K.; Grob, G. *J. High Resolut. Chromatogr. Chromatogr. Commun.* **1979**, *2,* 109–117.
21. Budzikiewicz, H.; Djerassi, C.; Williams, D. H. *Mass Spectrometry of Organic Compounds*; Holden-Day: San Francisco, CA, 1967.
22. McLafferty, F. W.; Turecek, F. *Interpretation of Mass Spectra*, 4th ed.; University Science Books: Sausalito, CA, 1993.
23. Keith, L. H., Ed. *Identification and Analysis of Organic Pollutants in Water*; Ann Arbor Science: Ann Arbor, MI, 1976.
24. Wise, S. A.; Chesler, S. N.; Hertz, H. S.; Hilpert, L. R.; May, W. E. *Anal. Chem.* **1977**, *49*, 2306–2310.
25. Schulz, W. *Chem. Eng. News* **1998**, *76*, 14.
26. Swackhamer, D. L.; Charles, M. J.; Hites, R. A. *Anal. Chem.* **1987**, *59*, 913–917.
27. Coleman, W. E.; Melton, R. G.; Kopfler, F. C.; Barone, K. A.; Aurand, T. A.; Jellison, M. G. *Environ. Sci. Technol.* **1980**, *14*, 576–588.
28. Munch, D. J.; Graves, R. L.; Maxey, R. A.; Engel, T. M. *Environ. Sci. Technol.* **1990**, *24*, 1446–1451.
29. Eichelberger, J. W.; Bellar, T. A.; Donnelly, J. P.; Budde, W. L. *J. Chromatogr. Sci.* **1990**, *28*, 460–467.
30. Ho, J. S. *J. Chromatogr. Sci.* **1989**, *27*, 91–98.
31. Wilkins, C. L.; Giss, G. N.; Brissey, G. M.; Steiner, S. *Anal. Chem.* **1981**, *53*, 113–117.
32. Gurka, D. F.; Titus, R. *Anal. Chem.* **1986**, *58*, 2189–2194.
33. Kuehl, D. W. *Anal. Chem.* **1977**, *49*, 521–522.
34. Hites, R. A.; Biemann, K. *Anal. Chem.* **1967**, *39*, 965–970.
35. Hites, R. A.; Biemann, K. *Anal. Chem.* **1968**, *40*, 1217–1221.
36. Hites, R. A.; Biemann, K. *Anal. Chem.* **1970**, *42*, 855–860.
37. Biller, J. E.; Biemann, K. *Anal. Lett.*, **1974**, *7*, 515.
38. Shackelford, W. M.; Cline, D. M.; Faas, L.; Kurth, G. *Anal. Chim. Acta* **1983**, *146*, 15–27.
39. Shackelford, W. M.; Cline, D. M. *Anal. Chim. Acta* **1984**, *164*, 251–256.
40. Colby, B. N. *J. Am. Soc. Mass Spectrom.* **1992**, *3*, 558–562.

41. Herron, N. R.; Donnelly, J. R.; Sovocool, G. W. *J. Am. Soc. Mass Spectrom.* **1996,** *7*, 598–604.
42. Price, P.; *J. Am. Soc. Mass Spectrom.* **1991**, *2*, 336–348.
43. Oswald, E. O.; Albro, P. W.; McKinney, J. D. *J. Chromatogr.* **1974**, *98*, 363–448.
44. Heller, S. R.; McGuire, J. M.; Budde, W. L. *Environ. Sci. Technol.* **1975**, *9*, 210–213.
45. Fenselau, C. *Anal. Chem.* **1977**, *49*, 563A–570A.
46. Gohlke, R. S.; McLafferty, F. W. *J. Am. Soc. Mass Spectrom.* **1993**, *4*, 367–371.
47. Kearns, G. L. *Anal. Chem.* **1964**, *36*, 1402–1403.
48. Junk, G. A.; Svec, H. J. *Anal. Chem.* **1965**, *37*, 1629–1630.
49. Rake, A. T.; Miller, J. M. *J. Chem. Soc. A* **1970**, 1881–1888.
50. *NIST/EPA/NIH Mass Spectral Database*; National Institute for Standards and Technology: Gaithersburg, MD, 1992.
51. Eichelberger, J. W.; Harris, L. E.; Budde, W. L. *Anal. Chem.* **1975**, *47*, 995–1000.
52. Stafford, G. C., Jr.; Kelley, P. E.; Syka, J. E. P.; Reynolds, W. E.; Todd, J. F. J. *Int. J. Mass Spectrom. Ion Process.* **1984**, *60*, 85.
53. Eichelberger, J. W.; Budde, W. L. *Biomed. Environ. Mass Spectrom.* **1987**, *14*, 357–362.
54. Eichelberger, J. W.; Budde, W. L.; Slivon, L. E. *Anal. Chem.* **1987**, *59*, 2730–2732.
55. McLuckey, S. A.; Glish, G. L.; Asano, K. G.; Van Berkel, G. J. *Anal. Chem.* **1988**, *60*, 2312–2314.
56. Ong, V. S.; Hites, R. A. *Mass Spectrom. Rev.* **1994**, *13*, 259–283.
57. Donnelly, J. R.; Sovocool, G. W.; Mitchum, R. K. *J. Assoc. Off. Anal. Chem.* **1988**, *71*, 434–439.
58. *Annual Book of ASTM Standards*; American Society for Testing and Materials: W. Conshohocken, PA, 1974; Parts 17 and 18.
59. Method 625, *Federal Register* 3 December, **1979**, *44*, 69464; *Title 40 Code of Federal Regulations Part 136.*
60. Method 624, *Federal Register* 26 October, **1984**, 49, 43234–43439; *Title 40 Code of Federal Regulations Part 136.*
61. Personal communication, Robert E. Finnigan, ca. 1979.
62. Hunt, D. F.; Crow, F. W. *Anal. Chem.* **1978**, *50*, 1781–1784.
63. Ong, V. S.; Hites, R. A. *J. Am. Soc. Mass Spectrom.* **1993**, *4*, 270–277.
64. Eichelberger, J. W.; Kerns, E. H.; Olynyk, P.; Budde, W. L. *Anal. Chem.* **1983**, *55*, 1471–1479.
65. Abramson, F. P. *Anal. Chem.* **1975**, *47*, 45.
66. Sphon, J. A. *J. Assoc. Off. Anal. Chem.* **1978**, *61*, 1247–1252.
67. Cai, Z.; Ramanujam, V. M. S.; Giblin, D. E.; Gross, M. L.; Spalding, R. F. *Anal. Chem.* **1993**, *65*, 21–26.
68. Stein, S. E.; Ausloos, P.; Lias, S. G. *J. Am. Soc. Mass Spectrom.* **1991,** *2*, 441–443.
69. Stein, S. E.; Scott, D. R. *J. Am. Soc. Mass Spectrom.* **1994**, *5*, 859–866.
70. Sparkman, O. D. *J. Am. Soc. Mass Spectrom.* **1996**, *7*, 313–318.
71. Swallow, K. C.; Shifrin, N. S.; Doherty, P. J. *Environ. Sci. Technol.* **1988**, *22*, 136–142.
72. Richardson, S. D.; Thruston, A. D.; Collette, T. W.; Patterson, K. S.; Lykins, B. W.; Majetich, G.; Zhang, Y. *Environ. Sci. Technol.* **1994**, *28*, 592–599.
73. Glaser, J. A.; Foerst, D. L.; McKee, G. D.; Quave, S. A.; Budde, W. L. *Environ. Sci. Technol.* **1981**, *15*, 1426–1435.
74. Taylor, J. K. *Quality Assurance of Chemical Measurements*; Lewis: Chelsea, MI, 1987; p 79.

3

Organic Compounds Amenable to Gas Chromatography

Compounds amenable to gas chromatography (GC) are thermally stable and have sufficient vapor pressures at the injection port and column oven temperatures to achieve adequate separations in a reasonable analysis time. Furthermore, these compounds are not so polar, acidic, basic, or reactive that they adhere to acidic or basic sites on surfaces or react with other substances. These requirements were discussed in more detail in Chapter 2 in the section "Limitations of Mass Spectrometry". In spite of these limitations, a very large number of compounds are amenable to GC especially the lower molecular weight members of many classes of compounds:

- Saturated and unsaturated aliphatic and alicyclic hydrocarbons, aromatic hydrocarbons, and some polycyclic aromatic hydrocarbons.
- Halogenated hydrocarbons of all types.
- Oxygenated hydrocarbons including many alcohols, phenols, aldehydes, ketones, ethers, epoxides, acetals, ketals, esters, lactones, anhydrides, heterocyclic compounds, and a few carboxylic acids.
- Sulfur analogs of oxygenated compounds including thiols, thiophenols, thioethers (sulfides), and disulfides.
- Nitrogen compounds including primary, secondary, and tertiary amines, imines, nitriles, isonitriles, and heterocyclic compounds.
- Oxygen- and sulfur-containing compounds including thioesters sulfoxides, and sulfones.
- Oxygen- and nitrogen-containing compounds including *N*-oxides, amides, lactams, nitro compounds, and some *N*-nitrosamines.
- Silicon compounds especially trimethylsilyl derivatives of alcohols and phenols.

- Phosphorus compounds especially phosphines, and esters of phosphorous, phosphoric, and phosphonic acids.
- Some organometallic compounds such as metallocenes and polyalkyl- or polyaryl-metals.

The very large number of compounds amenable to GC accounts, in part, for the widespread utilization of this separation technique and the availability of superb commercial instrumentation. Excellent resolving power for a large number of compounds is readily attained with a variety of high-quality commercial GC columns. The manufacture's current literature should be consulted for many examples of the separations that are possible with these columns. Commercial GC/MS instruments are usually tightly integrated instrument systems with highly developed computer software to control instrument functions and acquire, store, reduce, and display analytical data (Chapter 2, "Conventional Gas Chromatography/Mass Spectrometry"). Conventional GC/MS is the technique specified in all the analytical methods discussed in this chapter and in Chapter 4. Conventional GC/MS is the *killer application* of MS in environmental chemical analysis, and the issues, analytes, and methods discussed in this chapter and Chapter 4 are the major reasons for this status. Alternative ionization and other techniques are used with GC/MS for some environmental and other analyses and these are discussed in Chapter 5.

While these powerful analytical capabilities are very supportive of the broad spectrum analytical strategy, specialists in GC separations and measurements generally have a strong tendency toward the adoption of the target analyte strategy. This tendency is supported by the availability of many commercial GC columns and detectors that are optimized for specific classes of compounds. Analytical chemists may also favor the target analyte strategy because of the potential presence of many compounds amenable to GC in many different types of samples. Analyses are greatly simplified by focusing on a few compounds of special interest while ignoring the other components of the sample.

The target analyte strategy dominates environmental analyses in the United States because federal regulations promulgated between the late 1970s and early 1990s specified GC with standard detectors and conventional GC/MS for a number of specific analytes. The compounds discussed in this chapter and in Chapter 4 are generally target analytes in environmental analyses and they are organized into groups that have similar properties. However the broad spectrum analytical strategy was made practical by GC/MS and this valuable approach should not be neglected because of the emphasis on target analytes. The many target compounds discussed in this chapter and in Chapter 4 and the analytical methods devised for their determination should be considered model compounds and strategies for broad spectrum analyses whenever this strategy is appropriate.

Target Analytes—The Priority Pollutants

As described in Chapter 1, a major environmental and analytical chemistry event of the 1970s was the consent decree of 1976. Under the terms of this order the

U.S. Environmental Protection Agency (USEPA) was compelled to implement certain provisions of the Federal Water Pollution Control Act Amendments of 1972.[1] In order to accomplish this, a substantial analytical program was initiated by the USEPA to identify and measure toxic substances in wastewater effluents from industries in 21 categories. While broad spectrum chemical analyses were being practiced in a few laboratories at the time, the USEPA believed that a specific list of target analytes was necessary to manage a large-scale chemical analysis program. The USEPA and the environmental interest groups, who were the plaintiffs in the consent decree, agreed on 129 priority pollutants which included 105 specific organic compounds, eight commercial products which were mixtures of organic compounds, 13 elements, and fibrous asbestos. The other two priority pollutants were actually method-defined tests for total inorganic cyanides and a class of phenols. A target analyte strategy had been selected for the consent decree analyses, and this strategy would dominate environmental analyses for regulatory purposes in the United States for many years. The priority pollutant list was incorporated into the Clean Water Act of 1977 and had a major impact on the development of conventional GC/MS and some other analytical technologies.

Selection Criteria for the Priority Pollutants

The reasons for the selection of the specific 129 priority pollutants are described by Larry Keith and William Telliard, who were involved in the process in 1976.[1] The consent decree included a list of 65 compounds and classes of substances, which was called the toxic pollutant list. About 32 of these are specific organic compounds and the other 33 are classes of substances, for example, phthalate esters and halomethanes. The classes of substances included all compounds of Ag, As, Be, Cd, Cr, Cu, Hg, Ni, Pb, Sb, Se, Tl, and Zn as well as the elements themselves. Many of the specific compounds, elements, and classes of substances on this list were suspected or known to have adverse impacts on wildlife or were toxic or carcinogenic. Many also had been reported as environmental pollutants in various types of samples. For example, the pesticide 1,1-bis(4-chlorophenyl)-2,2,2-trichloroethane (4,4′-DDT) was known to interfere with the reproduction of some birds at the top of the food chain,[2] and benzidine, 3,3′-dichlorobenzidine, and some *N*-nitrosamines were known human or animal carcinogens.[3] The 33 classes of substances, including all compounds of the 13 elements, were a major analytical problem because thousands of individual compounds could be included in some of these classes. A broad spectrum analytical strategy for these was judged by the USEPA to be impractical as well as very costly.

In order to simplify the analytical problem, the target list of 129 specific priority pollutants was established.[1] All the specific organic compounds on the consent decree toxic pollutant list were selected for the priority pollutant list. Most of the other organic priority pollutants were chosen to represent the classes of compounds on the toxic pollutant list. Criteria for selection of these representatives included availability of commercial samples for use as standards in chemical analyses, reports in the literature indicating some compounds were potentially significant water pollutants, and chemical production data. Four compounds, chloroform, bromodichloromethane, chlorodibromomethane, and bromoform, were known chlorine disinfection by-products in drinking water and it was reasonable to assume these would also be produced in chlorinated wastewaters. One additional compound and five commercial mixtures of organic compounds were added to the list at the request of the environmental interest groups, who were the plaintiffs in the consent decree. The requirement to measure all compounds of the 13 elements was addressed by measuring the total amount of each element after suitable digestion of the compounds in the sample. The elemental analyses under the priority pollutant program is briefly described in Chapter 7.

Two other selection criteria for the organic priority pollutants are clearly implied by the chemical and physical properties of the compounds on the list. Essentially, all of the organic compounds selected were known to be amenable to separation by GC. Furthermore, almost all the organic compounds were known to be amenable to separation from water by at least one of several techniques. Not as much analytical information was available for only a few compounds on the list, for example, acrolein, acrylonitrile, the two benzidines, and the three *N*-nitroso compounds. However, these few were either specifically named on the consent decree toxic pollutant list or were the only viable representatives of classes of substances on the list. It was reasonable to assume that they would be amenable to the same extraction and GC techniques.

In effect, analytical considerations played a major role in the selection of the priority pollutants, which have been the basis of many other regulatory and nonregulatory target compound lists prepared during the last 20 years. Some of the priority pollutants were selected because they were known environmental contaminants, and they were known because analytical techniques were available to separate and identify them. Potentially widespread environmental contaminants that were known to be difficult to extract from water or unsuitable for GC were not on the priority pollutant list. Compounds not amenable to GC, or not easily converted into derivatives amenable to GC, were usually not known to be implicated in environmental problems because there were no analytical techniques comparable to GC/MS to separate and identify them. After some time only a few of the priority pollutants were found to be difficult to extract from water or separate by GC, and these were either deleted from target compound lists or determined by other analytical techniques.

Volatile and Semivolatile Organic Compounds

The 105 specific organic compounds and eight commercial product mixtures selected as priority pollutants were divided into two broad groups based on their chemical and physical properties. These two groups quickly became known as the volatile organic compounds (VOCs) and the semivolatile organic compounds.

The VOCs were initially defined as compounds with low water solubilities that are efficiently transferred into the vapor phase during purging of a water sample with an inert gas at ambient temperature. The original consent decree priority pollutant (PP) VOCs are indicated by a "+" in the PP column in Table 3.1. Many other VOCs were subsequently discovered as environmental pollutants and many of these are also included in Table 3.1. Some VOCs were discovered in drinking water, and maximum contaminant levels (MCLs) or monitoring requirements were established in the United States for some of these (Table 3.1). Many VOCs were found in ground water and soil near solid and hazardous waste disposal sites. The Clean Air Act Amendments of 1990 contains a list of 189 hazardous air pollutants (HAPs) and 107 of these are VOCs listed in Table 3.1 and identified by a "+" in the HAP column. Some of the VOCs are common solvents and/or high-volume industrial intermediates or products.

The VOCs are generally C_1–C_8 with molecular weights usually less than 200. Only about 6% of the VOCs in Table 3.1 have molecular weights greater than 200 and these few always contain at least two Br or four Cl atoms. Many VOCs contain only one or two carbons and one or more Br or Cl atoms. The presence of Br and/or Cl in almost 47% of the VOCs in Table 3.1 is indicated in their mass spectra by the patterns of abundances in clusters of ions caused by the statistical distribution of naturally occurring isotopes (see Table 2.4). The structures of the 143 VOCs in Table 3.1 are generally either familiar to most analytical chemists or easily deduced from their systematic names. Since about 94% of the compounds in Table 3.1 have molecular weights less than 200, their electron ionization (EI) mass spectra usually have only a few ions and are not difficult to understand and interpret. Molecular ions are observed in the EI spectra of about 80% of the compounds in Table 3.1. The structures and nominal 70-eV EI mass spectra of these VOCs are contained in the NIST/EPA/NIH database, which is available for personal computers.[4]

The second group of organic consent decree priority pollutants are less volatile than the VOCs with generally higher molecular weights in the 100–500 amu range. They are also amenable to GC and are called semivolatiles. Generally, these compounds partition between water and a mildly polar water-immiscible organic solvent such as methylene chloride. The distribution favors the organic phase for most compounds although this is pH dependent for phenols, nitrogen bases, and other acidic or basic compounds. When the immiscible solvent is carefully evaporated to concentrate the extract, compounds in this group are generally retained in the liquid phase. The original semivolatile priority pollutants are identified and discussed in Chapter 4. Many other semivolatile compounds were discovered as environmental pollutants and these are also discussed in Chapter 4 along with analytical methods for their determination.

Volatile Organic Compounds in Water

In order to obtain information about the presence and concentration of the priority pollutant VOCs (Table 3.1) in industrial wastewaters, and address the consent decree requirements, an analytical method was required. By the mid-

Table 3.1 Volatile organic compounds amenable to GC that are consent decree priority pollutants (PP) and/or hazardous air pollutants (HAP) listed in the Clean Air Act Amendments of 1990 and/or have maximum contaminant levels (MCL) or monitoring requirements in drinking water or are candidates for future regulation

Compounds	PP[1]	HAP	GC/MS (air methods)	MCL[5] (μg/L)	GC/MS (water methods)
Acetaldehyde		+	TO-15		Water soluble
Acetone					524.2
Acetonitrile		+	TO-15		Water soluble
Acetophenone		+	TO-15		
Acrolein	+	+	TO-15		Unstable
Acrylamide		+	TO-15		Water soluble
Acrylic acid		+	TO-15		Water soluble
Acrylonitrile	+	+	TO-15		524.2
Aniline		+	TO-15		Water soluble
Benzene	+	+	TO-14A, TO-15	5	524.2, D5790
Benzyl chloride		+	TO-14A, TO-15		Reacts with H_2O
Bromobenzene				[a]	524.2, D5790
Bromochloromethane				[a]	524.2, D5790
Bromodichloromethane	+			[b]	524.2, D5790
Bromoform	+	+	TO-15	[b]	524.2, D5790
Bromomethane	+	+	TO-14A, TO-15	[a]	524.2, D5790
1,3-Butadiene		+	TO-15		
Butanone		+	TO-15		524.2
n-Butylbenzene				[a]	524.2, D5790
sec-Butylbenzene				[a]	524.2, D5790
tert-Butylbenzene				[a]	524.2, D5790
Carbon disulfide		+	TO-15		524.2, D5790
Carbon tetrachloride	+	+	TO-14A, TO-15	5	524.2, D5790
Carbonyl sulfide		+	TO-15		
Chloroacetic acid		+	TO-15		Water soluble
Chloroacetonitrile					524.2
1-Chlorobutane					524.2
Chlorobenzene	+	+	TO-14A, TO-15	100	524.2, D5790
2-Chloro-1,3-butadiene		+	TO-15		
1-Chloro-2,3-epoxypropane		+	TO-15		
Chloroethane	+	+	TO-14A, TO-15	[a]	524.2, D5790
Bis(2-chloroethyl) ether	+	+	TO-15		Does not purge
2-Chloroethyl vinyl ether	+				624
Chloroform	+	+	TO-14A, TO-15	[b]	524.2, D5790
Chloromethane	+	+	TO-14A, TO-15	[a]	524.2, D5790
Bis(chloromethyl) ether		+	TO-15		Reacts with H_2O
Chloromethyl methyl ether		+	TO-15		Reacts with H_2O
3-Chloro-1-propene		+	TO-15		524.2
2-Chlorotoluene				[a]	524.2, D5790
4-Chlorotoluene				[a]	524.2, D5790
Diazomethane		+	TO-15		Reacts with H_2O
Dibromochloromethane	+			[b]	524.2, D5790
1,2-Dibromo-3-chloropropane		+	TO-15	0.2	524.2, D5790
1,2-Dibromoethane		+	TO-14A, TO-15	0.05	524.2, D5790
Dibromomethane				[a]	524.2, D5790
1,2-Dichlorobenzene	+		TO-14A	600	524.2, D5790
1,3-Dichlorobenzene	+		TO-14A	[a]	524.2, D5790

(*continued*)

Table 3.1 (*continued*)

Compounds	PP[1]	HAP	GC/MS (air methods)	MCL[5] (μg/L)	GC/MS (water methods)
1,4-Dichlorobenzene	+	+	TO-14A, TO-15	75	524.2, D5790
trans-1,4-Dichloro-2-butene					524.2, D5790
Dichlorodifluoromethane			TO-14A	[a]	524.2, D5790
1,1-Dichloroethane	+	+	TO-14A, TO-15	[a]	524.2, D5790
1,2-Dichloroethane	+	+	TO-14A, TO-15	5	524.2, D5790
1,1-Dichloroethene	+	+	TO-14A, TO-15	7	524.2, D5790
cis-1,2-Dichloroethene			TO-14A	70	524.2, D5790
trans-1,2-Dichloroethene	+			100	524.2, D5790
Dichloromethane	+	+	TO-14A, TO-15	5	524.2, D5790
1,2-Dichloropropane	+	+	TO-14A, TO-15	5	524.2, D5790
1,3-Dichloropropane				[a]	524.2, D5790
2,2-Dichloropropane				[a]	524.2, D5790
1,1-Dichloropropanone					524.2
1,1-Dichloropropene				[a]	524.2, D5790
cis-1,3-Dichloropropene	+	+	TO-14A, TO-15	[a, c]	524.2, D5790
trans-1,3-Dichloropropene	+		TO-14A	[a, c]	524.2, D5790
1,2-Dichloro-1,1,2,2-tetrafluoroethane			TO-14A		
Diethyl ether					524.2
Diethyl sulfate		+	TO-15		Reacts with H_2O
N,N-Dimethylaniline		+	TO-15		
Dimethylcarbamyl chloride		+	TO-15		Reacts with H_2O
Dimethylformamide		+	TO-15		Water soluble
1,1-Dimethylhydrazine		+	TO-15		Water soluble
Dimethyl sulfate		+	TO-15		Reacts with H_2O
1,4-Dioxane		+	TO-15		Water soluble
1,2-Epoxybutane		+	TO-15		
Ethyl acrylate		+	TO-15		Water soluble
Ethylbenzene	+	+	TO-14A, TO-15	700	524.2, D5790
Ethyl carbamate		+	TO-15		
Ethyleneimine (aziridine)		+	TO-15		Water soluble
Ethylene oxide (oxirane)		+	TO-15		Water soluble
Ethyl methacrylate					524.2
Formaldehyde		+	TO-15		Water soluble
Hexachloro-1,3-butadiene	+	+	TO-14A, TO-15	[a]	524.2, D5790
Hexachloroethane	+	+	TO-15		524.2, D5790
Hexane		+	TO-15		
2-Hexanone					524.2
2-Hydroxyphenol		+	TO-15		Water soluble
Iodomethane		+	TO-15		524.2
Isophorone	+	+	TO-15		Does not purge
Isopropylbenzene (cumene)		+	TO-15	[a]	524.2, D5790
4-Isopropyltoluene				[a]	524.2, D5790
Methacrylonitrile					524.2
Methanol		+	TO-15		Water soluble
Methyl acrylate					524.2
Methyl *tert*-butyl ether		+	TO-15		524.2, D5790
Methyl hydrazine		+	TO-15		Water soluble
Methyl isobutyl ketone		+	TO-15		524.2, D5790
Methyl isocyanate		+	TO-15		Reacts with H_2O

(*continued*)

Table 3.1 (*continued*)

Compounds	PP[1]	HAP	GC/MS (air methods)	MCL[5] (μg/L)	GC/MS (water methods)
Methyl methacrylate		+	TO-15		524.2
2-Methylphenol (*o*-cresol)		+	TO-15		Water soluble
3-Methylphenol (*m*-cresol)		+	TO-15		Water soluble
4-Methylphenol (*p*-cresol)		+	TO-15		Water soluble
Naphthalene	+			[a]	524.2, D5790
Nitrobenzene	+	+	TO-15		524.2
2-Nitropropane		+	TO-15		524.2
N-Nitrosodimethylamine	+	+	TO-15		Water soluble
N-Nitrosomorpholine		+	TO-15		
N-Nitroso-*N*-methylurea		+	TO-15		
Pentachloroethane					524.2
Phenol	+	+	TO-15		Water soluble
Phosgene		+	TO-15		Reacts with H_2O
1,3-Propranesultone		+	TO-15		
β-Propiolactone		+	TO-15		Water soluble
Propionaldehyde		+	TO-15		Water soluble
Propionitrile					524.2
M-Propylbenzene				[a]	524.2, D5790
1,2-Propyleneimine		+	TO-15		
Propylene oxide		+	TO-15		Reacts with H_2O
Styrene		+	TO-14A, TO-15	100	524.2, D5790
Styrene oxide		+	TO-15		
1,2,3,4-Tetrachlorobenzene					D5790
1,2,4,5-Tetrachlorobenzene					D5790
1,1,1,2-Tetrachloroethane				[a]	524.2, D5790
1,1,2,2,-Tetrachloroethane	+	+	TO-14A, TO-15	[a]	524.2, D5790
Tetrachloroethene	+	+	TO-14A, TO-15	5	524.2, D5790
Tetrahydrofuran					524.2
Toluene	+	+	TO-14A, TO-15	1000	524.2, D5790
1,2,3-Trichlorobenzene				[a]	524.2, D5790
1,2,4-Trichlorobenzene	+	+	TO-14A, TO-15	70	524.2, D5790
1,1,1-Trichloroethane	+	+	TO-14A, TO-15	200	524.2, D5790
1,1,2-Trichloroethane	+	+	TO-14A, TO-15	5	524.2, D5790
Trichloroethene	+	+	TO-14A, TO-15	5	524.2, D5790
1,1,2-Trichloro-1,2,2-trifluoroethane (Freon 113)			TO-14A		
Trichlorofluoromethane	+		TO-14A	[a]	524.2, D5790
1,2,3-Trichloropropane				[a]	524.2, D5790
Triethylamine		+	TO-15		Water soluble
1,2,4-Trimethylbenzene			TO-14A	[a]	524.2, D5790
1,3,5-Trimethylbenzene			TO-14A	[a]	524.2, D5790
2,2,4-Trimethylpentane		+	TO-15		
Vinyl acetate		+	TO-15		Unstable
Vinyl bromide		+	TO-15		
Vinyl chloride	+	+	TO-14A, TO-15	2	524.2, D5790
o-Xylene		+	TO-14A, TO-15	[d]	524.2, D5790
m-Xylene		+	TO-14A, TO-15	[d]	524.2, D5790
p-Xylene		+	TO-14A, TO-15	[d]	524.2, D5790

[a]Drinking water monitoring required by the Code of Federal Regulations (Ref. 6).
[b]The sum of four trihalomethanes shall not exceed 80 μg/L (Ref. 7).
[c]Monitoring required, but the *cis* and *trans* isomers are not differentiated in the regulation (Ref. 6).
[d]The sum of three xylenes shall not exceed 10 mg/L (Ref. 5).

1970s many techniques had been investigated for the determination of members of this class of substances in water and other condensed-phase sample matrices. Some of these laboratory techniques are briefly reviewed in this section and the extraction technique selected by the USEPA is described. Sometimes it is advantageous to conduct analyses of water at the sampling site or in a processing plant. Field and continuous mass spectrometric techniques for VOCs in water are considered in Chapter 8.

The priority pollutant VOC initiative resulted in the discovery of many VOCs in many kinds of aqueous systems including drinking waters,[8] industrial waste-waters,[1] ground waters near solid and hazardous waste disposal sites,[9,10] and commercial products such as bottled water.[11] Versatile and very reliable conventional GC/MS analytical methods for VOCs in water are widely used, documented in detail, and validated in multilaboratory studies. These methods have also been applied in many other areas of investigation including forensic studies, petroleum exploration, and analyses of beverages, blood, flavors, food, and recycled water used in space exploration.

Extraction Techniques for VOCs in Water

Perhaps the most obvious and simple technique for the determination of VOCs in water is injection of an aliquot of a water sample into the gas chromatograph. This technique has been used for many years, but suffers from several limitations and problems.[12] During vaporization of the water and other volatiles, inorganic salts and other nonvolatile material present in natural waters and wastewaters are deposited in the GC injection port liner and the first part of the GC column. These deposits can shorten the lifetime of these components, impact the separation efficiency of the column, and increase the need for maintenance. Since it is somewhat difficult to concentrate an aqueous sample and not lose the VOCs, the quantities of analytes in a few microliters of water sample may not be sufficient to identify or measure the analytes. Therefore, aqueous injection has limited utility although it may be the technique of choice for some polar water soluble but volatile analytes. The determination of these substances by aqueous injection GC/MS is discussed later in this chapter.

Liquid–liquid extraction of water and other samples followed by GC separation of the VOCs has been explored, but is of limited value. Generally, VOCs have vapor pressures at ambient temperatures similar to the vapor pressures of common immiscible solvents used to extract them from water. Some of the VOCs of environmental interest in Table 3.1 are common laboratory solvents. Therefore, concentration of the solvent extract by evaporation in order to present sufficient quantities of analytes to the GC/MS system is generally precluded because this would result in losses of VOCs. The maximum concentration factor that can be obtained with liquid–liquid extraction without solvent evaporation is determined by the equilibrium distribution of the analytes in the two phases. Another limitation of this approach is that some VOCs elute from the column with the solvent and cannot be readily separated from the often large solvent peak. Some success with some VOCs has been obtained by using a very low-boiling organic solvent, for example, *n*-pentane (bp 36 °C), to extract higher boiling VOCs from water.[13] This technique is feasible if the analyte concentrations in the extract are sufficiently high, or the spectrometer sensitivity is sufficiently high, that concentration of the solvent extract is not necessary. An alternative to a low-boiling solvent is a much higher boiling solvent, for example, hexadecane, to partition the VOCs from water. With this approach the VOCs elute from the column well before the high-boiling solvent. However, evaporative concentration of the extract is again precluded and high-boiling solvents produce more persistent background contamination in GC injection systems, GC columns, and the mass spectrometer ion source.

Headspace Analysis. A desirable approach to the determination of VOCs is the elimination of the solvent altogether, and one important solventless technique is equilibrium or static headspace analysis.[14] With this technique, which has been widely explored and used, the volatile components of a sample are allowed to equilibrate at constant temperature between the condensed phase and the confined vapor space, or headspace, above the condensed phase. The sample can be aqueous, nonaqueous, or even a solid phase. An aliquot of the vapors in the headspace is injected into the gas chromatograph, and the components of the vapor are separated and measured. Sampling of the vapor phase temporarily disturbs the equilibrium concentrations in the condensed and vapor phases and this can introduce errors which are larger for larger aliquots. The concentrations of analytes in the condensed phase are calculated from their measured concentrations in the vapor phase and a calibration which relates measured concentrations in the vapor phase to known concentrations in the liquid phase in a series of calibration solutions.

The distribution of a VOC between the vapor phase and the liquid phase depends on its Henry's law constant H in Eq. 3.1, where P is the partial pressure of the VOC vapor and C is its concentration in the liquid phase:

$$P = HC \tag{3.1}$$

Henry's law constants are known for many VOCs in water and reported values at 25 °C range from 0.12 kPa m^3/mol for 1,1,2-trichloroethane, which is somewhat water soluble (4.42 g/L), to 13.32 for 1,1-dichloroethene, which is more than a factor of 10 less soluble (0.400 g/L).[15] Therefore, a difference of more than a factor of 100 in H indicates that low micrograms per liter concentrations of less soluble VOCs may exert sufficient vapor pressures for measurement, but more soluble VOCs may not be detected. In addition to temperature, Henry's law constants are dependent on the nature of the aqueous phase including the concentrations of dissolved salts, the presence of miscible organic solvents, and the presence of other substances that affect the solubility of the VOCs in the condensed phase.

The advantage of the static headspace technique is that it is simple and does not require sample processing equipment other than suitable sample containers fitted with leak-free septum seals. The impact of some sample variables can be minimized by using internal concentration calibration standards (Chapter 2,

"Quantitative Analysis"). Calibration solutions should have identical volumes in the same size containers as the sample solutions and should have the same general composition as the sample matrices. This requirement may not be easily accomplished in a laboratory which must analyze a wide variety of aqueous sample types from drinking water to wastewater. The low vapor pressures of some analytes can be overcome somewhat by warming the sample, but this requires calibration with standards at the same temperature and additional time for equilibration to occur. Regardless of these limitations, the equilibrium static headspace technique is a valuable approach to VOC analyses by GC/MS.

Purge and Trap Technique. This technique was invented by USEPA chemist Thomas A. Bellar, who had conducted research to develop a method for VOCs in air, using the Tenax-GC™ adsorbent. When the air pollution research program was moved from the Taft Research Center in Cincinnati, OH, to Research Triangle Park, NC, in the early 1970s, Bellar chose to stay in Cincinnati and joined the USEPA's water research program. Figure 3.1 is a diagram showing the original design of an all-glass purging vessel for a 5 mL aqueous sample.[16] The water sample is placed in the vessel above the medium-porosity sintered glass disk. The volume above the sample is minimized to eliminate dead volume

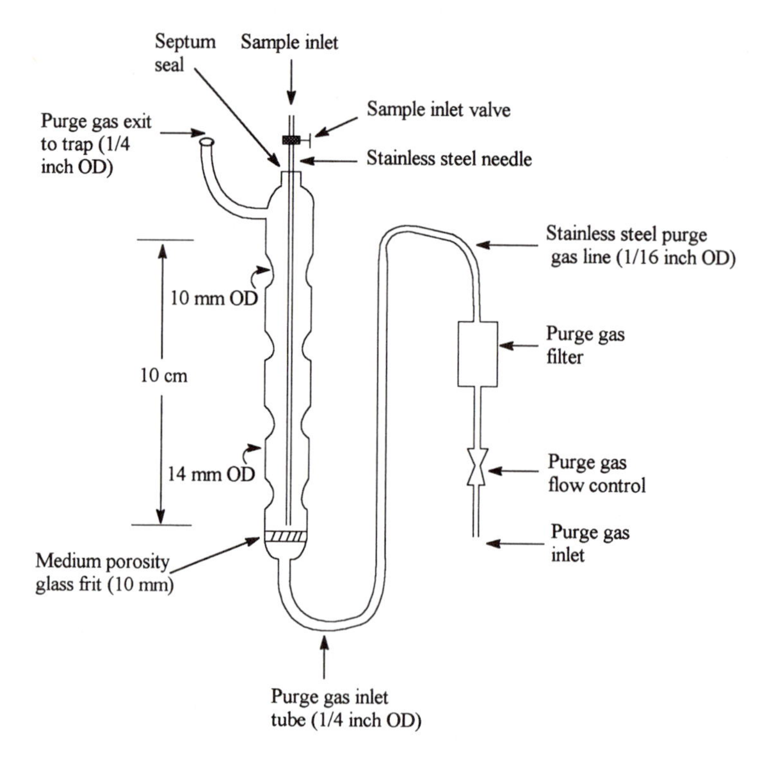

Figure 3.1 Diagram of the original design of an all-glass purging vessel for a 5 mL aqueous sample.

effects, yet sufficient space is allowed to permit most foams from wastewaters to disperse. The helium purge gas passes through the disk, and the finely divided helium bubbles agitate the water sample. The VOCs in the sample rapidly equilibrate between the condensed and vapor phases, and the vapors are entrained in the stream of gas bubbles which emerges from the liquid. The exit of the purging vessel is connected to a ~2.7 mm (ID) × 25 cm metal tube containing a porous adsorbent or several adsorbents in series. The helium and most of the water vapor are vented, but the VOCs are trapped on the adsorbents.

The adsorbent used initially was the organic polymer Tenax-GC™, which was effective for compounds boiling above about 30 °C.[8,16] Later, silica gel was placed in series after the Tenax to trap lower boiling compounds.[17] Still later, a third adsorbent, coconut charcoal, was optionally placed last in series to trap very volatile compounds, for example, dichlorodifluoromethane.[18] A major advantage of the Tenax polymer is its hydrophobic aromatic hydrocarbon structure (Chart 3.1), which minimizes the adsorption of water vapor. This polymer is also thermally stable to >200 °C and available in pure form with low background contamination. After completion of the sample purge, which is about 11 min for a 5 mL sample, the adsorbed compounds are thermally desorbed in a stream of helium in the reverse direction and introduced on to the GC column. The small water sample and short purge time are designed to minimize or eliminate breakthrough of analytes from the trap while achieving purging efficiencies of >70% for most compounds in just a few minutes. Larger sample volumes, for example, 25 mL, are used but purging efficiencies are diminished with larger volumes because of redissolution of VOCs in the sample. Purging of water is normally conducted at ambient temperature to minimize the production of water vapor, but some analyses are conducted at slightly elevated temperatures. Some water vapor is invariably collected, especially in the second and third adsorbents, and several techniques were later developed to minimize the presence of water vapor in the desorbed vapors.

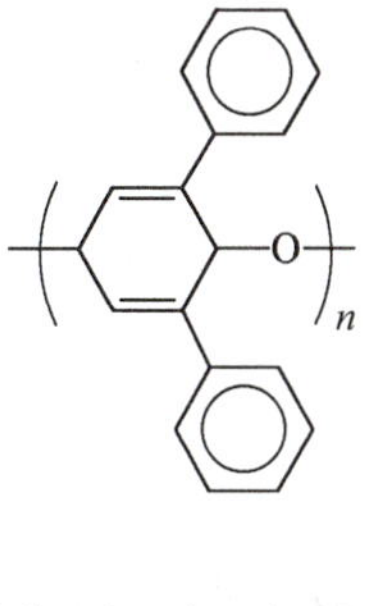

O
n

2,6-Diphenyl-*p*-phenylene oxide polymer
(Tenax-GC)

Chart 3.1

Method detection limits for the purge and trap technique combined with contemporary GC/MS are generally in the region of 1 μg/L or less.[18] Compounds that have small Henry's law constants, for example, the three dichlorobenzene isomers, and therefore relatively high detection limits with the static headspace technique, are purged and trapped with reasonable efficiencies. However, purging efficiencies for individual compounds can vary widely and depend on several factors including the purge gas flow rate, purge time, sample temperature, sample matrix, and desorption time. These operating parameters have been studied in great detail, and standard operating conditions have been established to minimize losses of all analytes, especially very volatile compounds.[18]

In order to minimize variations caused by fluctuations in operating conditions, internal concentration calibration standards are used with the mass spectrometric methods. Response factors are determined by purging calibration solutions containing internal standards and analytes (Chapter 2, "Quantitative Analysis"). With this calibration technique the measured concentrations in the sample are the true concentrations corrected to 100% purging efficiencies. The only assumption is that the purging efficiencies of all analytes and ISs are not significantly different in the samples and the calibration solutions. This may not be true if calibration standards are prepared in reagent water and samples contain miscible organic solvents or other substances that modify purging efficiencies.

Closed Loop Stripping. An alternative gas purging technique was developed by Kurt Grob in Europe in the early 1970s[19,20] This technique is similar to that invented by Bellar,[16] but a pump is used to recirculate the purge gas many times through the sample and a charcoal adsorbent. This extraction technique is known as closed loop stripping (CLS) or Grob closed loop stripping (GCLS) or sometimes as exhaustive gas stripping. Because the charcoal and other activated carbon adsorbents strongly retain many analytes, a solvent, usually carbon disulfide or methylene chloride, is used to elute the analytes from the adsorbent. The major advantage of CLS is lower detection limits in the low nanogram per liter range especially with large water samples, for example, 4–5 L. However, some volatile compounds may not be detected because they are continuously desorbed from the carbon surface and tend to remain in the vapor phase or they elute from the column with the solvent. Other more polar compounds are often strongly adsorbed on the carbon surface and are not eluted by the solvent. The CLS technique is most effective with low to intermediate molecular weight nonpolar compounds such as the chlorinated hydrocarbons and alkylated benzenes in Table 3.1.

Regulatory Adoption of Purge and Trap GC/MS

The purge and trap technique was used by the USEPA during the second half of the 1970s to determine VOCs in wastewater for the consent decree program.[1] Many of these analyses were conducted using packed-column GC/MS, but some potentially very important quantitative analyses used GC with standard detectors.[21] The packed GC columns used with these standard detectors did not have sufficient resolving power to separate fully the more than 30 priority pollutant

VOCs in calibration mixtures. Obviously, coeluting interfering compounds were much more likely in industrial wastewater samples and this deficiency was a serious source of error. The problem was partially addressed for standard detectors by employing several selective detectors for different subgroups of VOCs. These detectors included the electrolytic conductivity or microcoulometric detector for halogenated compounds and the photoionization detector for aromatic hydrocarbons. However, some coelutions were undetectable, especially when unexpected nontarget compounds were present in the sample (Chapter 2, "Alternatives to Conventional GC/MS"). With packed-column GC/MS the probability of coelution was just as great, but the continuous measurement of spectra data acquisition strategy made recognition of coeluting VOCs rather easy (Chapter 2, "Data Acquisition Strategies"). The extracted ion chromatogram technique, also described in Chapter 2, was widely used for identifications and measurements of coeluting VOCs.

The Detector Debate. During the period from the mid-1970s to the mid-1980s there was considerable controversy and debate over the issue of using GC/MS or GC with standard detectors for priority pollutant analyses.[21] Some investigators experienced in organic environmental analyses, especially those who measured chlorinated hydrocarbon pesticide residues, were staunch supporters of GC with standard detectors for regulatory analyses. Their opinions were that GC/MS was too complex and costly, that insufficient trained personnel and laboratory capacity were available, and that the nation could not afford this approach. They also contended that GC/MS was not as sensitive as standard GC detectors for some important chlorinated compounds. The counter argument was that GC/MS was far more reliable for compound identifications, more versatile, not as costly as it would first appear, and that instrument companies and commercial laboratories would provide the needed capacity if the market developed. Also, commercial GC/MS instrument systems were being improved rapidly with much more reliable and lower noise components. Some of these arguments are summarized in Chapter 2 in the section "Alternatives to Conventional GC/MS" and some were published in the scientific literature.[21–24]

Method 624. As described in Chapter 1, the USEPA was required by the Federal Water Pollution Control Act Amendments of 1972, and subsequently the Clean Water Act of 1977, to "promulgate guidelines establishing test procedures for the analysis of pollutants". Standard and validated analytical methods were considered essential to allow comparisons of analytical data obtained in different laboratories from the same samples and to minimize disputes between various personnel and laboratories.[25] After consideration of the techniques available for the extraction of VOCs from water, the USEPA selected the purge and trap technique for the first promulgated methods for VOCs in wastewater. However, instead of choosing either GC with standard detectors or GC/MS, the USEPA compromised and proposed on 3 December 1979 a suite of methods that used both GC with standard detectors and conventional GC/MS.[26] Methods 601–603 use the purge and trap extraction technique and packed-column GC with several standard GC detectors to determine 34 priority pollutant VOCs in wastewater. Method 624 uses the purge and trap extraction technique, packed-column GC, and continuous measurement of spectra to determine 31 priority pollutant VOCs in wastewater.

The 3 December 1979 proposal was very significant and had a major impact on the practice of environmental analytical chemistry and the development of commercial GC/MS instruments. While the wastewater compromise was probably a political necessity at the time, it was clear then and now that the methods employing GC/standard detectors and GC/MS are not equivalent for the reasons given in Chapter 2 in the section "Alternatives to Conventional GC/MS". However, the promulgation of both types of analytical methods did have some positive effects. Laboratories had a choice of several analytical technologies and could conduct environmental analyses, and compete in the emerging environmental laboratory business, regardless of their size and financial resources. Several sectors of the analytical instrument industry enjoyed substantially increased sales as many laboratories began using the proposed analytical methods for wastewater analyses.

Purge and Trap Analytes. Not all the compounds designated as VOC priority pollutants in Table 3.1 are viable purge and trap analytes. Some compounds, for example, *N*-nitrosodimethylamine and phenol, are not amenable to extraction from water by the purge and trap technique because of water solubility or chemical reactivity. They are included in Table 3.1, along with many analogous compounds such as chloroacetic acid and diazomethane, because they are VOC analytes in the air methods discussed later in this chapter. Some of the other VOC priority pollutants in Table 3.1, for example, naphthalene and nitrobenzene, are not practical analytes in the 600 series purge and trap methods because they are strongly retained on the specified packed GC columns. Even the three isomeric dichlorobenzenes, which are designated purge and trap analytes, have retention times of over 30 min, which makes them marginal analytes. However, these compounds, are amenable to extraction from water by purge and trap and can be readily separated using the fused silica capillary GC columns discussed later in this chapter. Some of the VOC priority pollutants in Table 3.1 are also designated semivolatiles and are analytes in methods for that class of compound, which are discussed in Chapter 4.

Innovations in the 600 Series of Analytical Methods

The USEPA's 600 series of analytical methods for wastewater proposed on 3 December 1979, including Methods 601–603 and 624, were innovative in several respects (other 600 series methods for semivolatiles are described in Chapter 4). The 600 series methods contained in single documents the sampling and sample preservation techniques, shipment and storage instructions, the extraction and sample preparation procedures, the instrument calibration procedure, the details of the GC separation and measurement, and mandatory quality control requirements. Published analytical methods of that time were not comprehensive or structured in this way. This documentation strategy was adopted so that virtually everything a reasonably trained analyst needed to know to conduct an analysis was included in a single organized document.

The integration of a complete set of techniques and procedures in an analytical method also made sense because It was known that parts of an analytical method are often interactive. Decisions made about sample preparation and calibration can have a significant impact on the results obtained with the separation and measurement procedures. While sometimes these procedures are separable with minimum impact, often they are not. The precision, accuracy, detection limits, and other figures of merit for analytical measurements should reflect the total analytical method including the sample preparation and all other processes used to obtain the final results.

Sampling and Sample Preservation Techniques. The sampling and sample preservation techniques used to collect water samples for determination of VOCs are crucial to the successful application of the analytical methods. These techniques have been studied in great detail and are described in standard analytical method descriptions.[18] Water samples for VOC analyses are collected in 40-mL or larger screw-cap vials fitted with perfluoroethene-faced silicone septa. The vials are filled to overflowing with minimum agitation of the sample and without allowing air bubbles to pass through the sample as the container is filled. If residual chlorine is present in a drinking water sample, a reducing agent—usually ascorbic acid—is added to the vial before filling to stop formation of disinfection by-products during shipment and storage prior to analysis. Acidification is also required to prevent microbiological degradation of some analytes. The vials are sealed without a headspace to prevent transfer of VOCs to the gas phase and they are shipped and stored at 4 °C or below.

Analytical Quality Control. The issue of analytical quality control was not new in the 600 series of methods, but had been pioneered by investigators working for predecessor agencies of the USEPA in the late 1960s.[25] They recognized that credible enforcement of environmental regulations required accurate and precise analytical measurements and documentation to prove that measurement systems were in control (Chapter 2, "Analytical Method Validation and Analytical Quality Control"). In May of 1979 USEPA Administrator Douglas M. Costle announced in an internal policy memorandum a mandatory quality assurance program for all USEPA supported or required monitoring activities. The purpose of this program was to ensure that valid analytical data are available to support decisions that could have a significant financial and social impact on the country. The incorporation of mandatory quality control provisions in the USEPA 600 series analytical methods was pioneering and sparked another intense round of debate, which lasted for many years, over the cost and value of analytical quality control.

The quality control requirements in the 1984 version of Method 624 state that the mass spectrometer must be tuned to produce an EI mass spectrum of the performance test compound 4-bromofluorobenzene (BFB) that meets the cri-

Table 3.2 Original BFB key ions and relative abundance criteria

m/z	Relative abundance criteria	Purpose of checkpoint
50	15–40% of mass 95	Low-mass relative abundance
75	30–60% of mass 95[a]	Low-mass relative abundance
95	Base peak, 100% relative abundance	
96	5–9% of mass 95	Low-mass resolving power and isotope ratio
173	< 2% of mass 174	Mid-mass resolving power
174	> 50% of mass 95	Mid-mass resolving power and relative abundance
175	5–9% of mass 174	Mid-mass resolving power and isotope ratio
176	> 95%, but < 101% of mass 174	Mid-mass resolving power and isotope ratio
177	5–9% of mass 176	Mid-mass resolving power and isotope ratio

[a]This acceptance range has been widened to 30–80% of mass 95 in USEPA Method 524.2.

teria in Table 3.2.[27] This compound was selected to serve as an alternative to decafluorotriphenylphosphine (DFTPP), which was described in Chapter 2 in the section "Standardization of Conventional GC/MS". The compound DFTPP is not a VOC and cannot be separated from water by purge and trap. Therefore, it is inconvenient to use DFTPP for performance tests on GC/MS instruments dedicated to VOC measurements. The compound BFB is a VOC, performs well with Method 624 and similar methods, and is unlikely to be present in an environmental sample. Therefore, it can be included in calibration solutions, or as a surrogate analyte in real environmental samples, and it provides a frequent quality control check on the performance of the instrument system.

The BFB criteria in Table 3.2 are designed to ensure that the mass spectrometer is calibrated correctly, that is, it gives the correct *m/z* values of measured ions. The relative abundance (RA) criteria are intended to encourage the design and tuning of mass spectrometers to produce GC/MS EI mass spectra that are at least similar to those recorded in the MS literature and databases of standard mass spectra. In addition, RAs of adjacent ions that meet the criteria in Table 3.2 are an indication that the mass spectrometer's resolving power is correctly adjusted. In order to lower detection limits, there is a tendency among some operators to use insufficient resolving power that does not adequately separate ions of adjacent *m/z*. This is sometimes done to avoid maintenance, for example, cleaning the ion source or quadrupole rods, which is required to achieve optimum signal/noise. The result of tuning for maximum signal with minimum concern for correct resolving power is often incorrect mass spectra that are not supportive of analyte identifications. Some operators will resist making correct resolving power adjustments to give standard mass spectra and refer to this as *detuning* the instrument.

Method 624 Status and Derivatives

As of 1999, Method 624 had not been revised to allow use of improved trap materials, open tubular GC column technology, or other improvements in analytical techniques. The mass spectral performance criteria for BFB in Method 624 (Table 3.2) have not changed although slightly revised criteria are used for other GC/MS methods for VOCs (Table 3.2, footnote "a"). Therefore, technically, but not legally, Method 624 is obsolete. Users of Method 624 for regulatory purposes usually have approval from appropriate authorities to use alternative traps, capillary GC columns, and other improvements discussed in the next section. Method 624 is historically important because of the innovations in its content and the precedents established for other analytical methods used for regulatory programs in the United States.

An analytical method very similar to Method 624 was used for many years in the USEPA's contract laboratory program which analyzed ground water and surface water samples taken from abandoned hazardous waste sites. Another version of Method 624, Method 1624, was included in the 1984 wastewater regulations to describe isotope dilution calibration for those samples where matrix effects cause significant differences in purging efficiencies of analytes in calibration and sample solutions (Chapter 2, "Quantitative Analysis").[27]

Method 8240 is a derivative of Method 624 that was published by the USEPA's Office of Solid Waste for the determination of VOCs in a variety of nonaqueous liquids, semisolids, solid wastes, and ground water.[28] For information on the techniques used for solid samples see the section of this chapter entitled "Volatile Organic Compounds in Sediments, Soil, Solid Wastes, and Tissue".

Purge and Trap with Fused Silica Capillary GC Columns

As described in Chapter 1, the development and commercial availability of high-resolution fused silica capillary GC columns in the early 1980s had a profound impact on the practice of organic environmental analysis including the determination of VOCs in all types of samples. With the resolving power limitations of packed GC columns removed and with the commercial availability of GC instruments with multiramp oven-temperature programming capabilities, the scope of the purge and trap method was significantly expanded. Many compounds amenable to the purge and trap extraction technique, for example, naphthalene, nitrobenzene, and the trichlorobenzene isomers, could be separated by GC in a reasonable total analysis time.

The connection of the exit of the purge and trap thermal desorption system to the entrance of a narrow-bore fused silica capillary GC column, for example, 0.32 mm ID, requires a technique to reduce the helium gas flow from the trap. One approach is a deactivated metal tube, or a section of uncoated but deactivated fused silica capillary, connected to the exit of the trap and cooled with liquid nitrogen. This cryogenic trap condenses the desorbed sample components in a narrow band, but allows the helium to be vented. After thermal desorption is complete, the cryogenic trap is connected to the injection port of the GC column, and the trap is rapidly heated with an appropriate flow of helium to transfer the analytes onto the column. This technique has the advantage of physically focusing the mixed analytes as they are injected into the chromatograph and thereby not limiting the resolving power possible with the GC column. However, precise focusing of the mixed analytes and optimum GC resolving power are far more critical with standard GC detectors than with capillary column GC/MS. A frequent problem with this approach is plugging of the cryogenic trap with ice from the small amount of water vapor desorbed from the trap. Water management techniques used to overcome this problem include purging the trap with dry helium for several minutes before thermal desorption to remove adsorbed water and the use of moderately cooled surfaces before the cryogenic trap to selectively remove water vapor.

An alternative to the cryogenic trap, if extremely low nanograms per liter detection limits are not required, is a standard gas stream splitter in the GC injection port to divert most of the gas flow from the GC column. Various types of gas stream splitters have been used including the open-split and all-glass jet separators shown in Figures 2.2 and 2.3. Sample components can be focused on the first part of the GC column by cooling the column oven to subambient temperature with cold nitrogen or carbon dioxide. Temperature programming of the GC oven, and GC, begin when desorption is complete.

Wider bore GC columns have the capacity to accept standard gas flows from the trap during thermal desorption, and chromatography can begin immediately or after cold trapping the analytes in the first part of the GC column. However, wider bore columns may not have the resolving power necessary for all types of samples. Depending on the pumping capacity of the mass spectrometer, an interface between the end of the GC column and the spectrometer inlet may be required. The chromatogram in Figure 2.5 was obtained using a wider bore 30 m × 0.53 mm (ID) fused silica GC column, multiramp linear temperature programming, and an open-split interface to the mass spectrometer.[29]

Drinking Water. Some of the stimulus for expansion of the purge and trap GC/MS method to other target analytes came from the discovery of chlorine disinfection by-products and other VOC contaminants in drinking water.[30] Thomas A. Bellar of the USEPA discovered the chlorine disinfection by-products chloroform, bromodichloromethane, chlorodibromomethane, and bromoform in Cincinnati drinking water in 1973 using the newly developed purge and trap extraction system.[30] These four VOCs became known as the *trihalomethanes* (THMs) and by the mid-1970s the widespread occurrence of these and other compounds in chlorine-disinfected drinking water was well established.[8] In 1979 the USEPA set a maximum contaminant level (MCL) in drinking water of 100 μg/L for the sum of the four THMs.[31] This regulatory limit was probably based more on practical analytical issues, that is, the availability of analytical methods capable of measuring the THMs with relatively low-cost equipment, than on an assessment of human health risks. A GC/MS analytical method for compliance monitoring of the THMs in drinking water was not specified in the regulation. During the late 1980s the MCL became more of a human health risk-based value and, therefore, an indication of the known potential for adverse effects from exposure to these compounds. In 1998 the USEPA lowered the MCL for the sum of the four THMs to 80 μg/L.[7]

In 1986 the Congress passed and President Reagan signed amendments to the Safe Drinking Water Act, which mandated the USEPA to establish national standards within three years for 83 drinking water contaminants that were named in the Act. The law also mandated national standards for 25 additional contaminants every three years beginning in 1990. In 1987 the USEPA completed the first of several new phases of regulation of contaminants in the nation's drinking water supplies. Maximum contaminant levels were established for eight VOCs and monitoring was required for an additional 51 VOCs.[32] During 1991 and 1992, MCLs were established for 14 VOCs in drinking water.[33,34] These MCLs are included in Table 3.1; footnote "a" to this table designates VOCs with monitoring requirements, but no MCL, in drinking water in the United States.

Method 524.2. The 1987 drinking water regulation included promulgation of USEPA Method 524.2, which uses the purge and trap extraction technique and capillary column GC/MS.[32] Method 524.2 requires the GC/MS system to meet the BFB specifications in Table 3.2 including the slight modification in footnote "a". However, Method 524.2, revision 4.1, allows flexibility in the construction of the purge and trap device, the materials and GC columns used, and some other operations.[18] For example, needle spargers and alternative adsorbents may be used subject to certain conditions and contingent on meeting specified accuracy, precision, method detection limit, and other quality control requirements. Several manufacturers produce GC columns suitable for Method 524.2, and their commercial literature should be consulted for examples of the conditions used and separations that are readily achieved. Instrument manufacturers offer automated purge and trap GC/MS systems to accomplish calibration, multiple analyses, and complete reporting of data with minimal operator intervention.

Method 524.2 Analytes. Revision 4.1 of Method 524.2 has been at least single-laboratory validated for 84 analytes, which are identified in Table 3.1 by the method number 524.2 in the far right column.[18] While no environmental samples are expected to contain all 84 validated analytes, calibration mixtures containing all analytes may be analyzed and any reasonable trade-off between analysis time and chromatographic resolution may leave some analytes unresolved. The GC conditions selected to separate 61 VOCs in about 30 min gave the chromatogram depicted in Figure 2.5 and left 12 pairs of compounds unresolved.[29] However, as described in Chapter 2 in the section "Alternatives to Conventional GC/MS", each member of the coeluting pairs has a sufficiently different mass spectrum, which permits identification of the coeluting compounds and accurate measurements using different quantitation ions.

Hundreds or thousands of other VOCs probably have physical and chemical properties that make them amenable to purge and trap extraction from water and separation under the GC conditions used in Method 524.2. Inspection of the compounds listed in Table 3.1 reveals that many have isomers, congeners, or analogs that would very likely be amenable to this analytical method. For example, 2- and 4-chlorotoluenes are validated Method 524.2 analytes and 3-

chlorotoluene is not, but surely it is a viable method analyte. While only a few hydrocarbons, mostly benzene derivatives, are validated Method 524.2 analytes, it is very reasonable to assume that numerous volatile aliphatic and alicyclic saturated and unsaturated hydrocarbons present in gasoline and other petroleum products are amenable to this method. Only a few of the many possible chlorofluorocarbons are in Table 3.1, but there are no chlorofluorohydrocarbons, fluorohydrocarbons, or perfluorocarbons and essentially all of these should be amenable to this analytical method. In any real environmental or other sample one or more of thousands of compounds could be present, which makes the number of potential Method 524.2 analytes very large. The validated method analytes in Table 3.1 should be viewed as model compounds for an analytical strategy that encompasses many analogous substances.

Method 524.2 Performance. Method 524.2, revision 4.1, contains suggested quantitation ions, retention data on several GC columns, and accuracy, precision, and method detection limit data for the 84 established method analytes.[18] The method performance obtained with a 30 m × 0.53 mm (ID) fused silica capillary column and an ion trap mass spectrometer (Figure 2.5) has been published.[29] With a 5-mL water sample and the analytes at 2 µg/L, the grand mean measurement accuracy for 54 compounds was 95% of the true value with a mean relative standard deviation (RSD) of 4%. With a 5-mL water sample and the analytes at 0.2 µg/L, the grand mean measurement accuracy for 52 compounds was 95% of the true value with a mean RSD of 3%. Two compounds, chloromethane and dichlorodifluoromethane, were not detected at 0.2 µg/L. The RSDs from repetitive determinations of dichlorodifluoromethane and another ambient-temperature gas, bromomethane, were 25 and 27%, respectively, at 2 µg/L. This variability is indication of the difficulty of measuring extremely volatile substances at very low concentrations with Method 524.2. Nevertheless, Method 524.2 is probably the most accurate, precise, and reliable of all the general purpose broad spectrum analytical methods available for organic compounds in water samples.

Multilaboratory Validation of Method 524.2 in Several Water Matrices. During 1992 the USEPA and the American Society for Testing and Materials (ASTM) Committee D-19 on Water jointly conducted a multilaboratory study of the ASTM derivative of Method 524.2, which is designated ASTM standard D5790. Over 40 volunteer laboratories participated in the study to characterize the performance of D5790 with fortified reagent water, ground water, industrial wastewater, drinking water, and a simulated hazardous waste site aqueous leachate. Table 3.3 shows the types of sample matrices analyzed, the sample sizes used, the number of participating laboratories, and the number of target analytes determined in each sample matrix. In this study the fortified analyte concentrations ranged from 0.2 to 80 µg/L. The summary report gives, for each analyte and each sample size in each matrix, the true concentration, the number of measurements, the mean recovery, the percentage recovery, the multilaboratory standard deviation, and the estimated single-analyst standard deviation.[35]

Table 3.3 Summary of the multilaboratory validation of ASTM Standard D5790 (USEPA Method 524.2)[35]

Water matrix	Sample size (mL)	Number of laboratories	Number of analytes
Reagent water	5	15	66
Reagent water	25	23	66
Ground water	5	10	65
Ground water	25	10	66
Wastewater	5	8	62
Drinking water	25	14	65
Synthetic leachate	5	3–6	9

The ASTM validated analytes are indicated in Table 3.1 by the designation D5790 in the far right column.

Method 524.2 Derivatives. Method 8260 is a derivative of Method 524.2 that was published by the USEPA's Office of Solid Waste for the determination of VOCs in a variety of nonaqueous liquids, semisolids, solid wastes, and ground water.[28] For information on the techniques used for solid samples see the section of this chapter entitled "Volatile Organic Compounds in Sediments, Soil, Solid Wastes, and Tissue". A method similar to Methods 524.2 and 8260 was used in the USEPA's contract laboratory program, which analyzed ground water and surface water samples taken from abandoned solid and hazardous waste sites.

Alternative Extraction Method for VOCs in Water

Because of the environmental importance of VOCs, and the strong interest in this class of substances in other areas of investigation, research has continued to develop alternative techniques for the extraction of these compounds from water and other fluids. Fused silica fibers coated with various polymeric substances have been used as microextraction devices for VOCs.[36] The coated fiber is placed in contact with the sample, and the analytes are partitioned into the stationary phase on the surface of the fiber. The fiber is then placed in the GC injection port, using a syringe-like device, and the analytes are thermally desorbed into the GC column. This technique has been evaluated with a range of compounds including some nonpolar VOCs and semivolatile compounds that are discussed in Chapter 4. It has several advantages including the use of inexpensive materials and the potential for application in the field, using a portable GC/MS instrument.

Polar Water Soluble Volatile Organic Compounds

Eighty-seven of the 143 VOCs listed in Table 3.1 are validated analytes with the purge and trap GC/MS water methods.[18,35] However, some of these validated analytes are measured with somewhat poorer precision than most analytes and consequently have method detection limits (MDLs) higher than those of most analytes (Chapter 2, "Detection Limits"). Some of the analytes measured with poorer precision are the more polar and water soluble which have low purging efficiencies.[37] For example, the method analytes acetone, methyl acrylate, methyl methacrylate, nitrobenzene, and tetrahydrofuran are measured with RSDs in the 5.7–18% range and have MDLs in the 0.28–1.6 μg/L range.

The 56 VOCs listed in Table 3.1 that are not validated analytes with the purge and trap GC/MS water methods, along with many of the validated analytes in water, are designated hazardous air pollutants (HAPs) and are discussed in the next section. However, many of these 56 compounds could also be of interest in aqueous media and these can be divided into four groups. One group consists of insoluble and unreactive compounds, for example, hexane and 2,2,4-trimethylpentane, that are reasonably expected to be amenable to the purge and trap methods although actual performance data may not be available. A second group consists of compounds, for example, benzyl chloride and diazomethane, that may be viable air pollutants, especially in dry air where collisions with water molecules are rare, but react with water and are converted into other compounds. A third group of compounds are not reactive in water, but are too water soluble to be amenable to extraction by purge and trap. However, the VOCs in this third group are reasonably amenable to liquid–liquid or liquid–solid extraction and are also classified as semivolatiles, which are discussed in Chapter 4. Compounds in this class include bis(2-chloroethyl)ether, isophorone, *N*-nitrosodimethylamine, and phenol. Also in this class is chloroacetic acid, which can be extracted from water at low pH or determined without extraction by using one of the condensed-phase separation techniques described in Chapter 6.

The fourth group consists of compounds that are not purgeable and do not react with water, but are water soluble and are not efficiently partitioned into an organic solvent or on to a solid phase adsorbent. These compounds are known as polar water-soluble VOCs and an alternative analytical approach is required for their determination in water. Compounds in this category in Table 3.1 include the common industrial and laboratory solvents acetonitrile, dimethylformamide, 1,4-dioxane, and methanol. Others not listed in Table 3.1 but in the same category include a broad range of low molecular weight C_1–C_6 alcohols, aldehydes, ketones, ethers, acids, esters, and amines, for example, ethanol, 1,2-dimethoxyethane, acetic acid, ethyl acetate, and pyridine. Even if some of these could be partitioned into an organic solvent or on to a solid-phase adsorbent, concentration of the extract or eluate by evaporation of the solvent may be precluded because of the similar volatilities of the analytes and the solvent. Similarly, separation of the analytes from the extraction or elution solvent on the GC column would be hindered by the similar vapor pressures. While some compounds in this fourth group may not be of great environmental interest because of rapid biodegradation, some are high-volume industrial solvents which could be of great interest as contaminants in water. Three techniques are useful for some members of this class of compounds and further research is needed to develop alternative techniques.

Aqueous Injection GC/MS. Aqueous injection GC/MS is attractive as a simple and straightforward technique with the potential of producing information quickly and without the need for sample preparation. However, aqueous injection GC has the disadvantage of potentially inadequate detection limits because the volume injected into a capillary GC column is limited to a few microliters and it is somewhat difficult to concentrate a water sample without loss of some of the VOCs. However, concentration of some analytes in water is possible by careful distillation or lyophilization.[38] Caution is also required with aqueous injection GC/MS because minerals and other nonvolatile material in water samples will contaminate the injection port liner and perhaps the first part of the GC column and probably shorten the lifetime of these components.

Aqueous injection packed-column GC/MS was evaluated during the early 1970s and found useful for some polar water-soluble VOCs.[12] Concentrations of 1–50 mg/L, which may be present in some environmental or other samples, were required to obtain reasonable EI mass spectra. Aqueous injection GC/MS was used for the determination of acrolein and acrylonitrile in early priority pollutant analyses.[1] A substantial improvement in performance has been obtained with a fused silica capillary column and an ion-trap mass spectrometer.[39] Detection limits for acetonitrile, acrylonitrile, propionitrile, butanone, ethyl acetate, 1,4-dioxane, and 3-pentanone were reported to be 4 μg/L in distilled water. This technique does not appear to be widely used, which may be related to concerns over the effects of water on the GC column and mass spectrometer. Detailed studies are needed to determine the accuracy and precision of quantitative measurements as well as the long-term impact of aqueous injection on GC/MS instruments.

Membrane Introduction Mass Spectrometry. Membrane introduction is the use of a semipermeable membrane to separate low molecular weight compounds from water and introduce them into a mass spectrometer. Essentially, all reported applications of this technique are concerned with field or real-time analyses which usually do not include a GC separation. The techniques and equipment used for field or real-time analyses with membrane introduction MS are described in Chapter 8. One study has been conducted to determine whether polar and water-soluble VOCs can be separated from water with a semipermeable membrane, trapped on a solid-phase adsorbent, and determined by capillary column GC/MS. This *permeate and trap* technique was applied to several compounds including acrylonitrile, 1,4-dioxane, 2-methyl-1-propanol, *N*-nitrosodimethylamine, propylene oxide, 2-propanol, and pyridine. Preliminary results were promising, but indicated the need for improved membrane materials and solid-phase adsorbents.[40]

Aqueous Phase Chemical Derivatization. Some compounds in this class are amenable to aqueous phase chemical reactions that produce derivatives that can be partitioned into an immiscible organic solvent or adsorbed on a solid-phase adsorbent. Aqueous phase derivatization of aldehydes and ketones is discussed under *Impingers* in the section of this chapter entitled "Sampling Techniques for VOCs in Air". Aqueous phase derivatization of carboxylic acids to esters may be feasible if the esters are not themselves polar water-soluble compounds.

Volatile Organic Compounds in Sediments, Soil, Solid Wastes, and Tissue

The VOCs listed in Table 3.1 are also of interest in other condensed-phase materials. Sediments are solids deposited at the bottoms of lakes, rivers, and streams. They are of environmental interest because water contaminants from intentional discharges or accidental spills often infiltrate and contaminate sediments which are an important part of natural ecosystems. The VOC contaminants may be adsorbed on the surfaces of sediment particles, dissolved in the associated water, or occluded in vacancies in the particle structures. The transfer of VOCs between these phases may be promoted by agitation from fast moving water, storms, and wave action. The two-phase nature of sediments complicates the sampling techniques and analytical methods used to analyze these materials. The VOCs are common soil contaminants in dry or moist soils from solid or hazardous waste disposal sites. These compounds are also found in agricultural soils where soil fumigants such as bromomethane have been used. Solid and semisolid wastes also contain VOCs often at concentrations much higher than found in sediments and soils.

Variations on the purge and trap method for water samples are commonly used to determine VOCs in sediments, soils, and solid wastes.[28,41] In one approach which has been widely used, samples collected in the field are placed in a wide-mouth container with little or no headspace, and the container is tightly sealed and shipped to the laboratory. An aliquot of the sample is placed in a purging vessel along with a small quantity of reagent water and the mixture is warmed to a moderate temperature, for example, 40 °C, to promote desorption of VOCs from the solids. The analytes are purged from the water with an inert gas, trapped on the same adsorbents used for water analyses, and thermally desorbed into a GC/MS system. This technique is not recommended because of probable large losses of VOCs both when the sample container is filled in the field and when it is opened in the laboratory to place an aliquot in the purging vessel.

Another commonly used technique is extraction of VOCs from a sample with a water-miscible solvent such as methanol.[28,41] The sample is collected and shipped to the laboratory as described above. An aliquot is transferred in the laboratory to a screw-cap vial containing a small volume of methanol and fitted with a perfluoroethene-faced silicone septum. The vial is sealed and the mixture is agitated in an ultrasonic bath at a low temperature to release the adsorbed and occluded VOCs from the particles. A small aliquot of the methanol extract is mixed with reagent water in a purging vessel, and the solution is analyzed by purge and trap GC/MS. This technique is also subject to significant losses of VOCs as the sample container is filled in the field and when it is opened in the laboratory to place an aliquot in the methanol.

Significantly improved results are obtained when a small sample of sediment or soil, for example, a few grams, is placed in the screw-cap vial with methanol in the field and the vial is quickly sealed and shipped at reduced temperature to the laboratory for analysis.[41] After sonication and centrifuging, an aliquot of the methanol solution is diluted with reagent water and the water is analyzed by the purge and trap method. With this technique, losses are still observed but at greatly reduced levels with mean recoveries of representative analytes in the 74–80% range. The referenced study was conducted using a conventional GC detector and external concentration calibration standards. Compensation for some or most of the observed losses may be possible with an internal concentration calibration standard added to the methanol in the field or to the extract before an aliquot is taken in the laboratory. The ideal sample preparation technique does not expose the sample to the atmosphere after the field sampling.

A vacuum distillation technique has been reported for the separation of VOCs from soil, oil, fish tissue, and high-organic water samples.[42] With this technique the sample is distilled under a moderate vacuum supplied by a mechanical pump (~ 1 Torr) and water vapor is condensed at −5 °C. Organic vapors, which pass through the water condenser, are condensed with liquid nitrogen at −196 °C in a cryogenic trap and subsequently thermally desorbed directly into a GC/MS system. This technique has been tested with over 35 VOCs and should be very useful for many types of complex and high-background condensed phases.

Volatile Organic Compounds in Air

Volatile organic compounds (VOC) in air have been identified and measured by environmental scientists for many years.[43] These analyses are of considerable importance because of the high potential for direct exposure to VOCs by all forms of life and by the impact of VOCs on valuable resources ranging from fine art to historic buildings. Atmospheric photochemical oxidation of some VOCs is responsible for the production and accumulation of air pollutants, for example, ozone, that have known adverse human health effects.[44] Research in the 1960s and earlier focused on hydrocarbons from automobile exhaust and industrial sources, and later, on pesticides from aerial spraying operations. These studies utilized early GC techniques and influenced the development of GC instrumentation including the flame ionization detector, which is widely used for hydrocarbon analyses. Interest in developing GC/MS methods for a broad spectrum of VOCs in ambient air began after passage of the Clean Air Act Amendments of 1970 and 1977. The issues surrounding the determination of VOCs in air are much different from those concerning this class of analytes in water and related media:

- The VOCs in air are defined as compounds having vapor pressures greater than 0.1 Torr at 25 °C and 760 mm Hg. The VOCs in water are defined as compounds amenable to extraction from a 5 mL water sample at ambient temperature by inert gas purging with an efficiency generally greater than 50%. Both vapor pressure and water solubility are crucial in determining whether a compound is amenable to extraction from water by inert gas purging.

- Sampling techniques and sample processing before GC/MS are more complex with air samples than with water and other condensed-phase samples. Sampling ambient air or emissions from automobiles and industrial stacks generally requires equipment and techniques more specialized than those required to obtain a water or other condensed-phase sample.
- Organic and inorganic compounds are present in air in both the vapor phase and as particulate matter (aerosols) of various dimensions. In water these substances are usually dissolved, associated with sediment particles, or present as an immiscible liquid on the surface or bottom of the body of water.
- Some VOCs are viable analytes in air, especially dry air, but react rapidly in water to form other substances that are usually much less hazardous and more biodegradable. Table 3.1 contains examples of compounds in this class including benzyl chloride, bis(chloromethyl)-ether, diazomethane, methyl isocyanate, and phosgene. These and other compounds can persist in dry air for some time because collisions with water molecules are infrequent.
- Photochemical and other reactions in air can rapidly convert some VOCs into other compounds, whereas hydrolysis and microbiological degradation are important reactions in water.
- Federal air legislation and regulatory requirements before 1990 focused on controlling emissions of particulate matter, inorganic gases, and hydrocarbons from stationary (industrial) and mobile (automobile) sources. There was no federal regulatory requirement to monitor VOCs in ambient air in 1999. In contrast, federal regulations that required monitoring of four trihalomethanes in drinking water were promulgated in 1979 and expanded to many other VOCs during the 1980s and 1990s.
- Because of impending federal regulations for wastewater and drinking water, more emphasis was given to the development of GC/MS methods for a wide variety of VOCs in water in the 1970s and early 1980s than to the development of analogous methods for VOCs in air.
- Analytical methods for VOCs in water are standardized, multi-laboratory validated, and mandated in some federal regulations. No single sampling or analytical method for VOCs in air is a standard or reference method or is required by a federal regulation.

Clean Air Act Amendments (CAAA) of 1990

The CAAA provided major stimulation to the further development and evaluation of sampling, separation, and measurement techniques for VOCs in ambient air. Title III of the act contains a list of 189 hazardous air pollutants (HAPs) and

107 of these are listed in Table 3.1 and identified by a "+" in the HAP column. The CAAA require the USEPA to issue maximum achievable control technology standards designed to reduce emissions of HAPs from industrial and other sources. Title II of the CAAA also requires reduced emissions of VOCs from mobile sources through the use of reformulated gasoline. Title VI requires the complete phase out of chlorofluorocarbons and other VOCs, including carbon tetrachloride and 1,1,1-trichloroethane, that cause depletion of the stratospheric ozone layer. Under the CAAA, the USEPA is required to conduct a research program, which must include monitoring of ambient air for HAPs in a representative number of urban areas, and other programs.

Sampling Techniques for VOCs in Air

Both field and laboratory methods are used for the determination of VOCs in air. Sometimes it is advantageous to conduct analyses of air in the field or in a processing plant. Field and continuous mass spectrometric techniques for VOCs in air are considered in Chapter 8. There are two general approaches to sampling air for laboratory identification and measurement of VOCs. Bulk air samples can be taken in the field in various containers and transported to the laboratory for analysis. Alternatively, the VOCs can be separated from the main components of air in the field and just the VOCs and their collection device are transported to the laboratory. Both of these strategies are used to obtain samples for laboratory analyses. Regardless of the general approach, the air entering the sampling system is often filtered to remove particulates that may contain organic compounds. The filtered particulate matter may be analyzed for semivolatile compounds as described in Chapter 4 and/or submitted to elemental analysis as described in Chapter 7.

Containers used for bulk air samples include flexible poly(vinyl fluoride) (Tedlar^{TM}) bags, evacuated glass or metal reservoirs, and thermally insulated cryogenic collection vessels. Plastic bags are generally limited to air samples that can be analyzed within a short time after sample collection. This limitation is due to potential losses of analytes by surface adsorption and surface chemical reactions. Furthermore, transportation of inflated bags over long distances to a laboratory is cumbersome and can result in significant losses of samples due to punctures and other accidents. Plastic bags are used in some laboratory operations, for example, to collect automobile exhaust and vaporized fuel, and for samples that can be conveniently transported to a field laboratory.[45] Evacuated metal canisters are very important and widely used air sample collection devices and these are discussed in more detail later in this section. Condensation of an entire air sample with liquid nitrogen or liquid helium is a strategy that has been used for many studies.[46] However, this technique is expensive to implement and requires specialized portable equipment for handling cryogenic fluids in the field.

Cold Trapping of VOCs in the Field. Cold trapping of VOCs is a technique that is used to separate VOCs from the main components of air in the field.[46] Air is drawn by a pump through an inert, often nickel, metal tube immersed in a fluid at a very low temperature, for example, −150 °C. The tube may be packed with some inert material such as Pyrex™ glass beads, and the low temperature is sufficient to condense the VOCs but not sufficient to condense oxygen or nitrogen. However, at this temperature, or the lower temperatures used to condense bulk air, many air components besides the VOCs are condensed including carbon dioxide, nitrogen oxides, ozone, sulfur dioxide, and water. Plugging of the condensation tube with ice or other solids is a major problem when sampling large volumes of humid air. This problem is addressed by the use of air dryers which trap moisture but allow the nonpolar VOCs to pass into the cold stage of the trapping system. However, the more polar and water-soluble VOCs are also removed by efficient air-drying systems. Another potential problem is that some VOCs may react during subsequent processing with trapped ozone, nitrogen oxides, or other substances present in the air. A practical limitation of the cold trapping technique is the requirement for liquid nitrogen or liquid argon in the field during extended sampling periods. This sampling technique is described in detail in USEPA Method TO-3 (Table 3.4) for VOCs in ambient air.[47] Although separation of VOCs from air by cold trapping in the field is somewhat complex and costly, this technique is important when used in the laboratory in connection with other air sampling techniques.

Impingers. Impingers are used to extract various substances, including some VOCs, from air. An impinger is a closed glass or metal vessel with an air inlet tube that extends to near the bottom of a liquid and an air outlet tube well above the surface of the liquid. The impinger may contain various aqueous or non-aqueous liquids including solutions of derivatizing agents. Air is drawn by a pump into the inlet tube and bubbled through the liquid, which dissolves soluble

Table 3.4 USEPA analytical methods for volatile organic compounds in ambient air

Method	Target analytes	Sampling technique	Separation and measurement	Reference
TO-1	VOCs	Tenax-GC™	GC/MS	47
TO-2	VOCs	Carbon molecular sieve	GC/MS	47
TO-3	VOCs	Cryogenic trap	GC/standard detectors	47
TO-5	Aldehydes and ketones	Impinger/DNPH	HPLC/UV	47
TO-7	*N*-Nitrosodimethylamine	Thermosorb/N adsorbent	GC/MS	47
TO-8	Phenols and cresols	Impinger/NaOH	HPLC	47
TO-11A	Aldehydes and ketones	Solid support with DNPH	HPLC/UV	62, 68
TO-14	VOCs (same as 14A)	Canister	GC/MS	47
TO-14A	VOCs (Table 3.1)	Canister	GC/MS	62, 69
TO-15	VOCs (Table 3.1)	Canister	GC/MS	62
TO-17	VOCs	Multiple solid adsorbents	GC/MS	62, 70

compounds, or compounds in the air react with reagents in solution to form soluble compounds. The liquid may also condense various substances, including water vapor, and collect fine particulate material that passes through a coarse filter or another separation device. A important advantage of an impinger compared to some bulk air sampling techniques is that hundreds of liters of air can be drawn through the device over several hours. Sampling trains with multiple collection devices in series are used to collect different fractions of an air sample from stationary sources. Components of sampling trains can include particulate filters of several types and sizes, particle-size separation devices, multiple impingers, and solid-phase adsorbents.

Impingers containing water are not efficient traps for most of the less polar VOCs listed in Table 3.1 including the hydrocarbons and most of the halogenated hydrocarbons. Generally, these compounds are too volatile and insoluble in water and are literally purged out of the impinger by the air bubbles. An impinger containing a nonaqueous solvent with a low vapor pressure would allow collection of some of these VOCs, especially at a reduced temperature, but concentration of the extract by evaporation of the solvent would not be possible. Impingers are most effective for the more polar VOCs in Table 3.1 that are readily soluble in water or other solvents, and for some semivolatile compounds associated with aerosols that are discussed in Chapter 4. For example, impingers containing distilled water are used to extract products of the photochemical oxidation of unsaturated hydrocarbons from smog chamber air.[48] These products, which include formaldehyde and acetaldehyde (Table 3.1), are also components of automobile exhaust and industrial emissions.

Polar VOCs and semivolatile compounds collected in impinger water can be determined using the methods for semivolatile compounds in water described in Chapter 4. The analytes are partitioned into an organic solvent, or on to a solid-phase adsorbent, and the organic solvent extract or adsorbent eluate is analyzed. However, many of the polar VOCs are too soluble in water and are not efficiently partitioned into an organic solvent or on to a solid-phase adsorbent. Other VOCs may be extracted or adsorbed, but cannot be concentrated in the extract or eluate by solvent evaporation because their vapor pressures are similar to the solvents used to extract or elute them. This group of VOCs also cannot be easily separated from organic solvents by GC because of similar vapor pressures and elution behavior. These are the same compounds discussed in the section of this chapter entitled "Polar Water Soluble Volatile Organic Compounds" under "Volatile Organic Compounds in Water". One of the techniques described in that section or an alternative sampling technique is required for the determination of these compounds.

Aqueous phase derivatization is used for polar, water-soluble aldehydes, ketones, or polyfunctional oxidation products that contain carbonyl groups. In one technique, air is drawn into an impinger containing an aqueous acidic solution of 2,4-dinitrophenylhydrazine (DNPH) in contact with an immiscible C_6–C_8 hydrocarbon solvent.[49] Carbonyl compounds react with DNPH to form 2,4-dinitrophenylhydrazones which dissolve in the hydrocarbon solvent. Figure 5.10 shows the structure and the EI mass spectrum of the 2,4-dinitrophenylhydrazone of *n*-hexanal. Some properties of these derivatives are described in Chapter 5 (under "Chemical Derivatization"). The 2,4-dinitrophenylhydrazones are not amenable to separation by GC, but they can be separated by using one of the techniques described in Chapter 6. This sampling technique is described in detail in USEPA Method TO-5 (Table 3.4) for aldehydes and ketones in ambient air.[47] Collection efficiencies $>80\%$ have been reported for formaldehyde, acetaldehyde, and benzaldehyde under some ambient air conditions.[50]

A technique that gives derivatives amenable to separation by GC is collection of the hydrocarbon oxidation products in impinger water followed by an aqueous phase reaction with the reagent *O*-(2,3,4,5,6-pentafluorobenzyl)hydroxylamine (PFBHA).[51] Aldehydes and ketones react with PFBHA to form the corresponding aldehyde and ketone oximes (Chart 3.2). The oximes are partitioned into an organic solvent, and the extract is analyzed by GC/MS. With formaldehyde and symmetrical ketones one oxime is formed, but other aldehydes and unsymmetrical ketones produce two isomers which may be separated by high-resolution GC. The EI spectra of the saturated aldehyde oximes contain either very low abundance $M^{+\cdot}$ ions or no $M^{+\cdot}$ ions, but oximes of unsaturated aldehydes and most ketones contain useful $M^{+\cdot}$ ion abundances. All the oxime EI spectra have base peaks at *m*/*z* 181, which is the pentafluorotropylium ion (Chart 3.2). While this ion is not indicative of the structure of the original aldehyde or ketone, it signals the presence of a derivatizable carbonyl group in the original substance. Some $(M–OH)^+$, $(M–NO)^+$, and other fragment ions are present in the EI spectra of the oximes and these are supportive of the structure of the original aldehyde or ketone. Additional information may be obtained from the positive-ion chemical ionization spectra and the electron capture ionization spectra of these derivatives (Chapter 5).

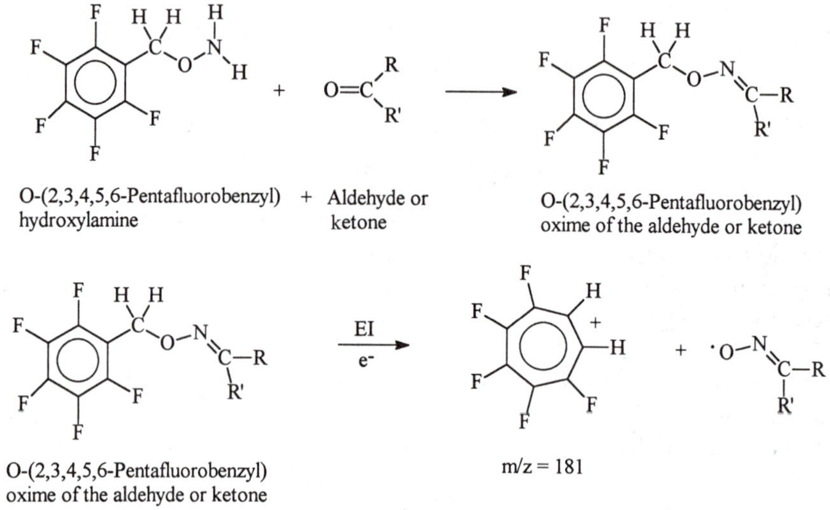

O-(2,3,4,5,6-Pentafluorobenzyl) hydroxylamine + Aldehyde or ketone → O-(2,3,4,5,6-Pentafluorobenzyl) oxime of the aldehyde or ketone

O-(2,3,4,5,6-Pentafluorobenzyl) oxime of the aldehyde or ketone —EI, e⁻→ m/z = 181 + ·O–N=C(R)R'

Chart 3.2

Impingers containing water also collect carboxylic acids and phenols produced by the photochemical oxidation of hydrocarbons in air.[52] These acids and phenols are partitioned into an organic solvent and derivatized with pentafluorobenzyl bromide to produce pentafluorobenzyl esters and ethers that are amenable to separation by GC. This derivatization reaction is discussed in Chapter 5 in the section "Chemical Derivatization" (Chart 5.5) and in Chapter 6 in another section entitled "Chemical Derivatization". An impinger sampling technique is described in detail in USEPA Method TO-8 (Table 3.4) for phenols and cresols in ambient air. Method TO-8 specifies an impinger containing aqueous sodium hydroxide, and the phenoxides formed are acidified, separated by liquid chromatography, and measured with a standard detector. Since phenols and cresols are amenable to extraction into an organic solvent, they could also be determined by using one of the methods for semivolatile compounds described in Chapter 4.

An impinger containing the organic solvent acetonitrile was used to collect a variety of photochemical oxidation products containing hydroxyl, carbonyl, and carboxylic acid groups.[53] The carbonyl containing compounds were converted to oximes with PFBHA (Chart 3.2) and analyzed by GC/MS. While impingers have been used to collect various polar water-soluble VOCs from ambient air, automobile exhaust, and stack emissions to allow identification of various oxidation products, collection efficiencies and quantitative analyses are not widely reported. Substantial efforts are required to assess fully the quantitative aspects of this technique.

Solid-Phase Adsorbents. One of the most attractive sampling strategies for VOCs in air is a solid-phase adsorbent that efficiently separates the VOCs from the principal components of air in the field. This potentially simple and inexpensive approach has been thoroughly investigated using a variety of candidate adsorbents. Typically, a porous solid-phase adsorbent is placed in a glass or metal tube which is taken to the field where air is drawn through the adsorbent to trap the VOCs.[43] The tube is then sealed and returned to the laboratory for analysis. Many types of solid-phase adsorbents were tested to identify those with the most favorable properties for the separation of VOCs from air.[54] Substances evaluated in numerous studies include alumina; activated carbons of various types; charcoal; graphitized carbon black (Carbopack™ B and C); carbon molecular sieves; GC packing materials such as Chromosorb™ 101 and 102; ethylvinylbenzene–divinylbenzene copolymer (Porapak Q™); styrene–ethylvinylbenzene–divinylbenzene terpolymer (Porapak P); silica gel; 2,6-diphenyl-*p*-phenylene oxide (Tenax-GC™—Chart 3.1); styrene–divinylbenzene copolymers (XAD-1™, XAD-2™, and XAD-4™); and acrylic ester polymers (XAD-7™ and XAD-8™). Adsorbent trapping was found to be applicable to a wide variety of nonpolar and some polar VOCs in Table 3.1 and to many analogous compounds.

Activated charcoal, which is widely used in occupational health studies, and other carbonaceous materials are generally strong adsorbents which tenaciously hold many compounds. With these adsorbents a solvent, typically carbon disulfide or methylene chloride, is used to desorb the VOCs. However, some compounds are adsorbed so strongly that poor recoveries are not uncommon. Elution with a solvent also has the effect of diluting the VOCs and therefore negatively affecting detection limits. Activated charcoal also may contain a significant background of VOCs which makes it best suited for the higher concentrations, about 1–100 $\mu g/m^3$, found in some workplace environments.

Many of the adsorbents that were evaluated have favorable properties for various VOCs, but during the early 1980s the porous organic polymer Tenax-GC (Chart 3.1) emerged as the adsorbent of choice of most investigators.[55] This development was not surprising because, by then, the favorable properties of Tenax-GC had been demonstrated as the primary adsorbent for VOCs purged from water. Among the attractive features of Tenax-GC are availability in reasonably pure form, a moderate adsorbing power that does not bind nonpolar and slightly polar molecules too strongly, a low tendency to adsorb water, and thermal stability above 200 °C, which allows nondiluting thermal desorption of adsorbed VOCs. These properties permit the determination of VOCs in ambient air at concentrations in the sub-$\mu g/m^3$ range. Tenax-GC was used in many studies of VOCs in ambient air, including indoor and outdoor air in the vicinity of the infamous Love Canal hazardous waste disposal site.[9,10] A Tenax-GC sampling technique is described in detail in USEPA Method TO-1 (Table 3.4) for VOCs in ambient air.[47]

The environmental studies of VOCs in air did reveal some limitations of Tenax-GC. One problem was controlling breakthrough of the more volatile polar and nonpolar VOCs from the adsorbent during long ambient air sampling periods.[55] Breakthrough of VOCs from Tenax-GC with purged water samples is controlled by using small (5 mL) samples and short purge times, typically 11 min. Also, two additional adsorbents, silica gel and charcoal, are often used after the Tenax-GC to trap the most volatile compounds. However, with ambient air, longer sampling times are often required because of generally lower concentrations, and breakthrough is more of a problem. In one study the breakthrough volumes of methanol and vinyl chloride were 2 L or less at 10 °C, but the breakthrough volumes of 1,2-dichlorobenzene and ethylbenzene were over 1000 L.[55] In later studies, additional adsorbents were placed after the Tenax-GC or new formulations of adsorbents were developed to address the breakthrough problem without losing the advantages of thermal desorption. There was a resurgence of interest in graphitized carbon blacks (Carbotrap™ and Carbotrap C™), carbon molecular sieves (Carbosieve S-III™ and Carboxen 569™), and Tenax-GR, which is the 2,6-diphenyl-*p*-phenylene oxide adsorbent mixed with 30% graphitized carbon.[56,57] While these strategies are generally useful, the carbonaceous materials are more likely to collect water vapor or adsorb polar organics more tenaciously than would the Tenax-GC. A sampling technique using carbon molecular sieves is described in USEPA Method TO-2 (Table 3.4) for VOCs in ambient air.[47]

Another limitation of solid-phase adsorbents was revealed by the identification of products of chemical reactions between VOCs and reactive components in air including nitrogen oxides and ozone.[58,59] Long air sampling times also risk oxidation of the adsorbent and the production of background VOCs and other artifacts. These reactions are not observed in water analyses because helium is used to purge water samples, and residual chlorine and other strong oxidizing agents in drinking water are destroyed by chemical reduction when samples are taken. The potential for oxidation during sampling contributed to the reduced attractiveness of Tenax-GC and related materials for collection of VOCs from ambient air. However, Tenax-GC alone or in combination with other materials continues to be used for sampling VOCs in ambient air and in emissions from automotive and stationary industrial sources.

Just as with the adsorbent trap used for VOCs in water, the connection of the exit of the thermal desorption system to the entrance of a narrow-bore fused silica capillary GC column requires a technique to reduce the high helium gas flow from the trap. The techniques described in the previous section "Purge and Trap with Fused Silica Capillary GC Columns" are also used for VOCs in air. However, cryogenic focusing in a transfer line before the GC column is much more prevalent in ambient air analyses than in water analyses.[46] This is probably because of recent strong interest in the large number of VOCs found at very low concentrations in ambient air and the perceived need for optimum GC resolving power especially with standard GC detectors. However, plugging of the trap with ice is a serious problem especially when large volumes of humid air are sampled with multisorbent traps. Various techniques are used by investigators and commercial system suppliers to control moisture. These include purging the primary trap with dry helium for several minutes before thermal desorption to reduce adsorbed water and the use of moderately cooled surfaces before the cryogenic trap to selectively remove water vapor.[57] More aggressive drying techniques are more effective at removing water, but they also remove polar water-soluble VOCs that are of interest.

A special type of solid-phase adsorbent is used to trap volatile aldehydes and ketones in air. The adsorbent consists of an inert solid support with a surface coating of a reagent that forms derivatives of the airborne carbonyl compounds. A typical inert support is a reverse-phase liquid chromatography packing material such as finely divided silica with surface bonded C_8, C_{18}, phenyl, or other nonpolar organic groups. A standard reagent is 2,4-dinitrophenylhydrazine (DNPH), which reacts with aldehydes and ketones to form the corresponding 2,4-dinitrophenylhydrazones.[60] The hydrazones are eluted with an organic solvent and the eluate is analyzed. Figure 5.10 shows the structure and the EI mass spectrum of the 2,4-dinitrophenylhydrazone of *n*-hexanal. However, 2,4-dinitrophenylhydrazones are not amenable to separation by GC, but they can be separated by liquid chromatography and detected by using one of the techniques described in Chapter 6. This sampling technique has been reported to provide collection efficiencies $>85\%$ for formaldehyde and acetaldehyde under some ambient air conditions.[50] A solid support coated with DNPH and contained in a cartridge is used in USEPA Method TO-11A (Table 3.4) for aldehydes and ketones in ambient air. As previously discussed under "Impingers", the reagent PFBHA reacts with carbonyl compounds to produce derivatives that are amenable to separation by GC (Chart 3.2). This reagent may also be a viable surface derivatizing reagent for aldehydes and ketones in ambient air.

Metal Canisters. The limitations of solid-phase adsorbents, and the other techniques to separate and collect VOCs in the field, refocused attention on bulk air samples during the second half of the 1980s and the 1990s. Deactivated metal canisters were thoroughly investigated as an alternative to the inconvenient and limited capability plastic bags mentioned previously.[61] The use of stainless steel canisters was supported by manufacturing techniques to produce canisters with smooth and inert internal surfaces and few or no active sites that adsorb VOCs or catalyze chemical reactions. Small (1–6 L) canisters that are easily transported and stored became practical with the development of very sensitive measurement techniques, especially the ion-trap mass spectrometer. A common type of canister is an internally electropolished stainless steel sphere equipped with a sampling valve and a flow rate controller. This type of canister is specified in USEPA Methods TO-14, TO-14A, and TO-15 (Table 3.4).[47,62]

Canisters are leak tested and cleaned in the laboratory, evacuated to about 5×10^{-2} Torr or less, and transported to the sampling site where samples are taken by opening the sampling valve. Composite samples can be taken over time and/or space and an in-line pump can be used to pressurize the container with either additional sample air or pure air if sample dilution is required. Pressurized samples are useful when longer term composite samples are taken or when larger samples are needed to lower detection limits. At the laboratory, the VOCs in the canister are concentrated and focused in a cryogenic trap or an adsorbent trap, then thermally vaporized into the capillary column GC/MS system.[46,63]

Summary of Sampling Techniques for VOCs in Air. In the United States the deactivated metal canister is probably the most widely used sampling technique for the determination of a variety of nonpolar and polar VOCs in ambient air. However, solid-phase adsorbents, impingers, plastic bags, and the cryogenic collection of both whole air and separated VOCs are used for various general and special purposes. Details of sampling techniques for VOCs in ambient air are included in the TO series of analytical methods (Table 3.4).[47,62] Sampling strategies and additional information on sampling techniques are included in the second edition of *Principles of Environmental Sampling*, which was published by the American Chemical Society in 1996, and in other reference books.[64,65] Some of these sampling techniques are general and applicable to a broad range of studies other than environmental monitoring, including the determination of VOCs produced by bacteria and human breath analysis.[66,67]

Toxic Organic (TO) Series of Analytical Methods

The TO series of analytical methods was developed by the USEPA to provide guidance on the determination of selected toxic organic compounds in ambient air.[47,62] Unlike the analytical methods for drinking water and wastewater, the TO methods are not incorporated into federal regulations as approved or required methods for regulatory monitoring. However, several of these methods are incorporated into American Society for Testing and Materials (ASTM) standards, and multilaboratory validation data are available.[68–70] Table 3.4 contains summary information about the TO series of methods for VOCs and references to USEPA methods and ASTM standards. While only some of these methods specify GC/MS as the analytical technique, and some specify standard GC detectors, all the TO methods for GC amenable analytes can and should be implemented with GC/MS for the reasons given in the section of Chapter 2 entitled "Alternatives to Conventional GC/MS".

Method TO-14 is a general method for nonpolar and slightly polar VOCs in ambient air and specifies a particular type of deactivated stainless steel canister for containment of the bulk air sample.[47] The sampling procedure used in Method TO-14 has options for final canister pressures both below and above atmospheric pressure. The method lists 40 analytes that are preserved in the vapor phase for several days or up to 30 days in both subambient pressure and pressurized canisters of the type specified in the method. Many other VOCs are likely analytes with this method, and testing and validation data will likely be available in the future. In the laboratory the air sample is pumped through a Nafion™ permeable membrane dryer to reduce water vapor, and the VOCs are concentrated and focused in a cryogenic trap. This fairly aggressive drying system is effective at removing water, but also removes some of the more polar and water-soluble VOCs. The focused analytes are rapidly thermally vaporized and determined by capillary column GC/MS. Method TO-14 requires the mass spectrometer to be tuned to produce an EI mass spectrum of the performance test compound 4-bromofluorobenzene (BFB) that meets the criteria in Table 3.2.

Method TO-14A is a 1999 version of TO-14 which contains more general specifications for the deactivated metal canister, but has been evaluated with the same 40 target analytes as Method TO-14.[62] These analytes are indicated in Table 3.1 by the entry *TO-14A*. The target analytes can be measured in 0.3–0.5 L samples at the 0.1–1 part per 10^9 (by volume) level in ambient air with Method TO-14A.

Method TO-15 uses a sampling technique very similar to that of Method TO-14A, but specifies a modified approach to reducing water vapor in the air sample in order to encompass a broader variety of polar water-soluble VOCs.[62] Method TO-15 lists the 97 HAPs in Table 3.1 as target analytes, but does not include 10 of the Method TO-14A analytes that are not designated as HAPs. However, these 10 would surely be detected by Method TO-15. A major change in Method TO-15 is elimination of the Nafion™ permeable membrane dryer used in Methods TO-14 and TO-14A to reduce water vapor in the air sample before cryogenic focusing. Method TO-15 provides several options to reduce water vapor and the flexibility to use other strategies that may be available or that may be developed in the future. These options include VOC concentration and focusing on a hydrophobic solid adsorbent, for example, Tenax-GC, or a multisorbent trap as discussed in the section on solid-phase adsorbents. Water vapor that accumulates on the adsorbent is reduced as much as possible by purging the trap with dry helium before thermal desorption of the analytes on to the GC column. Alternatively, cryogenic focusing is used as in Methods TO-14 and TO-14A, but the sample size is reduced to minimize the risk of plugging the trap with ice. Method TO-15 also incorporates the IS calibration procedure as described in Chapter 2 in the section "Quantitative Analysis". Method TO-15 represents the current state-of-the-science for the reliable determination of VOCs in ambient air.[62]

Method TO-17 uses a multiadsorbent sampling technique and thermal desorption into the GC/MS system.[62]

Summary

Volatile organic compounds, as defined and described in this chapter, are the most widely dispersed of all classes of environmental pollutants. These compounds are also constituents of beverages, foods, perfumes, many other consumer products, industrial products, and waste streams. They contribute to flavors, odors, and tastes and are determined in drinking water, breath, blood, sweat, urine, and recycled water used in space exploration. They provide important clues in forensic studies, petroleum exploration, and many other areas of investigation. The VOCs are also among the most important of all classes of compounds because all forms of life, as well as inanimate matter, are so readily and frequently exposed to these substances which have a wide variety of effects. The GC/MS methods for their extraction from various environmental media, concentration, and determination are among the most highly developed and reliable of all analytical methods. These analytical methods are also applied, or are applicable with modest modifications, to the determination of VOCs in a variety of other types of samples and areas of research.

REFERENCES

1. Keith, L. H.; Telliard, W. A. *Environ. Sci. Technol.* **1979**, *13*, 416–423.
2. Carson, R. L. *Silent Spring*: Houghton Mifflin: New York, 1962.
3. Weisburger, J. H.; Weisburger, E. K. *Chem. Eng. News*, 7 February 1966.
4. *NIST/EPA/NIH Mass Spectral Database*; National Institute for Standards and Technology: Gaithersburg, MD, 1992.
5. Code of Federal Regulations, Title 40, Part 141.61.
6. Code of Federal Regulations, Title 40, Part 141.40.
7. Code of Federal Regulations, Title 40, Part 141.30.
8. Symons, J. M.; Bellar, T. A.; Carswell, J. K.; DeMarco, J.; Kropp, K. L.; Robeck, G. G.; Seeger, D. R.; Slocum, C. J.; Smith, B. L.; Stevens, A. A. *J. Am. Water Works Assoc.* **1975**, *67*, 634–647.
9. Deegan Jr., J. *Environ. Sci. Technol.* **1987**, *21*, 328–331.
10. Deegan Jr., J. *Environ. Sci. Technol.* **1987**, *21*, 421–426.
11. Borman, S. *Chem. Eng. News*, 19 February 1990, 5–6.
12. Harris, L. E.; Budde, W. L.; Eichelberger, J. W. *Anal. Chem.* **1974**, *46*, 1912–1917.

13. Henderson, J. E.; Peyton, G. R.; Glaze, W. H. In *Identification and Analysis of Organic Pollutants in Water*; Keith, L. H., Ed.; Ann Arbor Science Publishers: Ann Arbor, MI, 1976, pp. 105–111.
14. Drozd, J.; Novak, J. *J. Chromatogr.* **1979**, *165*, 141–165.
15. Kotzias, D.; Sparta, C. In *Chemistry and Analysis of Volatile Organic Compounds in the Environment*; Bloemen, H. J. T.; Burn, J., Eds; Blackie Academic & Professional:London, England, 1993.
16. Bellar, T. A.; Lichtenberg, J. J. *J. Am. Water Works Assoc.* **1974**, *66*, 739–744.
17. Bellar, T. A.; Budde, W. L.; Eichelberger, J. W. In *Monitoring Toxic Substances*; Schuetzle, D., Ed.; American Chemical Society Symposium Series No. 94; American Chemical Society: Washington, DC, 1979.
18. Method 524.2, Revision 4.1 In *Methods for the Determination of Organic Compounds in Drinking Water*, Suppl. III, USEPA Report EPA/600/R-95/131, August **1995**; URL http://www.epa.gov/nerlcwww/methmans.html.
19. Grob, K. *J. Chromatogr.* **1973**, *84*, 255–273.
20. Grob, K.; Zurcher, F. *J. Chromatogr.* **1976**, *117*, 285–294.
21. Kagel, R. O.; Stehl, R. H.; Crummett, W. B. *Anal. Chem.* **1979**, *51*, 223A.
22. Budde, W. L.; Eichelberger, J. W. *Anal. Chem.* **1979**, *51*, 567A–574A.
23. Finnigan, R. E.; Hoyt, D. W.; Smith, D. E. *Environ. Sci. Technol.* **1979**, *13*, 534–541.
24. Hites, R. A. *Environ. Sci. Technol.* **1981**, *15*, 974–975.
25. Ballinger, D. G. *Environ. Sci. Technol.* **1979**, *13*, 1362–1366.
26. *Guidelines Establishing Test Procedures for the Analysis of Pollutants; Proposed Regulations, Federal Register*; 3 December **1979**, Vol. 44, 69464–69575.
27. *Guidelines Establishing Test Procedures for the Analysis of Pollutants Under the Clean Water Act; Final Rule and Interim Final Rule and Proposed Rule, Federal Register*; 26 October **1984**, Vol. 49, 43234-43439; Title 40 Code of Federal Regulations Part 136.
28. *Test Methods for Evaluating Solid Waste, Physical/Chemical Methods*; USEPA Publication SW-846, 3rd ed. and Updates I, II, IIA, IIB, and III; URL http://www.e-pa.gov/epaoswer/hazwaste/test/8xxx.htm.
29. Eichelberger, J. W.; Bellar, T. A.; Donnelly, J. P.; Budde, W. L. *J. Chromatogr. Sci.* **1990**, *28*, 460–467.
30. Bellar, T. A.; Lichtenberg, J. J.; Kroner, R. C. *J. Am. Water Works Assoc.* **1974**, *66* 703–706.
31. *National Interim Primary Drinking Water Regulations; Control of Trihalomethanes in Drinking Water; Final Rule, Federal Register*; 29 November **1979**, Vol. 44, 68624–68707; Title 40 Code of Federal Regulations Part 141.30.
32. *National Primary Drinking Water Regulations; Synthetic Organic Chemicals; Monitoring for Unregulated Contaminants, Federal Register*; 8 July **1987**, Vol. 52, 25690–25717; Title 40 Code of Federal Regulations Part 141.
33. *National Primary Drinking Water Regulations*; *Final Rule Federal Register*; 30 January **1991**, Vol. 56, 3526–3597; Title 40 Code of Federal Regulations Part 141.
34. *National Primary Drinking Water Regulations; Synthetic Organic Chemicals and Inorganic Chemicals; Final Rule Federal Register*; 17 July **1992**, Vol. 57, 31776–31848; Title 40 Code of Federal Regulations Part 141.
35. *Standard D5790-95, Standard Test Method for Measurement of Purgeable Organic Compounds in Water by Capillary Column Gas Chromatography/Mass Spectrometry, Annual Book of ASTM Standards*; Vol. 11.02, American Society for Testing and Materials: W. Conshohocken, PA, **1998**.
36. Zhang, Z.; Yang, M. J.; Pawliszyn, J. *Anal. Chem.* **1994**, *66*, 844A–853A.
37. Munch, J. W.; Eichelberger, J. W. *J. Chromatogr. Sci.* **1992**, *30*, 471–477.

38. Peters, T. L. *Anal. Chem.* **1980**, *52*, 211–213.
39. Gurka, D. F.; Pyle, S. M.; Titus, R. *Anal. Chem.* **1992**, *64*, 1749–1754.
40. Shoemaker, J. A.; Bellar, T. A.; Eichelberger, J. W.; Budde, W. L. *J. Chromatogr. Sci.* **1993**, *31*, 279–284.
41. Liikkala, T. L.; Olsen, K. B.; Teel, S. S.; Lanigan, D. C. *Environ. Sci. Technol.* **1996**, *30*, 3441–3447.
42. Hiatt, M. H.; Youngman, D. R.; Donnelly, J. R. *Anal. Chem.* **1994**, *66*, 905–908.
43. Pellizzari, E. D.; Bunch, J. E.; Berkley, R. E.; McRae, J. *Anal. Chem.* **1976**, *48*, 803–807.
44. Penkett, S. A. *Environ. Sci. Technol.* **1991**, *25*, 631–635.
45. Schuetzle, D.; Jensen, T. E.; Nagy, D.; Prostak, A.; Hochhauser, A. *Anal. Chem.* **1991**, *63*, 1149A–1159A.
46. McClenny, W. A.; Pleil, J. D.; Holdren, M. W.; Smith, R. N. *Anal. Chem.* **1984**, *56*, 2947–2951.
47. *Compendium of Methods for the Determination of Toxic Organic Compounds in Ambient Air*; USEPA Report EPA/600/4-89/017, June **1988**.
48. Yu, J.; Jeffries, H. E.; Le Lacheur, R. M. *Environ. Sci. Technol.* **1995**, *29*, 1923–1932.
49. Grosjean, D. *Environ. Sci. Technol.* **1982**, *16*, 254–262.
50. Grosjean, D.; Fung, K. *Anal. Chem.* **1982**, *54*, 1221–1224.
51. Le Lacheur, R. M.; Sonnenberg, L. B.; Singer, P. C.; Christman, R. F.; Charles, M. J. *Environ. Sci. Technol.* **1993**, *27*, 2745–2753.
52. Chien, C.-J.; Charles, M. J.; Sexton, K. G.; Jeffries, H. E. *Environ. Sci. Technol.* **1998**, *32*, 299–309.
53. Yu, J.; Flagan, R. C.; Seinfeld, J. H. *Environ. Sci. Technol.* **1998**, *32*, 2357–2370.
54. Ciccioli, P. In *Chemistry and Analysis of Volatile Organic Compounds in the Environment*; Bloemen, H. J. T; Burn, J., Eds; Blackie Academic & Professional:London, **1993**.
55. Krost, K. J.; Pellizzari, E. D.; Walburn, S. G.; Hubbard, S. A. *Anal. Chem.* **1982**, *54*, 810–817.
56. Helmig, D.; Vierling, L. *Anal. Chem.* **1995**, *67*, 4380–4386.
57. Oliver, K. D.; Adams, J. R.; Daughtrey Jr., E. H.; McClenny, W. A.; Yoong, M. J.; Pardee, M. A.; Almasi, E. B.; Kirshen, N. A. *Environ. Sci. Technol.* **1996**, *30*, 1939–1945.
58. Pellizzari, E. D.; Krost, K. J. *Anal. Chem.* **1984**, *56*, 1813–1819.
59. Calogirou, A.; Larsen, B. R.; Brussol, C.; Duane, M.; Kotzias, D. *Anal. Chem.* **1996**, *68*, 1499–1506.
60. Grosjean, D.; Williams, E. L.; Grosjean, E.; Andino, J. M.; Seinfeld, J. H. *Environ. Sci. Technol.* **1993**, *27*, 2754–2758.
61. Gholson, A. R.; Jayanty,, R. K. M.; Storm, J. F. *Anal. Chem.* **1990**, *62*, 1899–1902.
62. *Compendium of Methods for the Determination of Toxic Organic Compounds in Ambient Air*; 2nd ed., USEPA Report EPA/625/R-96/010b, January **1999**.
63. Kelly, T. J.; Callahan, P. J.; Pleil, J.; Evans, G. F. *Environ. Sci. Technol.* **1993**, *27*, 1146–1153.
64. Keith, L. H., Ed. *Principles of Environmental Sampling*, 2nd ed.; American Chemical Society: Washington, DC, **1996**.
65. Bloemen, H. J. T.; Burn, J., Eds. *Chemistry and Analysis of Volatile Organic Compounds in the Environment*; Blackie Academic & Professional: London, **1993**.
66. Rivers, J. C.; Pleil, J. D.; Wiener, R. W. *J. Exposure Anal. Environ. Epidemiol., Suppl. 1* **1992**, 177–188.
67. Pleil, J. D.; Lindstrom, A. B. *J. Chromatogr. B* **1995**, *665*, 271–279.

68. *Standard D5197-97, Standard Test Method for Determination of Formaldehyde and other Carbonyl Compounds in Air (Active Sampler Methodology), Annual Book of ASTM Standards*; Vol. 11.03, American Society for Testing and Materials: W. Conshohocken, PA, **1998**.
69. *Standard D5466-95, Standard Test Method for Determination of Volatile Organic Chemicals in Atmospheres (Canister Sampling Methodology), Annual Book of ASTM Standards*; Vol. 11.03, American Society for Testing and Materials: W. Conshohocken, PA, **1998**.
70. *Standard D6196-97, Standard Practice for Selection of Sorbents and Pumped Sampling/Thermal Desorption Analysis Procedures for Volatile Organic Compounds in Air, Annual Book of ASTM Standards*; Vol. 11.03, American Society for Testing and Materials: W. Conshohocken, PA, **1998**.

4

Semivolatile Organic Compounds Amenable to Gas Chromatography

The semivolatile compounds were introduced and briefly described in the section of Chapter 3 entitled "Target Analytes—The Priority Pollutants". The *semivolatile* descriptor was coined in the mid-1970s when the USEPA was compelled by the consent decree of 1976 to implement certain provisions of the Federal Water Pollution Control Act Amendments of 1972.[1] This major environmental and analytical chemistry event was described in Chapter 1 and the section of Chapter 3 entitled "Target Analytes—The Priority Pollutants".

Semivolatile compounds are less volatile than the volatile organic compounds (VOCs) described in Chapter 3. They generally have molecular weights in the range of 100–500 amu and vapor pressures in the range 0.1–10^{-7} Torr at 25 °C and 760 Torr. Some compounds in this group have borderline vapor pressures and can be classified either as VOCs or semivolatiles depending on their other properties especially their water solubilities. The semivolatile compounds are extracted from air particulate filters, vapor adsorbents, sediment, soil, tissue, or water by a slightly polar water-immiscible organic solvent such as methylene chloride. The distribution favors the organic phase for most compounds although this is pH dependent in water for phenols, nitrogen bases, and some other acidic or basic compounds. When the immiscible solvent is carefully evaporated at slightly elevated temperatures, for example, ~50 °C, to concentrate the extract, semivolatiles are generally retained in the liquid phase.

Semivolatile compounds are thermally stable at temperatures required for gas chromatography (GC), many have low chemical reactivity, and they are amenable to separation by GC. However, some of the more chemically reactive, polar, acidic, and basic compounds are marginal analytes and may be more appropriately determined by using the techniques described in Chapter 6. The semivolatile compounds amenable to GC are often readily separated on a variety of commercial bonded-phase fused silica capillary GC columns. The manufacturer's literature should be consulted for many examples of these separations.

Table 4.1 lists 175 semivolatile compounds or mixtures of semivolatile compounds. These are a small fraction of the total number of semivolatiles amenable to separation by GC. Most of the compounds in Table 4.1 are industrial chemicals, by-products of industrial processes, pesticides, products of incomplete combustion, or other environmentally significant substances. Many are also representatives of classes of compounds that include many isomers, congeners, and chiral forms. Among the 129 priority pollutants selected for the 1976 consent decree analytical effort were 76 semivolatile compounds, and these are indicated by a "+" in the PP column of Table 4.1. Many other semivolatile compounds were later discovered as environmental pollutants in air, drinking water, soils, and other media and some of these are also included in Table 4.1. The Clean Air Act Amendments of 1990 contains a list of 189 hazardous air pollutants (HAPs) and 54 of these are listed in Table 4.1 and identified by a "+" in the HAP column. Since it would be almost impossible to list all the semivolatile compounds of potential interest, the compounds in this table should be considered models for the large number of potential analytes amenable to separation and measurement by GC/MS.

There are some general and important differences between the broad class VOCs described in Chapter 3 and the semivolatiles discussed in this chapter. Some of these differences impact the strategies and analytical methods used to determine these two classes of compounds in environmental media:

- The VOCs are often familiar compounds whose structures are easily deduced from their systematic names. The semivolatile compounds often have far more complex structures, less familiar common names, and very complex systematic names.
- Unlike the VOCs, the semivolatile compounds include a number of commercial products that are mixtures of isomers, congeners, or other closely related compounds. Examples in Table 4.1 include the Aroclors, toxaphene, and technical chlordane.
- The semivolatiles contain more atoms and have higher molecular weights than the VOCs and, therefore, have a significantly larger number of possible isomers, congeners, and chiral forms than the smaller VOCs.
- The larger and higher molecular weight semivolatiles generally have more complex electron ionization (EI) mass spectra than the smaller VOCs.
- While some VOCs have very long lifetimes in the environment, most are susceptible to oxidation, hydrolysis, or biodegradation in the atmosphere or in the aquatic environment. More semivolatiles, especially the polychlorinated compounds with no functional groups, are resistant to degradation and can persist in ice, soils, sediment, and tissue for decades or longer.

Table 4.1 Semivolatile organic compounds amenable to GC that are consent decree priority pollutants (PP) and/or hazardous air pollutants (HAP) listed in the Clean Air Act Amendments of 1990 and/or have maximum contaminant levels (MCL) or monitoring requirements in drinking water or are candidates for future regulation

Compounds	PP[1]	HAP	GC/MS (air methods)	MCL[a] (μg/L)	GC/MS (water methods)
Acenaphthene	+		TO-13A		Ref. 2
Acenaphthylene	+		TO-13A		525.2, Ref. 2
2-Acetylaminofluorene		+	Chapter 6		
Alachlor (acetanilide, Chart 4.2)			TO-4A, TO-10A	2	525.2
Aldrin (Chart 4.1)	+		TO-4A, TO-10A	[a]	525.2, Ref. 2
Ametryn (triazine)					525.2
4-Aminobiphenyl		+	Chapter 6		
Anthracene	+		TO-13A		525.2, Ref. 2
Aroclor-1016™	+	+		[b]	525.2
Aroclor-1221™	+	+		[b]	525.2, Ref. 2
Aroclor-1232™	+	+		[b]	525.2
Aroclor-1242™	+	+	TO-4A, TO-10A	[b]	525.2
Aroclor-1248™	+	+		[b]	525.2, Ref. 2
Aroclor-1254™	+	+	TO-4A, TO-10A	[b]	525.2, Ref. 2
Aroclor-1260™	+	+	TO-4A, TO-10A	[b]	525.2
Atraton (triazine)					525.2
Atrazine (triazine, Chart 4.2)			TO-4A, TO-10A	3	525.2
Benzidine (Chart 6.1)	+	+	Chapter 6		Ch. 6, Ref. 2
Benzo[*a*]anthracene (Chart 4.2)	+		TO-13A		525.2, Ref. 2
Benzo[*b*]fluoranthene	+		TO-13A		525.2, Ref. 2
Benzo[*k*]fluoranthene	+		TO-13A		525.2, Ref. 2
Benzo[*g,h,i*]perylene (Chart 2.1)	+		TO-13A		525.2, Ref. 2
Benzo[*a*]pyrene (Chart 2.1)	+		TO-13A	0.2	525.2, Ref. 2
Benzo[*e*]pyrene			TO-13A		
1,4-Benzoquinone		+			
Benzylbutyl phthalate	+				525.2, Ref. 2
Biphenyl (Chart 4.1)		+			
Bromocil					525.2
4-Bromophenyl phenyl ether	+				Ref. 2
Butachlor (acetanilide)				[a]	525.2
Butylate (thiocarbamate)					525.2
Carboximide (MGK 264)					525.2
Carboxin					525.2
Chlordane (technical mixture)	+	+	TO-4A, TO-10A	2	See isomers
α-Chlordane (Chart 4.1)					525.2
γ-Chlordane					525.2, Ref. 2
Chlorneb					525.2
Chlorobenzilate		+			525.2
2-Chlorobiphenyl					525.2
Bis(2-chloroethoxy)methane	+				Ref. 2
Bis(2-chloroethyl) ether	+				Ref. 2
Bis(2-chloroisopropyl) ether	+				Ref. 2
4-Chloro-3-methylphenol	+				Ref. 2
2-Chloronaphthalene	+				Ref. 2
2-Chlorophenol	+				Ref. 2
4-Chlorophenyl phenyl ether	+				Ref. 2
Chlorothalonil			TO-4A, TO-10A		525.2
Chlorpropham (carbamate)					525.2

(*continued*)

Table 4.1 (*continued*)

Compounds	PP[1]	HAP	GC/MS (air methods)	MCL[a] (μg/L)	GC/MS (water methods)
Chlorpyrifos (phosphate)			TO-4A, TO-10A		525.2
Chrysene (Chart 5.1)	+		TO-13A		525.2, Ref. 2
Coronene			TO-13A		
Cyanazine (triazine)					525.2
Cycloate (thiocarbamate)					525.2
1,1-Bis(4-chlorophenyl)-2,2-dichloroethane (4,4′-DDD)	+				525.2, Ref. 2
1,1-Bis(4-chlorophenyl)-2,2-dichloroethene (4,4′-DDE)	+	+	TO-4A, TO-10A		525.2, Ref. 2
1,1-Bis(4-chlorophenyl)-2,2,2-trichloroethane (4,4′-DDT)	+		TO-4A, TO-10A		525.2, Ref. 2
Diazinon (thiophosphate)			TO-4A, TO-10A		525.2
Dibenzo[*a*,*h*]anthracene	+		TO-13A		525.2, Ref. 2
4,4′-Dibromobiphenyl					Ref. 2
4,4′-Dibromooctafluorobiphenyl					Ref. 2
Di-*n*-butylphthalate	+	+			525.2, Ref. 2
3,3′-Dichlorobenzidine	+	+	Chapter 6		Ch. 6, Ref. 2
2,3-Dichlorobiphenyl					525.2
2,4-Dichlorophenol	+				Ref. 2
Dichlorvos (phosphate)		+	TO-10A		525.2
Dieldrin (Chart 4.1)	+		TO-4A, TO-10A	[a]	525.2, Ref. 2
Di(2-ethylhexyl) adipate				400	525.2
Di(2-ethylhexyl) phthalate	+	+		6	525.2, Ref. 2
Diethyl phthalate	+				525.2, Ref. 2
3,3′-Dimethoxybenzidine		+	Chapter 6		Ch. 6, Ref. 2
Dimethylaminoazobenzene		+			
3,3′-Dimethylbenzidine		+	Chapter 6		Ch. 6, Ref. 2
2,4-Dimethylphenol	+				Ref. 2
Dimethyl phthalate	+	+			525.2, Ref. 2
Dimethyl tetrachloroterephthalate (dacthal)			TO-4A, TO-10A		525.2
2,4-Dinitrophenol (Chart 6.1)	+	+			Ref. 2
2,4-Dinitrotoluene	+	+			525.2, Ref. 2
2,6-Dinitrotoluene	+				525.2, Ref. 2
Di-*n*-octylphthalate (Chart 4.2)	+				Ref. 2
Diphenamide					525.2
1,2-Diphenylhydrazine	+	+			
Disulfoton (thiophosphate)					525.2
Disulfoton sulfone					525.2
Disulfoton sulfoxide					525.2
Endosulfan I	+				525.2, Ref. 2
Endosulfan II	+				525.2, Ref. 2
Endosulfan sulfate	+				525.2, Ref. 2
Endrin	+			2	525.2, Ref. 2
Endrin aldehyde	+				525.2
Eptam (EPTC)					525.2
Ethprop					525.2
Etridiazole					525.2
Fenamiphos (phosphate)					525.2
Fenarimol					525.2

(*continued*)

Table 4.1 (*continued*)

Compounds	PP[1]	HAP	GC/MS (air methods)	MCL[a] (μg/L)	GC/MS (water methods)
Fluoranthene (Chart 4.2)	+		TO-13A		Ref. 2
Fluorene	+		TO-13A		525.2, Ref. 2
1-Fluoronaphthalene					Ref. 2
2-Fluorophenol					Ref. 2
Fluridone					525.2
Heptachlor (Chart 4.1)	+	+	TO-4A, TO-10A	0.4	525.2, Ref. 2
Heptachlor epoxide	+		TO-4A, TO-10A	0.2	525.2, Ref. 2
2,2′,3,3′,4,4′,6-Heptachlorobiphenyl					525.2
Hexachlorobenzene	+	+	TO-4A, TO-10A	1	525.2, Ref. 2
2,2′,4,4′,5,6′-Hexachlorobiphenyl					525.2
α-Hexachlorocyclohexane	+	+	TO-4A, TO-10A		525.2, Ref. 2
β-Hexachlorocyclohexane	+	+	TO-4A, TO-10A		525.2, Ref. 2
γ-Hexachlorocyclohexane (lindane—Chart 4.1)	+	+	TO-4A, TO-10A	0.2	525.2, Ref. 2
δ-Hexachlorocyclohexane	+	+			525.2, Ref. 2
Hexachlorocyclopentadiene	+	+		50	525.2, Ref. 2
Hexazinone					525.2
Indeno[1,2,3-*c*,*d*]pyrene	+		TO-13A		525.2, Ref. 2
Isophorone	+	+			525.2, Ref. 2
Maleic anhydride		+			
Merphos (trithiophosphite)					525.2
Methoxychlor		+	TO-4A, TO-10A	40	525.2
2-Methyl-4,6-dinitrophenol	+	+			Ref. 2
4,4′-Methylenebis(2-chloroaniline)		+	Chapter 6		
4,4′-Methylenedianiline		+	Chapter 6		
Methyl paraoxon (phosphate)					525.2
Metolachlor (acetanilide)			TO-4A, TO-10A	[a]	525.2
Metribuzin (unsym. triazine)				[a]	525.2
Mevinphos (phosphate)					525.2
Mirex (Chart 4.1)					
Molinate (thiocarbamate)					525.2
Naphthalene (Chart 4.2)	+	+	TO-13A		Ref. 2
Napropamide					525.2
4-Nitrobiphenyl		+			
2-Nitrophenol	+				Ref. 2
4-Nitrophenol	+	+			Ref. 2
N-Nitrosodimethylamine	+	+			Ref. 2
N-Nitrosodiphenylamine	+				Chapter 6
N-Nitrosodi-*n*-propylamine	+				Ref. 2
trans-Nonachlor			TO-4A, TO-10A		525.2
Norflurazon					525.2
2,2′,3,3′,4,5′,6,6′-Octachlorobiphenyl					525.2
Pebulate (thiocarbamate)					525.2
Pentachlorobenzene			TO-4A, TO-10A		Ref. 2
2,2′,3′,4,6-Pentachlorobiphenyl					525.2
Pentachlorophenol	+	+	TO-4A, TO-10A	1	525.2, Ref. 2
cis-Permethrin			TO-4A, TO-10A		525.2
trans-Permethrin			TO-4A, TO-10A		525.2

(*continued*)

Table 4.1 (*continued*)

Compounds	PP[1]	HAP	GC/MS (air methods)	MCL[a] (μg/L)	GC/MS (water methods)
Perylene			TO-13A		
Phenanthrene	+		TO-13A		525.2, Ref. 2
Phenol	+	+			Ref. 2
p-Phenylenediamine		+	Chapter 6		
Phthalic anhydride		+			
Prometon (triazine)					525.2
Prometryn (triazine)					525.2
Pronamide					525.2
Propachlor (acetanilide)				a	525.2
Propazine (triazine)			TO-4A, TO-10A		525.2
Pyrene	+		TO-13A		525.2, Ref. 2
Quinoline		+			
Simazine (triazine)			TO-4A, TO-10A	4	525.2
Simetryne (triazine)					525.2
Stirofos (phosphate)					525.2
Tebuthiuron (urea)			TO-4A, TO-10A		525.2
Terbacil					525.2
Terbufos (dithiophosphate)					525.2
Terbutryn (triazine)					525.2
1,2,3,4-Tetrachlorobenzene					Ref. 2
1,2,3,5-Tetrachlorobenzene					Ref. 2
2,2′,4,4′-Tetrachlorobiphenyl					525.2
2,3,7,8-Tetrachlorodibenzodioxin (Chart 4.1—parent)	+	+	TO-9A, 23	3×10^{-5}	8280, 8290, 1613
Toxaphene (Chart 4.2—parents)	+	+		3	525.2, Ref. 2
Triademefon					525.2
2,4,5-Trichlorobiphenyl					525.2
2,3,6-Trichlorophenol					Ref. 2
2,4,5-Trichlorophenol		+			
2,4,6-Trichlorophenol	+	+			Ref. 2
3,4,5-Trichlorophenol					Ref. 2
Tricyclazole					525.2
Trifluralin (aromatic amine)		+	TO-4A, TO-10A		525.2
2,4,6-Trimethylphenol					Ref. 2
Vernolate (thiocarbamate)					525.2

[a]Drinking water MCL or monitoring requirement.[3]
[b]Mixtures of polychlorinated biphenyls have a MCL of 0.5 μg/L measured as decachlorobiphenyl.

- The semivolatiles are far less elusive than the VOCs, and sampling techniques and sample handling procedures are simpler and less dependent on specialized equipment.

Classes of Semivolatile Organic Compounds

A comprehensive review of the chromatographic behavior and mass spectra of all the compounds in Table 4.1 is beyond the scope of this book. However, because the semivolatile class of compounds includes many less familiar, larger, and more complex structures than the VOCs, brief descriptions of some important classes are given in the following sections. References are provided in most sections to reviews and other publications, which provide more information about specific groups of compounds.

Aromatic Amines

Primary aromatic amines contain one or more $-NH_2$ groups attached to an aromatic ring system such as benzene, a polycyclic aromatic hydrocarbon, or a heterocyclic aromatic compound. Substitution of the amino hydrogens with one or two alkyl or aryl groups produces secondary and tertiary amines, respectively. Aromatic amines are widely used intermediates in the production of agricultural chemicals, azodyes, plastics, and pharmaceuticals. They are also generated in the production of gaseous and liquid synthetic fuels from coal, oil shale, and heavy petroleum fractions. Primary aromatic amines are of concern because some of these are known human carcinogens (4-aminobiphenyl, 2-aminonaphthalene, and benzidine) or reasonably anticipated to be human carcinogens (>10 others).[4]

Aromatic amines with three or fewer aromatic rings are generally thermally stable, have adequate vapor pressures at typical GC operating temperatures, and give abundant, often base peak, $M^{+\cdot}$ ions in their EI spectra. Furthermore, when carefully handled with clean and silanized glassware and equipment, GC of these compounds can be successful.[5] However, the buildup of residues from previous samples, development of active sites in injection port liners and GC columns, and the basicity of the aromatic amines can cause erratic and suddenly poor behavior in GC systems.[2] Amines, which are frequently strong complexing ligands for metals, adsorb on surfaces and acidic sites, which causes broad tailing GC peaks, variable peak areas, and poor analytical sensitivity. This series of compounds, especially the primary aromatic amines, should be considered marginal for GC/MS, and the strategies described in Chapter 6 are appropriate for many applications, particularly when long-term reliability and low detection limits are needed.

Benzidine (Chart 6.1), 3,3′-dichlorobenzidine, and 1,2-diphenylhydrazine, which have similar properties, are listed in Table 4.1 as original priority pollutants (PPs) and HAPs. The benzidine derivatives 3,3′-dimethybenzidine and 3,3′-dimethoxybenzidine are also listed as HAPs. Benzidine and its derivatives were widely used, and still may be, as intermediates in the production of azodyes and pigments. While these aromatic primary diamines can be successfully separated by GC, the problems discussed in the preceding paragraph are common.[5] These analytes are best determined using the techniques described in Chapter 6. The aromatic primary diamines *p*-phenylenediamine, 4,4′-methylenedianiline, and 4,4′-methylenebis(2-chloroaniline), which are listed in Table 4.1 as HAPs, are expected to behave similarly to benzidine. On the other hand, tertiary amines are often less reactive and more amenable to GC separation. The herbicide trifluralin (Table 4.1) is a tertiary aromatic amine containing both nitro and trifluoromethyl groups and it is a GC/MS analyte.

Benzene Derivatives

Benzene derivatives are compounds in which an element or a group of elements is substituted for one or more of the benzene hydrogens. The benzene derivatives considered in this section contain halogens, alkyl groups, alkyloxy groups, and combinations of these and analogous groups of elements. Substituted benzenes in this class are very widely used commercial products and industrial intermediates and some of them are natural products, for example, the alkylated benzenes found in petroleum. Some compounds in this broad category are by-products formed during the production of other compounds and they have been widely discharged into the environment especially at chemical waste disposal sites. Benzenes substituted by just one or a few halogens, a small alkyl group, or a small alkyloxy group, are generally VOCs and are listed in Table 3.1. These compounds include toluene, the three xylenes, ethylbenzene, chlorobenzene, the three dichlorobenzenes, the chlorotoluenes, and many others. They are usually amenable to the analytical methods for VOCs presented Chapter 3.

Benzenes which have more than two hydrogens substituted by, for example, halogens, alkyl groups, or alkyloxy groups are generally less volatile and are classified as semivolatile compounds. However, the nature and quantity of the substituents are important in determining whether a specific compound is a VOC or a semivolatile. In spite of six halogens, hexafluorobenzene is a VOC, but the two tetrachlorobenzenes in Table 4.1 are semivolatiles. Table 4.2 shows the number of congeners that could be formed by replacing one to six benzene hydrogens with the same substituent X. While there are only 12 possible chlorinated benzenes or brominated benzenes, a very large number of substituted benzenes could be formed using all combinations of, for example, F, Cl, Br, methyl, ethyl, and methoxy. Bromochloromethoxybenzenes are reported to be widespread environmental contaminants and there are 19 possible chloromethoxybenzenes, 19 bromomethoxy benzenes, and 96 bromochloromethoxybenzenes.[6] Pentachlorobenzene and hexachlorobenzene are included in Table 4.1 and the latter is an original PP, a HAP, and has a maximum contaminant level in drinking water in the United States. Generally, compounds in this group are amenable to separation by GC and they often produce EI mass spectra containing abundant $M^{+\cdot}$ ions and/or characteristic fragment ions.

Table 4.2 Distributions of possible congeners among levels of substitution of hydrogens by the substituent X in aromatic compounds

Aromatic systems	Total	X_1	X_2	X_3	X_4	X_5	X_6	X_7	X_8	X_9	X_{10}
Benzene	12	1	3	3	3	1	1				
Biphenyl	209	3	12	24	42	46	42	24	12	3	1
Dibenzo-*p*-dioxin	75	2	10	14	22	14	10	2	1		
Dibenzofuran	135	4	16	28	38	28	16	4	1		
Diphenyl ether	209	3	12	24	42	46	42	24	12	3	1
Naphthalene	75	2	10	14	22	14	10	2	1		

Biphenyl Derivatives

Biphenyl (left end of top row in Chart 4.1) is the parent compound of one of the most important and widely measured, studied, and discussed groups of environmental contaminants. Polybrominated biphenyls (PBBs) are a series of compounds in which bromine is substituted for some or most of the hydrogens of the biphenyl molecule. Some PBBs were manufactured and used as flame retardants in textiles and plastics in the United States during the early 1970s. Like the chlorinated biphenyls, which are described later in this section, there are 209 possible brominated biphenyl congeners ranging from the three possible monobromobiphenyl isomers to decabromobiphenyl (Table 4.2). One of these congeners, 4,4′-dibromobiphenyl, is listed in Table 4.1. The PBBs used as fire retardants contain only about nine major congeners and mainly 2,2′,4,4′,5,5′-hexabromobiphenyl.[7]

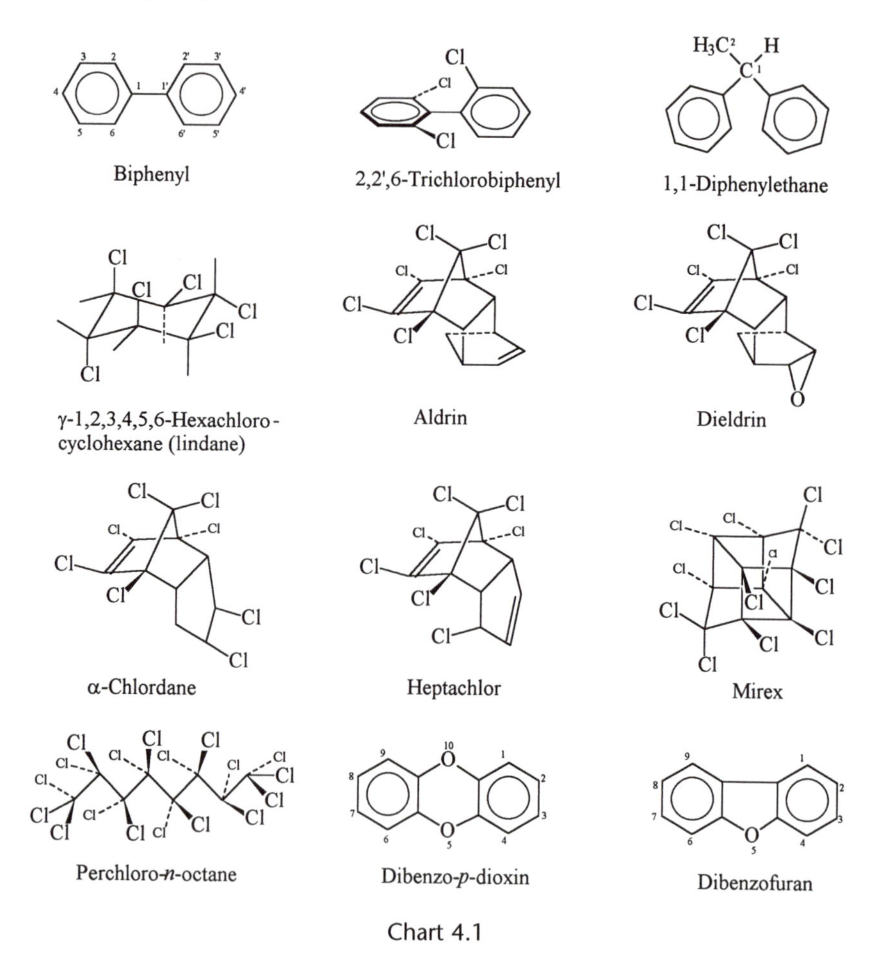

Chart 4.1

The PBBs became infamous environmental contaminants when a large quantity was accidentally included in a dairy cattle feed in 1973.[8] The fire retardant PBBs may be vaporized in emissions from solid waste incinerators and these PPBs are classified as reasonably anticipated to be human carcinogens.[4] The PBBs containing up to eight bromines have been separated by GC although this requires high oven temperatures, that is, 300 °C for over 30 min.[7] Because the masses of the tetra- to deca-bromobiphenyl $M^{+\cdot}$ ions are in the range 468–948 Da, accurate measurements of ion abundances require a mass spectrometer with adequate resolving power (Chapter 5, "High-Resolution Mass Spectrometry"). This level of resolving power may not be available on some quadrupole and other types of mass spectrometer. With EI the PBB congeners generally give moderate to very abundant $M^{+\cdot}$ ions, and their mass spectra contain fragment ions corresponding to the loss of one or more bromines from the $M^{+\cdot}$ ion.

The polychlorinated biphenyls (PCBs) are a series of compounds in which chlorine is substituted for some or all of the hydrogens of the biphenyl molecule. As shown in Table 4.2, there are 209 possible congeners which range in composition from mono- to deca-chlorobiphenyl. Only 206 of these are truly PCBs, but the three possible monochlorobiphenyls are commonly considered part of this class of compound. From 1929 to 1975 the PCBs were manufactured in the United States by chlorination of biphenyl and sold under several trade names. The largest producer, Monsanto Chemical Company, sold a series of PCB mixtures under the name Aroclor™ 12XX. The 12 in the Aroclor designation gives the number of carbons in the biphenyl molecule and the XX indicates the weight percentage of Cl in the mixture. Table 4.1 lists seven Aroclor commercial mixtures that range in composition from 21–60% Cl by weight. The Aroclor 1016 product, which was not named according to this convention, is a PCB with approximately 40% Cl by weight. Aroclor 1221 is mainly biphenyl and chlorobiphenyls, Aroclors with 32–48% Cl contain mono- to penta-chlorobiphenyls, and those Aroclors with 54–62% Cl contain significant quantities of hexa- to octa-chlorobiphenyls.[9] Only Aroclor 1268, which is not in Table 4.1, contains mostly nonachlorobiphenyls and decachlorobiphenyl. Except for Aroclor 1221, individual Aroclor mixtures contain approximately 30–125 congeners, but some are not favored in the chlorination reaction and are present in very small quantities or are not present at all. The Aroclor mixtures in Table 4.1 were designated PPs in 1976 and HAPs in 1990.

The commercial PCB mixtures were used in many products and applications where their thermal stability, low chemical reactivity, electrical properties, and other characteristics were important. They found wide application in the 1930s–1970s in electrical capacitors, transformers, and heat exchangers, and as plasticizers, hydraulic fluids, lubricants, paper additives, and other products.[10] They were sometimes blended with tri- and tetra-chlorobenzenes, and estimated Monsanto sales in the United States during this period were 2.8×10^9 kg.[10] Widespread environmental distribution resulted because very large quantities of PCBs were discarded in waste streams and landfills or were released in transformer fires and other accidents. These compounds, especially the more highly chlorinated, are resistant to chemical and biochemical degradation and tend to accumulate in nonpolar environments such as the fatty tissue of wildlife.[11] Their very low solubility in water results in adsorption on particulate matter and their accumulation in sediments.[12] Some PCBs also have sufficient vapor pressures at ambient temperatures in temperate regions of the Earth to allow atmospheric transport to colder regions where they were never manufactured or used.[13]

Adverse human health and ecological effects were attributed to PCBs and most manufacturing, processing, distribution, and uses were banned in the United States in 1979.[11] The PCBs are currently classified as reasonably anticipated to be human carcinogens.[4] In addition, a maximum contaminant level in drinking water is in effect in the United States (Table 4.1). The large quantities of PCBs in the environment and regulations in the United States designed to control their release into the environment resulted in the development of a number of analytical methods for PCBs in various matrices and the application of these methods in hundreds of thousands of chemical analyses.

The PCB congeners are generally very amenable to GC because of their high thermal stability and favorable vapor pressures. They produce highly characteristic EI mass spectra with $M^{+\cdot}$ and fragment ions that are strongly indicative of the number of chlorines in the ions (Chapter 2, "Identification Criteria"). However, PCB isomers at any level of chlorination produce virtually identical mass spectra and these isomers cannot generally be distinguished by MS (Chapter 2, "Limitations of Mass Spectrometry"). The analytical strategies for PCBs have evolved over the years with improvements in technology, and several significantly different approaches have been proposed and some of them are widely used.

Aroclors by Packed Column GC Patterns. Most analyses before about 1982 used packed GC columns for the identification and measurement of standard Aroclor mixtures in environmental and other samples. Separation of individual PCB congeners is not possible with a packed column, but the Aroclor mixtures give characteristic chromatograms that are used for identification and measurement. These chromatograms consist of 8–12 broad peaks with each peak consisting of 5–15 unresolved congeners.[14, 15] As expected, the lower molecular weight congeners generally elute before the more heavily chlorinated congeners.[9] However, structure also influences elution behavior and some more chlorinated congeners elute before some less chlorinated congeners.

The general patterns of packed-column GC peaks in Aroclor standards are compared with the patterns observed in environmental samples. If the samples contain only one environmentally unaltered Aroclor mixture, identification and measurement as an Aroclor gives a reasonable estimate of the total concentration of PCBs. Selected ion monitoring is used to minimize interferences from some coeluting substances.[14] This technique, which is used in some regulatory programs in the United States, is reasonably successful when analyzing, for example, used transformer oil which typically contains only one unaltered Aroclor. However, in the environment biological degradation or vaporization of less chlorinated species, and/or strong adsorption of some congeners in biota or on particulates, and/or mixing of several or more Aroclors makes this GC peak-matching strategy very difficult or impossible.

Isomer Groups by Level of Chlorination. With the development and widespread use of bonded-phase fused silica capillary columns (FSCCs) in the early 1980s, and the ability to separate most but not all PCB congeners, several alternative analytical strategies emerged. One approach was developed to provide a more realistic, but practical, regulatory approach than the packed-column Aroclor identification technique. This GC/MS approach employs a single pure isomer at each level of chlorination to calibrate the MS response for all isomers at each level of chlorination.[16] The calibration isomers selected have response factors (RFs) close to the mean RF of all isomers at each level of chlorination. With this technique, PCB concentrations are determined as isomer groups at each level of chlorination from monochlorobiphenyls to decachlorobiphenyl, and total PCBs are determined by summing the isomer group concentrations.

This strategy is a practical approach to the determination of environmentally altered Aroclors, mixed Aroclors, or incidentally generated PCBs.[17] The isomer group technique is less costly than a congener-specific analytical strategy and was supported by the development of software to automate data reduction.[18] This technique was also described in unpublished USEPA Method 680, but it did not attract much attention for regulatory or other analyses of PCBs. However, a similar strategy has been adopted by some investigators for the determination of chlorinated dibenzo-*p*-dioxin, dibenzofuran, and toxaphene isomer groups at various levels of chlorination. These compounds are discussed later in this chapter.

Measurements of Individual Congeners. By the mid-1980s all 209 chlorinated biphenyl congeners had been synthesized and their retention times measured on a FSCC.[19] This permitted the development of congener-specific analyses, which are now employed primarily in environmental research such as atmospheric transport and microbiological dechlorination studies.[13,20] Eight chlorinated biphenyl congeners are shown in Table 4.1 and these were used as model target analytes in USEPA Method 525.2, which is discussed later in this chapter.

Measurements of Planar Congeners. With the realization that certain planar PCBs with chlorines limited to the 3,4,5 and 3′,4′,5′ positions (biphenyl—left end of top row in Chart 4.1) are the most biologically active, many recent studies have concentrated on the identification and measurement of these planar congeners.[21] Substitution of hydrogens in the 2,6,2′, or 6′ positions with Cl tends to force the two phenyl rings into separate planes (center of top row in Chart 4.1) and these nonplanar PCBs are generally less biologically active. Measurements of some planar PCBs are difficult because they are not abundant in typical PCB mixtures and environmental samples.

The wide-ranging environmental contamination by PCBs and their environmental transport, persistence, microbiological degradation, human health effects, ecological impacts, and government regulation have stimulated literally thousands of studies over the last 30 years. A large number of analytical techniques have been used for the extraction of PCBs from environmental and other samples. The more highly chlorinated PCBs give abundant negative ions with electron capture ionization, which is described in Chapter 5. More information about extraction techniques and analytical methods is given in the last part of this chapter and in a recent comprehensive review.[22]

Chlorinated Hydrocarbon Pesticides

There are four principal groups of chlorinated hydrocarbon pesticides: (1) chlorinated and other derivatives of 1,1-diphenylethane (right end of top row in Chart 4.1); (2) a group of 1,2,3,4,5,6-hexachlorocyclohexane isomers represented by lindane (left end of the second row in Chart 4.1); (3) Diels–Alder condensation products of unsaturated hydrocarbons and hexachlorocyclopentadiene, which are sometimes further chlorinated and are represented by aldrin, dieldrin, α-chlordane, and heptachlor in Chart 4.1; and (4) a very large group of chlorinated camphene congeners called toxaphene, which is discussed in a separate section.

Chlorinated hydrocarbons were the first generation of the widely used and very effective synthetic organic compound pesticides. Some compounds in this group were the first recognized ecologically harmful organic environmental pollutants.[23] Their structures generally do not contain reactive functional groups and some of them are very slow to degrade in the environment. Because of their persistence, nonpolar structures, and hydrocarbon solubility, these compounds tend to accumulate in fatty tissue and concentrate in the food chain.[24] They are also found in vegetation and sediments.[25,26] Although many of these pesticides were banned from use in the United States and other countries over 20 years ago, some are still used in developed and undeveloped countries. They also continue to be dispersed in the environment by vaporization and atmospheric transport to colder climates where they are condensed.[27] Thus, these well-studied compounds are found in largely uninhabited nonagricultural areas where they have never been applied. This class of compound is among the most widely dispersed in the environment and among the most frequently monitored of all environmental pollutants.

Perhaps the most well known of the derivatives of 1,1-diphenylethane (right end of top row in Chart 4.1) is 1,1-bis(4-chlorophenyl)-2,2,2-trichloroethane (4,4′-DDT) in which the three hydrogens on C-2 of the ethane residue and the hydrogens on C-4 and C-4′ of the phenyl groups are replaced by chlorines (Table 4.1). Other isomers and congeners also have been used, for example, 4,4′-DDD (Table 4.1), and are present in the environment. Another derivative is methoxychlor (Table 4.1) in which the two phenyl ring chlorines of 4,4′-DDT are replaced by methoxy groups. The most widely dispersed environmental contaminant from this group is 1,1-bis(4-chlorophenyl)-2,2-dichloroethene (4,4′-DDE) which is formed by elimination of HCl (dehydrochlorination) from C-1 and C-2 of the ethane residue of 4,4′-DDT. The compounds 4,4′-DDD, 4,4′-DDT, and 4,4′-DDE are original PPs (Table 4.1); 4,4′-DDE is also listed as a HAP, and DDT is classified as reasonably anticipated to be a human carcinogen.[4] Another HAP, methoxychlor, has a maximum contaminant level in drinking water in the United States (Table 4.1).

The 1,2,3,4,5,6-hexachlorocyclohexane isomers are produced by the photochemical chlorination of benzene and are widely misnamed as benzene hexachlorides (BHC) in the chemical, environmental, and regulatory literature. Seven isomers are formed by positioning chlorines in axial and equitorial positions around the cyclohexane ring and these are designated the α, β, γ, δ, ϵ, η, and θ isomers. The α isomer exists as an enantiomeric pair, which provides another isomer.[28] The α, β, γ, and δ isomers are produced in the largest quantities, comprise nearly all of the technical mixtures, and are listed as original PPs and HAPs in Table 4.1. The β isomer with all six chlorines in equitorial positions is the most persistent. The γ isomer, known as lindane (left end of second row in Chart 4.1), is the most effective pesticide and has a maximum contaminant level in drinking water in the United States (Table 4.1). The hexachlorocyclohexane isomers are classified as reasonably anticipated to be human carcinogens.[4]

The final group of chlorinated hydrocarbon pesticides considered in this section is prepared by the condensation of hexachlorocyclopentadiene (HCCP) with several unsaturated hydrocarbons or with itself. Aldrin (center of second row in Chart 4.1) is formed by the Diels–Alder condensation of HCCP with bicyclo[2.2.1]heptadiene. Aldrin does not persist in the environment because it is readily epoxidized to dieldrin (right end of second row in Chart 4.1), which is persistent. Endrin (structure not shown) is a structural isomer of dieldrin. Technical chlordane is prepared by the Diels–Alder condensation of HCCP with cyclopentadiene followed by chlorination of the product to produce a complex mixture which is reported to have 147 components by FSCC GC.[29] The main components are α-chlordane (left end of third row in Chart 4.1), γ-chlordane (structure not shown) in which the two chlorines attached to the fused cyclopentane ring are *trans* to one another, heptachlor (center of third row in Chart 4.1), and *trans*-nonachlor (structure not shown) which combines the structural features of α-chlordane and heptachlor. Many other compounds, structural isomers, and congeners are also produced in the chlorination process, and many components are of unknown structure. Heptachlor epoxide is produced by the epoxidation of heptachlor analogous to the aldrin to dieldrin transformation. Among the most interesting of the HCCP condensation products is the pesticide mirex (right end of third row in Chart 4.1), which is a fusion of two HCCP molecules into a cage structure. Aldrin, dieldrin, endrin, and technical chlordane were very widely used pesticides and the persistent compounds in this group are widely dispersed in the environment.

Hexachlorocyclopentadiene, most of the HCCP derivatives mentioned in the previous paragraph, an endrin aldehyde oxidation product, and three endosulfans were designated original PPs (Table 4.1). Endosulfans I and II are isomers of the Diels–Alder condensation product of HCCP with an unsaturated cyclic sulfurous acid ester, and endosulfan sulfate is an oxidation product. Technical chlordane and one of its components, heptachlor, are also listed as HAPs. Chlordane, endrin, heptachlor, heptachlor epoxide, and HCCP have maximum contaminant levels in drinking water in the United States (Table 4.1). Mirex is classified as reasonably anticipated to be a human carcinogen.[4]

Generally, the chlorinated hydrocarbon pesticides are readily separated on contemporary GC columns and instruments, but some isomers may be separated by only a few seconds. With some instruments or experimental conditions, decomposition of some compounds on hot surfaces is not uncommon. Endrin and 4,4′-DDT, which are particularly sensitive to oxidation and dechlorination, respectively, are recommended to evaluate the performance of a GC column and instrument.[30] Nearly all the chlorinated hydrocarbon pesticides give highly fragmented EI mass spectra devoid of $M^{+\cdot}$ ions, but an exception is 4,4′-DDE which has an abundant $M^{+\cdot}$ ion. Some of these compounds also have highest *m/z* fragment ions that correspond to losses of two or more chlorines. With some compounds, for example aldrin, endrin, and dieldrin, base peaks are below *m/z* 100. While these spectra and the GC retention times generally provide reliable target compound identifications, the desire for lower detection limits and more definitive mass spectra has led many investigators to use GC/MS with electron capture ionization, which is described in Chapter 5.

Chlorinated Paraffins

Paraffin is the straight-chain saturated hydrocarbon fraction from petroleum. Industrial chlorination of paraffin in the C_{10}–C_{30} range produces a complex mixture of compounds in which chlorines replace some of the hydrogens of various hydrocarbons. In addition to substitution, some C–C bonds are broken during the chlorination process to produce shorter chain chlorinated hydrocarbons. An example of a chlorinated paraffin is perchloro-*n*-octane (left end of bottom row in Chart 4.1). Even with the assumption of no more than one chlorine attached to any one carbon, these mixtures could contain in excess of 6000 isomers and congeners. Mixtures of chlorinated paraffins are marketed in several carbon and chlorine content ranges and are used as fire retardants, plasticizers, metal cutting lubricants, and for other applications. Although these materials were commercial products as early as the 1930s, very little information is available on their environmental occurrence and ultimate fate. Some chlorinated paraffins have been determined in fish and sediments from the Detroit River and several other locations.[31] Chemical determinations of these mixtures is considered very difficult. The chlorinated paraffin mixture with 60% Cl is classified as reasonably anticipated to be a human carcinogen.[4]

The lower boiling C_{10}–C_{13} chlorinated congeners are amenable to GC at attainable oven temperatures. They generate very broad elution profiles containing six to ten unresolved main peaks. The $M^{+\cdot}$ ions are not observed with EI, which produces extensive fragmentation from losses of Cl, HCl, and multiple chlorines and HCl groups. The recommended techniques for these compounds are GC, electron capture ionization (Chapter 5), selected ion monitoring, and high-resolution MS (Chapter 5).[31] One of the techniques described in Chapter 6 may provide more information, especially for the C_{14}–C_{30} chlorinated congeners that are progressively less amenable to GC. However, capillary supercritical fluid chromatography with charge exchange ionization (Chapters 6 and 5, respectively) produced results similar to that obtained with FSCC GC.[32]

Dibenzo-p-dioxin and Dibenzofuran Derivatives

Dibenzo-*p*-dioxin and dibenzofuran (bottom row of Chart 4.1) are the parent compounds of a series of environmental pollutants in which chlorine, and occasionally bromine, replace some or all of the hydrogens of these molecules. There are 75 possible chlorinated dibenzo-*p*-dioxin congeners and 135 possible chlorinated dibenzofuran congeners, which range in composition from monochloro to octachloro (Table 4.2). Easily the most widely studied and infamous of this group of compounds is 2,3,7,8-tetrachlorodibenzo-*p*-dioxin (2378-TCDD), which is among the most controversial of all environmental pollutants.[33–35] This compound is listed in Table 4.1 as an original PP and a HAP. It is classified as reasonably anticipated to be a human carcinogen and has an extremely low maximum contaminant level in drinking water in the United States (Table 4.1).[4]

None of the chlorinated or brominated dibenzo-*p*-dioxins and dibenzofurans were deliberately manufactured as chemical products. The 2378-TCDD isomer was a by-product in the manufacture of 2,4,5-trichlorophenol, which was used to produce the herbicide 2,4,5-trichlorophenoxyacetic acid (2,4,5-T). During the 1960s, 2,4,5-T contaminated with 2378-TCDD was widely used as a defoliant in Vietnam. The 2378-TCDD isomer was also dispersed during the early 1970s in contaminated waste oils used to control road dust in eastern Missouri and during an explosion at a 2,4,5-trichlorophenol plant in Seveso, Italy, in 1976.[33–35] Later, chlorinated dibenzo-*p*-dioxins and dibenzofurans were discovered in atmospheric emissions and in residues from a variety of combustion processes, for example, solid waste incineration.[36, 37] These compounds are also formed during the bleaching of wood pulp with chlorine and they contaminate sediments and fish in streams near some pulp and paper manufacturing plants.[37–39] Some mixed bromochloro-dibenzo-*p*-dioxins and -dibenzofurans have been found in combustion residues, but these derivatives are not nearly as well studied as the chlorinated dibenzo-*p*-dioxins and dibenzofurans.[40] The chlorinated dibenzo-*p*-dioxins and dibenzofurans, like the PCBs and chlorinated hydrocarbon pesticides, are transported in the atmosphere from warmer to colder climate areas.[41] The polychlorinated congeners, particularly the Cl_4–Cl_8 compounds, are persistent in the environment and tend to accumulate in fatty tissues in the food chain.[39]

The thermally very stable chlorinated dibenzo-*p*-dioxins and dibenzofurans are readily amenable to GC. By the mid-1980s, GC columns and conditions had been developed for the separation of many of the chlorinated dibenzo-*p*-dioxin congeners.[42] However, at some levels of chlorination where a larger number of isomers are possible (Table 4.2), some but not all isomers are resolved. If isomer-specific measurements are needed, separation of the target isomers by GC is required because at each level of chlorination the mass spectra of all isomers are essentially identical (Chapter 2, "Limitations of Mass Spectrometry"). Fortunately, 2378-TCCD, the isomer of greatest interest in many studies, can be separated from the other tetrachlorodibenzo-*p*-dioxin isomers. The chlorinated dibenzofurans are more difficult to separate because of the significantly larger number of possible isomers at most levels of chlorination (Table 4.2), but conditions for the separation of almost all the tetrachloro isomers were reported by the late 1980s.[43,44] As information about the biological properties of various chlorinated dibenzo-*p*-dioxin and dibenzofuran congeners developed, it was concluded that compounds substituted in at least the 2, 3, 7, and 8 positions were of most concern. In the 1990s, attention has focused on 17 tetra-, penta-, hexa-, hepta-, and octa-chlorodibenzo-*p*-dioxin and dibenzofuran congeners with the 2, 3, 7, 8 substitution pattern.[45]

All the chlorinated dibenzo-*p*-dioxins and dibenzofurans give abundant, often base peak, $M^{+\cdot}$ ions in their EI mass spectra (e.g., see the spectrum of 2378-TCDD in Figure 5.1). One or more of the ions in the $M^{+\cdot}$ ion isotope group, and the corresponding ions in ^{13}C-labeled internal standards, are typically used for selected ion monitoring in GC/MS. An important fragment ion in these spectra is the $(M\text{–}63/65)^+$ ion, which corresponds to the loss of $CO^{35}Cl$ or $CO^{37}Cl$ from the $M^{+\cdot}$ ions. This fragment ion is used to verify the identification of these analytes. Lower abundance ions, corresponding to the loss of two COCl groups, are also observed in the spectra of the chlorinated dibenzo-*p*-dioxins.

Most environmental analyses in the 1970–1980s were concerned with soil, sediment, fly ash, and tissue extracts because both series compounds are very insoluble in water. Because of generally very low concentrations, often in the low nanogram/kilogram range or lower, the high chemical background at these levels, and the potential for interferences, preliminary separations and sample processing before GC/MS are crucial for successful analyses.[45] Nearly all investigators use selected ion monitoring (SIM) to minimize interferences and maximize selectivity. Commercial GC/MS systems allow sequential monitoring of different sets of selected ions to facilitate the simultaneous determination of many tetra- to octa-chlorinated dibenzo-*p*-dioxins and dibenzofurans as various congeners elute from the GC column.

Many investigators also use high-resolution MS (HRMS) for greater selectivity and lower detection limits, and this technique is described in Chapter 5 using 2378-TCDD as an example.[46] Four USEPA analytical methods which employ HRMS have been promulgated in regulations or published and these are summarized in the last section of this chapter. Electron capture ionization, which is described in Chapter 5, is also used for this class of compounds. It provides higher sensitivity for the more highly chlorinated congeners and is not sensitive to the myriad of nonelectron-capturing components in most environmental samples.[47] The technique of quantitation by isomer groups at various levels of chlorination, which was described in the section on biphenyl derivatives, is widely used in atmospheric transport and other studies where isomer specific analyses are not required.[47] Because of the importance of this class of environmental contaminants, the last section of this chapter is a review of the development of analytical methods and includes summaries of some of the published methods.

Diphenyl Ether Derivatives

Diphenyl ether (left end of top row in Chart 4.2) is the parent compound of a large number of possible derivatives (Table 4.2) in which bromine, chlorine, or other elements or groups of elements are substituted for some or all of the phenyl hydrogens of the molecule. The halogenated diphenyl ethers are represented in Table 4.1 by 4-chlorophenyl phenyl ether and 4-bromophenyl phenyl ether, which were selected as original PPs. The brominated diphenyl ethers are of special interest because they are used as fire retardants in textiles, plastics, and other products. It has been shown that brominated diphenyl ether fire retardants are converted into polybrominated dibenzo-*p*-dioxins, dibenzofurans, and other products at elevated temperatures that can exist in accidental fires or solid waste incinerators.[48] The dibenzo-*p*-dioxins and dibenzofurans are formed by oxidation at C-2 and C-2′ of the diphenyl ether with and without the incorporation of an oxygen atom (Chart 4.3). This reaction might also be catalyzed by hot metal surfaces in some equipment and by oxidative enzymes in some biological systems. The brominated diphenyl ethers are amenable to GC and give EI mass spectra with abundant $M^{+\cdot}$ ions and some fragment ions that correspond to losses of one or more bromines from the $M^{+\cdot}$ ion.[48]

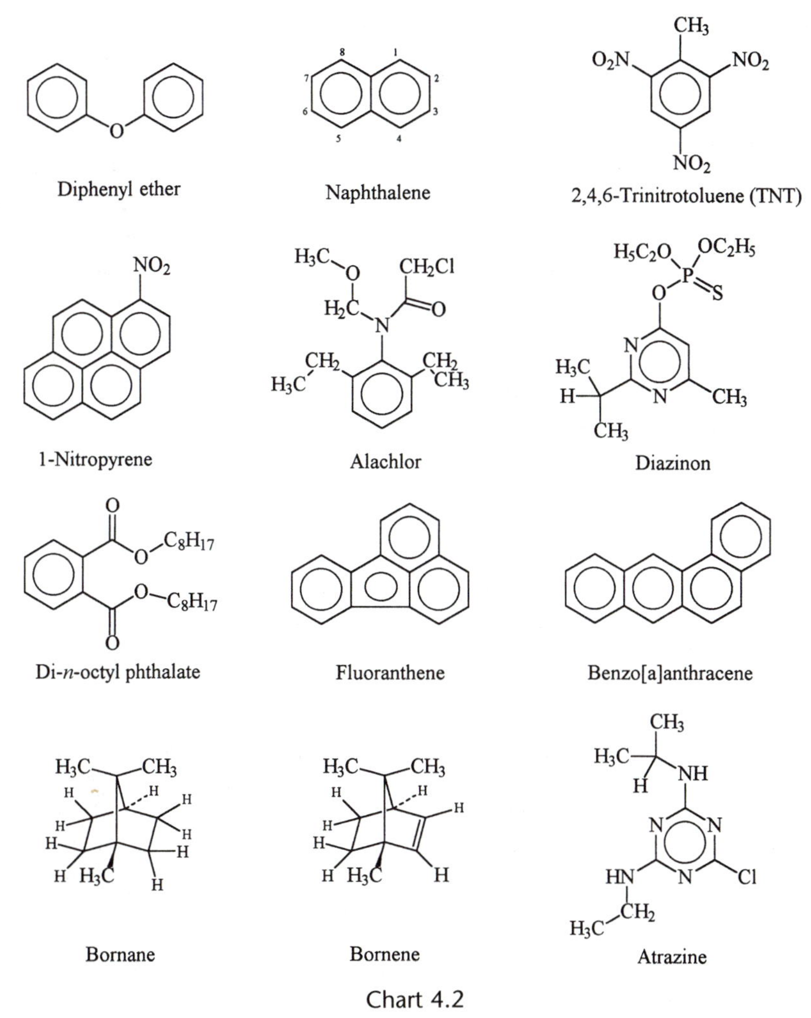

Chart 4.2

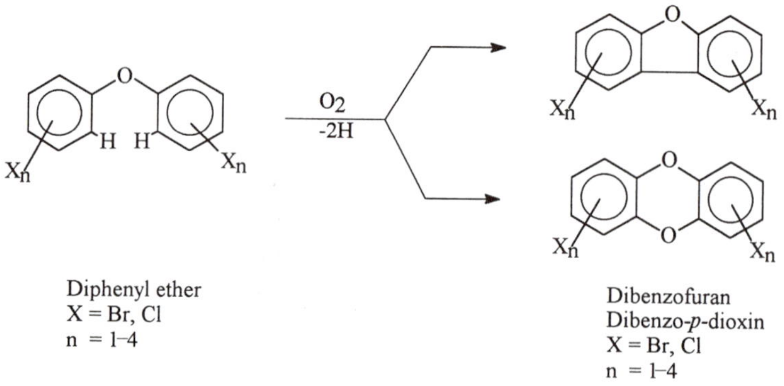

Diphenyl ether
X = Br, Cl
n = 1–4

O_2
-2H

Dibenzofuran
Dibenzo-*p*-dioxin
X = Br, Cl
n = 1–4

Chart 4.3

Herbicide Carboxylic Acid Esters

Many herbicides are carboxylic acids, for example, 2,4-dichlorophenoxyacetic acid (2,4-D) in Chart 6.1, and these compounds, which are not amenable to separation by GC, are discussed in Chapter 6. However, commercial herbicide formulations often contain methyl or other esters of these acids. For example, the infamous *agent orange*, which was used as a defoliant in Vietnam during the 1960s, was a mixture of the *n*-butyl esters of 2,4-D and 2,4,5-T. When herbicide carboxylic acids are target analytes in environmental analyses, aqueous samples are treated with a strong base to hydrolyze the esters before further processing and determination of the acids using the methods described in Chapter 6. However, if the hydrolysis reaction is omitted, and if the herbicide esters are stable to hydrolysis in the environment, these esters can be separated and determined by GC/MS.

Naphthalene Derivatives

Naphthalene (center of top row in Chart 4.2) and alkylated naphthalenes are found in petroleum and in products produced from coal. These compounds are the basis for a variety of industrial intermediates and other products in which bromine, chlorine, or other elements or groups of elements are substituted for the hydrogens of naphthalene. The compound 2-chloronaphthalene is representative of this group and is listed in Table 4.1 as an original priority pollutant. Mixtures of chlorinated naphthalenes were manufactured and sold under the name Halowax 10XX™ where the 10 indicates the number of carbons in the molecule and XX indicates the weight percentage of chlorine in the mixture. For example, two such mixtures are Halowax 1014 and 1051. As shown in Table 4.2, there are 75 possible chlorinated naphthalene congeners, but Halowax 1051 consists of mostly octachloronaphthalene and the two heptachloronaphthalene isomers.[49]

While there are many reports of the environmental occurrence of polychlorinated naphthalenes, these compounds were not nearly as widely used as the

PCBs and they have not been studied to the same extent as the PCBs, chlorinated dibenzo-*p*-dioxins, and chlorinated dibenzofurans. The chlorinated naphthalenes are amenable to GC and both 2-chloronaphthalene and octachloronaphthalene have base peak $M^{+\cdot}$ ions in their EI spectra.[50] Major fragment ions correspond to the loss of one chlorine from 2-chloronaphthalene and two chlorines from octachloronaphthalene. Other congeners have been synthesized and at least 32 of them have been identified in sediment samples.[49,51]

Nitroaromatic Compounds

Aromatic compounds in which one or more hydrogens are substituted by a nitro ($-NO_2$) group are important environmental compounds and some of them are also significant industrial products. Some explosives are polynitro aromatic or heterocyclic compounds, for example, 2,4,6-trinitrotoluene (TNT) (right end of top row in Chart 4.2). These explosives, and by-products of their manufacture such as 4-nitrotoluene, are environmental pollutants in soil and ground water in the vicinity of munitions manufacturing and processing plants. Nitroaromatics are also intermediates in the preparation of textile dyes, agricultural chemicals, and pharmaceuticals. Nitro compounds are subject to anaerobic reduction to produce nitroso compounds or aromatic amines, which are known or suspected carcinogens.[4] Many of these compounds are sufficiently thermally stable, volatile, and nonreactive to allow separation and measurement by GC/MS, but some are not and the methods described in Chapter 6 must be used. The compounds 2,4- and 2,6-dinitrotoluene are amenable to GC and are listed in Table 4.1 as original priority pollutants. Both of these compounds and TNT give very low abundance or no $M^{+\cdot}$ ions in their EI spectra, but they have base peak $(M-OH)^+$ ions. The strongly electron-withdrawing nitro groups in these polynitro compounds stabilize negative ions, and these analytes give abundant $M^{-\cdot}$ ions with electron capture ionization (Chapter 5).

Particulates in diesel engine exhaust and emissions from some other high temperature processes contain nitropolycyclic aromatic hydrocarbons (nitroPAHs), which are formed by the reaction of polycyclic aromatic hydrocarbons (PAHs) with hydroxyl radicals and oxides of nitrogen.[52] Many nitro- and dinitro-PAHs have sufficient vapor pressures at typical GC oven temperatures and can be separated and identified by GC/MS. In one study, 34 nitro-substituted anthracenes, biphenyls, fluoranthenes, fluorenes, naphthalenes, pyrenes, and other aromatics were identified in a diesel engine exhaust particulate extract.[52] Most of these had a single nitro group, for example, 1-nitropyrene (left end of second row in Chart 4.2), but some dinitroPAHs were identified. The compound 4-nitrobiphenyl, which was identified in this study, is listed as a HAP in Table 4.1. Some of these nitroPAHs are potent mutagens and they are classified as reasonably anticipated to be human carcinogens (1-nitropyrene, 4-nitropyrene, 6-nitrochrysene 1,6-dinitropyrene, and 1,8-dinitropyrene).[4] The mono- and dinitro-PAHs give abundant $M^{+\cdot}$, $(M-NO)^+$, and $(M-NO_2)^+$ ions in their EI mass spectra. It is likely that higher molecular weight nitroPAHs and polynitroPAHs will not be amenable to separation by GC, and the methods described in Chapter 6 will be required for these compounds. These nitroPAHs also give abundant $M^{-\cdot}$ ions with electron capture ionization, which is discussed in Chapter 5.

Nitrogen-Based Pesticides

Most second generation and later pesticides contain nitrogen in some form such as an amine, amide, carbamate, heterocycle, nitrile, urea, or thiourea. Some of these, typically the carbamates and ureas, are not amenable to GC because of thermal instability and are discussed in Chapter 6. The GC amenable triazine herbicides are described in a separate sections in this chapter. Among the other nitrogen-based pesticides, many are amenable to separation by GC including a few carbamates, the α-chloroacetanilides, and some other amides, imides, and thiocarbamates. The α-chloroacetanilides alachlor (center of second row in Chart 4.2), butachlor, metolachlor, and propachlor have the same basic structure, but with different substituents on the phenyl group and the amide nitrogen. These four compounds are listed in Table 4.1, and alachlor has a maximum contaminant level in drinking water in the United States. The EI spectra of the α-chloroacetanilide herbicides have either very low abundance or no $M^{+\cdot}$ ions, and fragment ions are used for identification and quantitative analysis.

The compounds butylate, cycloate, eptam, molinate, pebulate, and vernolate in Table 4.1 are thiocarbamates in which the noncarbonyl oxygen of the carbamate structure is substituted by a sulfur (the structure of a carbamate is illustrated by carbofuran in Chart 6.1). These compounds are amenable to separation by GC but have either very low abundance or no $M^{+\cdot}$ ions, and fragment ions are used for identification and quantitative analysis. The compounds carboxin, carboximide, diphenamide, napropamide, and pronamide in Table 4.1 are amides or imides amenable to GC.

N-*Nitroso Compounds*

Primary and secondary aliphatic and aromatic amines form *N*-nitroso compounds (R_2N–NO) by substitution of a nitric oxide (NO) molecule for a hydrogen on the nitrogen. This reaction occurs in industrial and other processes where amines are exposed to nitrogen oxides or nitrites and it may occur in biological systems. The compound *N*-nitrosodimethylamine, which is listed in Table 4.1 as a HAP, is produced in rocket engines that use unsymmetrical dimethylhydrazine fuel and liquid N_2O_4 oxidizer. Diphenylamine has been used for many years as a stabilizer in propellants where it reacts with nitrogen oxides, which form by slow decomposition of the propellants, to form *N*-nitrosodiphenylamine (Chart 6.1). These two *N*-nitroso compounds and *N*-nitrosodi-*n*-propylamine are listed as original priority pollutants in Table 4.1. Several *N*-nitrosoureas are known human carcinogens, and many *N*-nitroso compounds are classified as reasonably anticipated to be human carcinogens.[4]

Some of the lower molecular weight *N*-nitrosodialkylamines, including the two in Table 4.1, are amenable to separation by GC, but *N*-nitroso compounds in general are susceptible to thermal decomposition in a GC injection port or column oven. The N–NO bond in this class of compounds is weak and many compounds decompose readily on heating to produce nitric oxide and the $R_2N^{\cdot}$ free radical. For example, *N*-nitrosodiphenylamine decomposes easily on heating in a direct insertion probe or a GC injection port.[2] The NIST/EPA/NIH database spectrum of *N*-nitrosodiphenylamine does not have a $M^{+\cdot}$ ion at *m*/*z* 198, but has a base peak at *m*/*z* 168 which corresponds to the diphenylnitrogen cation.[50] In a GC injection port the initially formed diphenylnitrogen radical abstracts a hydrogen from a solvent molecule or some other hydrogen source and behaves exactly like diphenylamine. Therefore, it is impossible to distinguish between diphenylamine and *N*-nitrosodiphenylamine by conventional GC/MS.[2] The low-temperature condensed-phase separation methods described in Chapter 6 are required for the determination of most *N*-nitroso compounds.

Petroleum and Related Products

Crude petroleum consists of an enormous variety of saturated and unsaturated aliphatic and alicyclic hydrocarbons, aromatic hydrocarbons, and small quantities of polycyclic aromatic and heterocyclic compounds containing oxygen, nitrogen, or sulfur. Refined petroleum products such as gasoline, kerosene, and No. 2 fuel oil are hydrocarbon mixtures with standard boiling-point ranges. Crude oils produced from coal and oil shale have similar components, but often significantly greater amounts of polycyclic aromatic hydrocarbons (PAHs) and heterocyclic compounds. Environmental contamination from spills of crude petroleum and related products are not uncommon and large-scale disasters, such as the *Exxon Valdez* tanker accident in 1989, occur from time to time. Some natural seeps of crude petroleum are also long term and continuing sources of environmental pollution. The lower molecular weight hydrocarbons in crude petroleum, gasoline, and so on, are VOCs and, when present in, for example, air, water, and soils are amenable to the analytical techniques described in Chapter 3. The heavier hydrocarbons and heterocyclics are usually semivolatiles, and examples of these in Table 4.1 are the 20 PAHs, which are discussed in more detail a separate section. Alkylated aromatic hydrocarbons and alkylated heterocyclic compounds are also important semivolatile constituents of crude petroleum and some petroleum products. Biomarkers are often semivolatile hydrocarbons that provide information about the sources, migration, and maturity of petroleums.[53]

Most crude petroleum and refined product components are amenable to separation by GC, which has been widely used for the characterization and analysis of these complex mixtures. The hydrocarbons present in these mixtures are perhaps the most widely studied class of compounds by MS. The petroleum industry adopted MS early in its development in the 1940s to conduct rapid hydrocarbon analyses in support of refinery operations. Many significant developments in MS, including the discovery of chemical ionization (Chapter 5), occurred in petroleum industry research laboratories. The EI mass spectra of hydrocarbons vary widely from the highly fragmented but characteristic spectra of straight-chain saturated hydrocarbons to essentially no fragmentation of the $M^{+\cdot}$ and M^{2+} ions of PAHs.[50]

The GC/MS technique is widely used to determine biomarkers in petroleum, the sources of recently spilled and weathered crude petroleums, the impact of spilled oil on sediments and wildlife, and the pathways of degradation of hydrocarbons in the environment.[53–56] This technique is also used to support refinery operations and petroleum exploration. The determination of both volatile and semivolatile constituents of crude petroleum and related products can often be accomplished by direct analysis of the bulk material or fractions from a preliminary condensed-phase separation (Chapter 6). Techniques for the extraction of semivolatile petroleum hydrocarbons from environmental media are discussed in the last part of this chapter.

Phenols and Other Aromatic Hydroxy Compounds

Phenols are compounds with one or more hydroxy (–OH) groups attached to an aromatic ring system such as benzene, a PAH, or an aromatic heterocyclic compound. Phenol is also the name of a specific compound, hydroxybenzene. Phenol and some alkylated phenols are by-products of the manufacture of coke (roasted coal) used in steel production. Naphthols are the corresponding hydroxynaphthalenes that are also coke oven by-products. Creosote is a mixture of phenols, including the methylphenol isomers (cresols), obtained from wood and coal tars. Guaiacol (2-methoxyphenol) and related alkylated phenols are extracted from wood pulp, and similar compounds are produced in the degradation of humic substances and lignin.

Phenols are also major high-volume industrial intermediates used in the production of agricultural chemicals, pesticides, pharmaceuticals, plastics, surfactants, and many other products. Chlorinated phenols, for example, 2-chlorophenol and pentachlorophenol, are used in large volumes as wood preservatives. Phenols are a long-standing contamination problem for water utilities because they react with chlorine used for disinfection to produce chlorinated phenols that have undesirable tastes and odors in drinking water. Given the natural occurrences and industrial uses of various phenols, it is not surprising that these compounds are ubiquitous environmental pollutants. Table 4.1 contains 16 phenols including 11 original PPs, seven HAPs, and one compound, 2,4,6-trichlorophenol, that is classified as reasonably anticipated to be a human carcinogen.[4]

Phenol and the less acidic phenols such as the chlorophenols, dichlorophenols, trichlorophenols, chlorocresols, fluorophenols, nitrophenols, and alkylated phenols are thermally stable and amenable to GC/MS. They have abundant, usually base peak, $M^{+\cdot}$ ions in their EI spectra and often abundant fragment ions that correspond to the loss of a Cl or methyl group. However, the more acidic phenols with multiple electron-withdrawing nitro, fluoro, and chloro groups often give broad tailing GC peaks with variable peak areas and poor signal/noise. Pentachlorophenol and the dinitrophenols are difficult to separate and measure with accuracy, precision, and high signal/noise by GC/MS. This is probably a result of strong interactions between these acids and surfaces or basic sites in GC injection port liners and columns. The more acidic phenols are marginal GC/MS analytes, and the methods discussed in Chapter 6 may be required for these and the higher molecular weight substituted phenols, naphthols, and related compounds.

Phosphorus-Based Pesticides

Phosphorus-based pesticides are nearly always esters of phosphoric acid or thio analogs in which one or more of the oxygens bound to phosphorus are substituted by a sulfur. Some phosphorus pesticides are *N*-substituted amides of partially esterified phosphoric acid or thiophosphoric acid. In addition to the N, O, or S bound to the phosphorus, these and other elements are sometimes contained in functional groups remote from the phosphorus. Diazinon (right end of second row in Chart 4.2) is an example of one of these structures. The diversity of possible structures in this series of compounds is very large and the compounds have similarly diverse vapor pressures, thermal stabilities, and chemical reactivities. Some of these pesticides are quite volatile, for example, methyl parathion which is rapidly degraded by sunlight. Others are hydrolyzed in water, but some are sufficiently resistant to hydrolysis and can be measured in water samples. Table 4.1 contains 12 phosphorus-based pesticides, which are indicated as phosphates, thiophosphates, and so on, after their names. All of these have been at least single-laboratory validated as analytes in a drinking water analytical method (see Method 525.2 later in this chapter). These 12 and many other compounds in this group are amenable to separation by GC, and their EI mass spectra vary widely from having abundant $M^{+\cdot}$ ions to having only fragment ions.[50]

Phthalate Esters and Analogous Plasticizers

Esters of phthalic acid, for example, di-*n*-octyl phthalate (left end of third row in Chart 4.2), and several other dicarboxylic acid esters, are very widely used as plasticizers which give flexibility to a vast array of consumer products including fabrics, plastics, electrical wire insulation, and flooring. These compounds are also used as fillers and carriers in a wide variety of other products. They are extremely widely distributed in the environment, including fatty tissue, and it is almost impossible to avoid detecting one or more of them in a reagent blank (Chapter 2, "Analytical Method Validation and Analytical Quality Control") or an environmental sample.

Table 4.1 contains six dialkyl phthalates and one dialkyl adipate, and all six phthalate esters are original priority pollutants. Three of them are also HAPs and di(2-ethylhexyl) phthalate and di(2-ethylhexyl) adipate have maximum contaminant levels in drinking water in the United States. Di(2-ethylhexyl) phthalate is also classified as reasonably anticipated to be a human carcinogen.[4] One important related compound in Table 4.1, dimethyl tetrachloroterephthalate, is a herbicide derived from terephthalic acid which has the two methyl ester groups in the 1 and 4 positions (*para*) of the benzene ring. The compounds in this group of esters are generally amenable to separation by GC, and the EI mass spectra of phthalic acid esters typically contain very low abundance or no $M^{+\cdot}$ ions.[50] All phthalic acid esters, except dimethyl phthalate, give characteristic base peaks at m/z 149. This ion is protonated phthalic anhydride, which is formed by losses of both alkyl groups and one oxygen from the $M^{+\cdot}$ ion with a proton transfer from one of the alkyl groups. Dimethyl tetrachloroterephthalate gives an abundant $M^{+\cdot}$ ion in its EI spectrum.[57]

Polycyclic Aromatic Hydrocarbons and Related Compounds

Polycyclic aromatic hydrocarbons (PAHs) are planar molecules containing primarily fused six-member unsaturated carbon rings with a single hydrogen attached to each of the peripheral carbons. Examples of structures of PAH are benzo[*a*]pyrene and benzo[*g*,*h*,*i*] perylene in Chart 2.1; naphthalene, fluoranthene, and benzo[*a*]anthracene in Chart 4.2; and chrysene and triphenylene in Chart 5.1. Less frequently, a five-member unsaturated carbon ring is contained within these structures, for example, acenaphthylene (Table 4.1). These compounds have an even number of carbons and hydrogens and the number of possible isomers increases dramatically with an increasing number of fused rings. Some hydrocarbons such as fluorene and acenaphthene, which have one or two saturated carbons in a fused five-membered ring, and biphenyl (left end of top row in Chart 4.1) are not strictly PAHs, but they have similar properties and are usually considered along with the PAHs for analytical purposes. Alkylated PAHs containing one or a few small alkyl groups, for example, 1-methylpyrene, are also considered members of the same general class of compounds in analytical methods. Similarly, various PAHs containing heterocyclic O, N, or S, for example, dibenzofuran in Chart 4.1, have some similar properties and are found in some of the same types of samples.

The PAHs and related compounds are found in low concentrations in crude petroleum, petroleum products, carbon black, coal, coal tar, and shale oil. They are ubiquitous products of incomplete combustion in internal combustion engines, furnaces, cigarettes, solid waste incinerators, and fossil fuel burning electricity generating plants. They are produced by the pyrolysis of organic matter and are contained in residues from various elevated temperature distillation and other chemical processes. They are environmentally important because of their widespread occurrence, transport in the atmosphere, and because 15 PAHs, including six in Table 4.1, are classified as reasonably anticipated to be human carcinogens.[4] Fifteen PAHs are original PPs, several are HAPs, and benzo[*a*]pyrene has a maximum contaminant level in drinking water in the United States (Table 4.1).

Those PAHs with molecular weights of about 300 or less and with six or fewer fused rings are generally amenable to separation by GC. The 20 PAHs in Table 4.1 are analytes in GC/MS analytical methods. These products of incomplete combustion are very thermally stable and are generally not very polar or reactive, but the larger molecules often have low vapor pressures even at elevated temperatures. Some of these, for example, benzo[*a*]pyrene, benzo[*g*,*h*,*i*]perylene, and indeno[1,2,3-*c*,*d*]pyrene have retention times of 25–30 min or longer even when typical GC columns are heated to near maximum column temperatures. This can result in broad GC peaks with variable areas and poor signal/noise. Most PAHs with six or more fused rings, including several on the reasonably anticipated to be human carcinogens list, are not amenable to separation by GC in a reasonable time on standard columns at typical maximum column temperatures. The condensed-phase separation techniques described in Chapter 6 are required for these compounds.

The EI mass spectra of the PAHs consist principally of abundant $M^{+\cdot}$ ions, generally low abundance M^{2+} ions, and no fragment ions or just a few with very low abundances.[50] Therefore, isomers of PAHs generally cannot be distinguished by MS (however, see the section of Chapter 5 entitled "Charge Exchange Ionization"). Compounds with a partially saturated fused ring, for example, acenaphthene and fluorene, or alkyl groups attached to the ring structure, have abundant $(M-1)^{+}$ and $(M-2)^{+\cdot}$ ions. The PAH containing methyl groups have $(M-27)^{+}$ ions of moderate abundance from the loss of C_2H_3 groups. Compounds with larger alkyl groups have very abundant ions produced by homolytic fission of the bond between the first and second carbons of the alkyl group (β cleavage). Some PAHs with larger symmetrical ring systems, for example, coronene, have very abundant M^{2+} ions.

Terphenyl Derivatives

Terphenyls are biphenyls (left end of top row in Chart 4.1) with a phenyl group substituted for one of the hydrogens of the biphenyl molecule. Substitution can occur at the biphenyl 2, 3, or 4 positions, and three terphenyl isomers are possible. Chlorinated terphenyls were produced before the mid-1970s in the United States and Europe, but in quantities much lower than those of the PCBs. Mixtures of chlorinated terphenyls were sometimes blended with the PCBs and used for similar proposes. The chlorinated terphenyl products of the Monsanto Company in the United States were designated Aroclor 54XX where the XX is the weight percentage of chlorine, which is generally in the range 32–60%. This group of compounds has received considerably less attention than the PCBs, but their estimation in aquatic biota and sediments has been reported.[58] Gas chromatography of chlorinated congeners containing less than about 10 chlorines is possible, but requires 1 h or more at near maximum GC column temperatures. One of the condensed-phase separation techniques described in Chapter 6 may be useful for these compounds. Few pure chlorinated terphenyl congeners have been separated and few authentic standards are available. For the more chlorinated congeners, electron capture ionization (Chapter 5) provides analytical sensitivity greater than that of EI.

Toxaphene (Chlorinated Camphene)

Toxaphene is the name given to an extremely complex mixture of compounds produced by the chlorination of camphene, which is produced from the natural product α-pinene.[59] The compounds formed in this reaction are primarily derivatives of bornane and bornene (bottom row of Chart 4.2) in which chlorine is substituted for one or more hydrogens of these molecules. The total number of theoretically possible chlorinated bornane and bornene congeners exceeds 32,000 because at each level of chlorination there can exist a very large number of structural isomers including chiral pairs. The major components of toxaphene contain six to ten chlorines, and the mixture contains more than 600 individual compounds. Only a small number of congeners have known and confirmed chlorobornane or chlorobornene structures. Toxaphene was widely used as a pesticide for insects on cotton and a few other crops until it was banned in the United States and some other countries in the early 1980s. Like other chlorinated hydrocarbon pesticides discussed previously, some toxaphene components are persistent in the environment and have sufficient vapor pressures at ambient temperatures to allow atmospheric transport to cold regions of the earth. Toxaphene is listed as an original priority pollutant and has a maximum contaminant level in drinking water in the United States (Table 4.1). Toxaphene is also classified as reasonably anticipated to be a human carcinogen.[4]

The toxaphene components are amenable to GC, but few or none of the $\sim$ 100 observable GC peaks represent a single pure congener. The EI mass spectra of toxaphene components consist of a series of clusters of ions from below m/z 50 to $> m/z$ 300.[60,61] The $M^{+\cdot}$ ions are not observed and the EI spectra are difficult to interpret because of the likely coelution of similar but different compounds and the multiplicity of fragment ions corresponding to losses of one or more Cl, HCl, and other small groups. In early GC/MS analyses, toxaphene was identified and measured using SIM of a few characteristic EI ions with $m/z < 200$ and by comparing the patterns of GC peaks in standards and samples.[62] This approach is analogous to that used for the identification and measurement of Aroclor PCB mixtures. Conventional GC/MS is generally no longer used when toxaphene is a target analyte because SIM with electron capture ionization (Chapter 5) provides abundant $M^{-\cdot}$ ions of many components and much simpler and interpretable mass spectra.[60] The technique of quantitation by isomer groups at each level of chlorination, which was described for PCB analyses, has been adopted by some investigators.[59]

Triazine Pesticides and Related Compounds

Symmetrical triazine is a six-member aromatic ring containing alternating carbons and nitrogens. Most triazine pesticides are derivatives of symmetrical triazine as illustrated by atrazine (right end of bottom row in Chart 4.2). A few, for example, metribuzin (Table 4.1), have an unsymmetrical configuration of nitrogens in the ring. Table 4.1 contains 11 triazine pesticides (designated *triazine* after the common name) and they generally have various alkylamino groups substituted for one or two of the hydrogens on the ring carbons (see atrazine structure). Other groups substituted for hydrogen include alkyl, alkyloxy, alkylthio, amino, chloro (as in atrazine), and oxo. Some of these compounds, for example, atrazine and cyanazine, are among the most widely used herbicides on corn, cotton, and other crops.[63] Atrazine, cyanazine, simazine, and their degradation products are frequently found in ground and river waters in agricultural areas of the United States.[63,64] Atrazine and simazine, have maximum contaminant levels in drinking water in the United States (Table 4.1). The triazine pesticides and their dealkylated degradation products are generally amenable to separation by GC. Some of these compounds have EI mass spectra with abundant, even base peak, $M^{+\cdot}$ ions, and others have low abundance or no $M^{+\cdot}$ ions in their spectra. Most compounds do have abundant fragment ions produced by losses of small alkyl and other groups.

Analytical Methods, Sampling, and Solvent Selection

The development of broad spectrum GC/MS analytical methods for semivolatile organic compounds was strongly stimulated by the consent decree of 1976. This court ordered settlement is described in Chapter 1 and the first section of Chapter 3 entitled "Target Analytes—The Priority Pollutants". Briefly, the consent decree was concerned with discharges of industrial wastes into rivers and other bodies of water. It established a list of 76 semivolatile organic priority pollutants (PPs) which are indicated by a "+" in the PP column in Table 4.1. This was the first important list of semivolatile target analytes and it and subsequently expanded lists were used for many environmental monitoring and regulatory programs. Most of the original 76 PPs are included in one of the classes of semivolatiles described in the previous section of this chapter.

Several analytical methods were developed for the 76 semivolatile PPs in industrial wastewaters. These methods were used widely and they became models for later analytical methods for this class of compounds in other environmental matrices. Analytical methods for air, other types of waters, and solids were required for other regulatory and environmental research programs mandated by laws in the United States. These later methods utilized many of the innovative features in the consent decree industrial wastewater methods. They also included specialized sampling and sample preparation procedures for the target sample matrices. Table 4.3 lists the USEPA and a few other GC/MS analytical methods that were developed, the environmental matrices to which they apply, and the target analytes of these methods. The subsequent sections of this chapter contain detailed or brief summaries of the methods in Table 4.3. Interspersed within these summaries are discussions of some of the important analytical issues, strategies, and techniques that either were considered during the development of the methods or that have appeared as a result of experience using the methods for environmental analyses.

Table 4.3 Summary of USEPA and related GC/MS analytical methods for semivolatile organic compounds in various environmental matrices

Method designation	Environmental matrices	Target analytes
625	Industrial and municipal wastewater	PPs identified in Table 4.1 + others
1625	Industrial and municipal wastewater	Same as 625 + others
CLP—semivol.	Ground and surface waters, soil, and sediment	PPs identified in Table 4.1 + others
8250	Soil, solid wastes, and ground water	PPs identified in Table 4.1 + others
525, 525.1, 525.2	Drinking water and other low-particulate waters	Identified in Table 4.1
TO-4A	Ambient air particulates and vapors	Identified in Table 4.1
TO-10A	Ambient air particulates and vapors	Identified in Table 4.1
TO-13A	Ambient air particulates and vapors	Identified in Table 4.1
	Fish tissue[91]	Table 4.1 pesticides and PCBs + a few others
	Tree bark[92]	Some Table 4.1 chlorinated pesticides
8270C	Solid waste, soil, air particulates, and water	Most analytes in Table 4.1 + others
613	Industrial and municipal wastewater	2,3,7,8-Tetrachlorodibenzo-*p*-dioxin
CLP—dioxin	Soil, sediment, solid waste, others	2,3,7,8-Tetrachlorodibenzo-*p*-dioxin
	Fish tissue[113]	17 Dibenzo-*p*-dioxins and dibenzofurans[a]
8280A	Water, soil, sediment, fly ash, and chemical production wastes	17 Dibenzo-*p*-dioxins and dibenzofurans[a] + others
8290	Paper pulp, fish tissue, human adipose tissue + others in 8280A	17 Dibenzo-*p*-dioxins and dibenzofurans[a]
23	Air emissions from industrial boilers, furnaces, and municipal solid-waste incinerators	Total dibenzo-*p*-dioxins and dibenzofurans by level of chlorination
TO-9A	Ambient air	17 Dibenzo-*p*-dioxins and dibenzofurans[a] and other isomer groups
1613	Drinking water and wastewater	17 Dibenzo-*p*-dioxins and dibenzofurans[a] and other isomer groups
ECI[47,117]	Soil and sediment	Dibenzo-*p*-dioxins and dibenzofurans by level of chlorination

[a]Specific polychloro isomers and congeners are identified in Table 4.4.

Sampling Techniques

The sampling techniques and sample processing procedures used for VOCs, which were described in some detail in Chapter 3, are particularly important and are emphasized because of the elusive nature of these compounds. Many VOCs can easily be lost during sampling or subsequent analyte transfers, and specialized equipment and techniques are used to minimize or eliminate these losses. In contrast, the semivolatile analytes are less elusive than the VOCs, and sampling techniques and sample processing procedures are generally less complex. For example, semivolatiles in air are sampled using glass or quartz fiber filters and downstream adsorbents. The filters and adsorbents are extracted with an organic solvent to give a solution of analytes that can be further processed or immediately analyzed by GC/MS. The solvent extraction of semivolatiles from an air filter or adsorbent is not a complicated process and is similar to the solvent extraction of semivolatiles from a sediment, soil, solid waste, or tissue sample.

While there are certainly some important strategies, precautions, and techniques used in sampling water, sediment, soil, solid waste, and tissue for the determination of semivolatiles, the techniques used are, in general, less complex than those used for VOCs. Details of sampling equipment and techniques used for semivolatiles are included in the USEPA analytical methods that are summarized later in this chapter. Additional information on sampling strategies and techniques can be found in the second edition of *Principles of Environmental Sampling*, which was published by the American Chemical Society in 1996.[65]

Organic Solvent Selection

A significant issue in the development of analytical methods for semivolatiles in all media is the choice of an organic solvent to extract the analytes from the sample. In addition to its ability to dissolve the analytes, this solvent must be appropriate for GC or be easily exchanged by vaporization for a solvent that is appropriate. Analytical methods for some chlorinated hydrocarbon pesticides and PCBs in environmental media had been used since about the mid-1960s. These methods relied primarily on packed-column GC, the standard electron capture GC detector, and nonpolar solvents such as *n*-hexane, iso-octane, petroleum ether mixtures, and diethyl ether. The pesticides and PCBs are generally soluble in these solvents and, even when mixed with water-miscible acetone for a soil, sediment, or tissue extraction, they are transparent in the electron capture GC detector. However, for the analysis of industrial wastewaters, the broad range of structures and polarities of the 76 semivolatile PPs argued for a more versatile solvent. This solvent was needed for both liquid–liquid extractions of semivolatiles from the water samples and for GC calibration and other standard solutions.

Various solvents were studied during the 1970s including benzene, carbon tetrachloride, chloroform, diethyl ether, ethyl acetate, methylene chloride, toluene, and mixtures of these and others. A moderately polar solvent that is not miscible with water was needed for extractions of analytes from water and wet sediments. It was desirable, but not essential, for this solvent to be denser than water to allow somewhat more convenient draining of the organic layer from a standard laboratory separatory funnel. A similar polar solvent was needed for extractions of analytes from other media. It was important for the solvent to have a relatively low boiling point to facilitate concentration of extracts by evaporation without losses of semivolatile compounds. The solvent could not be toxic or irritating to laboratory workers and should be a good solvent for GC. Finally, its molecular weight should be as low as possible to permit efficient removal in a jet separator interface that was widely used to connect packed GC columns to a mass spectrometer (Figure 2.3).

Ethyl acetate is too soluble in water (1 mL in 10 mL of water) and does not efficiently partition a wide variety of organic compounds from water. Diethyl ether, which had been favored in classical qualitative organic chemical class separations, was discarded because it is a hazardous flammable solvent with too low a boiling point and a tendency to form peroxides. While peroxides and inhibitors can be removed to facilitate trace organic analyses, air oxidation of diethyl ether quickly produces more peroxides. Even shortly after purification, sufficient concentrations are present to react with trace organics and solvents and create unacceptable reagent blanks. Hydrocarbon solvents such as *n*-hexane, benzene, and toluene are too nonpolar, less dense than water, and flammable. In the early 1970s it was clear that carbon tetrachloride and chloroform were toxic and potentially carcinogenic. The boiling point of carbon tetrachloride is too high and neither carbon tetrachloride nor chloroform are efficiently removed by a jet separator interface to a mass spectrometer.

Methylene chloride emerged as the solvent of choice because it had the best combination of desirable physical and chemical properties. It was not suspected of being carcinogenic or toxic in the 1970s, had a convenient boiling point (41 °C), a density greater than water, low water solubility, and a polarity sufficient to extract a wide variety of compounds from water and other media. It was commercially available in pure form at a reasonable cost, contained no peroxides or peroxide inhibitors to react with analytes, and was a good solvent for GC. Methylene chloride was ultimately selected as the solvent of choice for most GC/MS methods for semivolatiles, but it was not selected for all methods. Benzene, in spite of being a known human carcinogen, was widely used for the extraction of chlorinated dibenzo-*p*-dioxins and dibenzofurans from solid samples. These analytes are very soluble in benzene and it efficiently extracts them from materials such as fly ash and particulate matter collected from ambient air. Some other solvents and solvent mixtures such as *n*-hexane–acetone, ethyl acetate, and supercritical carbon dioxide are used for some techniques that are described later in this chapter.

Specialized Standard Analytical Methods or Versatile and Flexible Methods

Analytical methods developed during the early stages of environmental regulatory programs in the United States were generally designed for a specific type of sample matrix, for example, drinking water, or a group of similar sample matrices such as industrial wastewaters (Table 4.3). These analytical methods give specific directions for sampling, sample preparation, and analysis with no provision for analyzing other sample matrices and only modest flexibility to make changes to the analytical techniques. Standard analytical methods were considered essential to allow comparisons of analytical data obtained in different laboratories from the same samples and to minimize disputes between various personnel and laboratories.[66] When standard analytical methods were not available, the regulated industries used this as a reason to delay enforcement of environmental regulations. A practical advantage of specialized standard methods is that they tended to be relatively short and generally analyst-friendly.

However, as many different sample matrices were presented for analysis, and some very complex matrices such as solid wastes became important, it was evident that it would be impossible to develop, test, and validate specialized standard methods for every type of sample matrix and every important analyte. Similarly, as newer analytical techniques were developed, some of which had important technical and economic advantages, some regulatory offices of the USEPA were very slow to incorporate improved techniques into standard methods. The standard methods had been used to build databases of environmental measurements that could be compromised by improved analytical techniques. Calls for less restrictive and more flexible regulatory analytical methods were published with an emphasis on performance-based methods.[67,68] Very slowly and with considerable internal and external debate the USEPA moved in the direction of less restrictive and more flexible regulatory analytical methods.

The trend from specialized standard analytical methods to less restrictive and more versatile regulatory methods is most apparent in the GC/MS methods for semivolatile organic compounds. One approach used to provide flexibility is to separate the sampling and sample preparation procedures from the GC/MS procedure. In some USEPA methods referenced in this chapter the GC/MS and quality control procedures are described, but the methods do not contain sampling or sample preparation procedures. These procedures are in separate numbered documents which are cited along with instructions to the analyst to utilize the one that is most appropriate.

Another strategy used to provide versatility is to expand the sampling and sample preparation sections of the method to include specialized procedures for several different sample matrices. Once the concentrated extract is obtained, the GC/MS procedure is the same for all extracts. This approach is employed in several USEPA methods which are notably longer, more complex, and much less analyst-friendly than the specialized standard methods. Another trend to provide flexibility is to specify either various standard GC detectors or GC/MS systems after sample collection and preparation. These methods may employ alternative solvents, for example, mixtures of diethyl ether and hydrocarbons, rather than methylene chloride to accommodate standard GC detectors. They usually include extract fractionation procedures that are required with standard GC detectors, but often not needed with GC/MS. Again these methods are usually longer, more complex, and less analyst-friendly than the specialized standard analytical methods.

Semivolatile Organic Compounds in Wastewater

A GC/MS method for semivolatile priority pollutants (PPs) in industrial wastewater evolved during the mid to late 1970s under pressure from the consent decree PP analytical program. While experiments continued with a few organic solvents, most investigators adopted the liquid–liquid extraction technique with a separatory funnel and methylene chloride to partition semivolatiles from industrial wastewater. The separatory funnel technique was familiar to many analysts, taught in many college chemistry laboratory courses, and widely used in industrial and other chemical analyses. Many VOCs in a sample are also extracted with this technique, but they are generally lost during subsequent concentration of the extract by solvent evaporation. Some industrial wastewaters form intractable emulsions during separatory funnel extractions and continuous extraction techniques were used for these samples.

The resolving power of the packed GC columns used for semivolatiles was not sufficient to separate more than small groups of semivolatiles. Even when the pesticides, phenols, commercial PCB mixtures, and toxaphene were disregarded, it was difficult to separate the remaining <40 semivolatile PPs in a calibration standard. Some investigators experimented with rigid glass capillary columns and achieved substantially improved results. However, there was no rush to employ these columns for environmental analyses because they were difficult to install in GC ovens and mass spectrometer interfaces. In addition, coated and ready-to-use columns were commercially available in the United States from only one manufacturer, the Perkin-Elmer Company, which held a restrictive patent until 1977. Just as with the VOCs, there was continuing controversy over the issue of using GC/MS and/or GC with standard detectors for analyses of wastewater (Chapter 3, "Regulatory Adoption of Purge and Trap GC/MS").[69] Some investigators favored the electron capture detector (ECD), which is extremely sensitive to highly chlorinated compounds, for the chlorinated hydrocarbon pesticides, PCBs, and other halogenated compounds.

When the USEPA was compelled to propose analytical methods for the determination of semivolatile compounds in wastewater for regulatory purposes, a compromise was made analogous to that described for the VOCs in Chapter 3. A suite of methods was proposed on 3 December 1979 that used packed GC columns with both standard GC detectors and the mass spectrometer.[70] Methods 604 and 606-612 describe liquid–liquid extractions with methylene chloride followed by GC of the concentrated extract. These methods specify a variety of standard GC detectors to determine 79 semivolatile and marginal VOC analytes. When an ECD or halogen-specific detector is specified, for example, Method 608 for pesticides and PCBs, the solvent is exchanged for hexane during extract concentration. The standard detector GC methods generally include a preliminary extract fractionation with open-column liquid chromatography, often called *clean-up*, to minimize interferences. Method 613 is the GC/MS method proposed for the determination of the single analyte 2,3,7,8-tetrachlorodibenzo-*p*-dioxin. The method is discussed in the last section of this chapter. Method 625 is the GC/MS method for 83 semivolatile and marginal VOC analytes. All the 600 series methods are integrated complete analytical methods that include sampling techniques, sample extraction, extract fractionation if needed, details of the GC separation and measurement, and quality control requirements.

The proposal of Method 625 for regulatory use was very significant and had a major impact on the subsequent practice of environmental analyses and the development of commercial GC/MS instrumentation. The prospect of using a single analytical method with no required extract fractionation for 83 analytes was attractive even though it required a more costly analytical instrument. The alternative was the implementation of eight different analytical methods with required extract fractionation and different GC columns and detectors. The economic implications were clear to the regulated industries and the emerging environmental testing industry (Chapter 2, "Alternatives to Conventional GC/MS"). Some manufacturers of GC/MS systems, especially the Finnigan Corporation, enjoyed very large increases in sales and they were able to use the new revenue to improve the capabilities of their products. It was clear then and now that the GC/standard detector and GC/MS methods are not equivalent for the reasons given in the section of Chapter 2 entitled "Alternatives to Conventional GC/MS". Nevertheless, there were some positive benefits from the compromise and these are noted in the section of Chapter 3 entitled "Method 624".

Method 625

Because of the limitations of packed-column GC, two different columns were required for the original semivolatile PPs (Table 4.1). One column was needed for the 11 phenols, and a different column was needed for the balance of the PPs that were either not ionized in water (neutrals including the pesticides and PCBs) or were weak bases in water, for example, benzidine. Therefore, a preliminary separation of the two groups of compounds was required before GC/MS analysis. This separation was accomplished by adjusting the pH of the aqueous sample first to about 11, extracting the bases and neutrals with methylene chloride, then reducing the sample pH to about 2, and extracting the phenols with fresh methylene chloride. The two extracts, which were called the base–neutral and acid fractions, were analyzed separately. Most of the methylene chloride was evaporated to concentrate the extracts and allow detection of the analytes at concentrations in the low microgram per liter range in the original sample. During extract concentration, even at ambient temperature with a gentle stream of dry nitrogen, most VOCs in the extracts were vaporized and lost. A concentration factor of about 1000 was attained by using a 1 L water sample and concentrating the extract to about 1 mL.

The two concentrated extracts were chromatographed on the separate GC columns, and the results from the separate fractions were combined for the analytical report. Data were acquired by continuous measurement of spectra as described in the section of Chapter 2 entitled "Data Acquisition Strategies". The resolving power of the packed column used for the phenols was sufficient to separate almost all of the 11 priority phenols in a calibration standard. The resolving power of the packed column used for the base–neutral fraction was not sufficient to separate the semivolatiles even when the pesticides, commercial PCB mixtures, and toxaphene were placed in separate calibration solutions. The limited resolving power of the GC columns was addressed by using a single quantitation ion for each analyte and avoiding the same quantitation ion for closely eluting or coeluting compounds (Chapter 2, "Quantitative Analysis"). Internal standards were included in the calibration solutions and added to the concentrated extracts for quantitative analysis. Nevertheless, the number of calibration solutions required a large number of GC injections for an adequate calibration of all analytes and this increased the time required and cost of the analyses. Often only a few calibration points were measured with each calibration solution and these analyses were considered to be semiquantitative.

A quality control provision in Method 625 is that mass spectra must meet the key ion and relative abundance criteria established for decafluorotriphenylphosphine (DFTPP). The rationale for these criteria is given in the section of Chapter 2 entitled "Standardization of Conventional GC/MS". This requirement was very controversial, but ultimately forced the manufacturers either to develop GC/MS instruments that produced EI spectra that met the performance criteria or lose some of the environmental market. The justification for the strong quality control requirement was that environmental regulations were too important to the nation to allow the acquisition of deficient or incorrect mass spectra that would lead to disputes among industry, government, and environmental testing laboratories. Total ion chromatograms from industrial wastewater samples were often extremely complex and required excellent laboratory techniques, instrumentation, and data interpretation strategies.

Fused Silica Capillary GC Columns. The very significant limitations of packed-column GC were clearly illustrated in the industrial wastewater analytical program with Method 625 and the other 600 series methods. The commercial introduction of high-resolution fused silica capillary columns (FSCCs) by the Hewlett-Packard Company in 1979 opened the way for vastly improved and much more efficient GC separations. The Office of Research and Development of the USEPA employed FSCCs with Method 625 during 1980–1981 to determine PPs and other semivolatiles in ground and surface waters at the infamous Love Canal hazardous waste site near Niagara Falls, NY. A modified Method 625 with a different sample preparation procedure was also used for soil and sediment analyses. Method 625 specified syringe injections of an aliquot of a concentrated extract in methylene chloride and was readily modified for FSCCs. The flexible FSCCs were easily installed in GC column ovens and readily threaded through the heated interface into the EI ion source of the mass spectrometer. The carrier gas flow in FSCCs was sufficiently low that no interface between the chromatograph and most mass spectrometers was needed. Pesticides, phenols, PCBs, polycyclic aromatic compounds, weak bases, and other classes of compounds all chromatographed well on the same FSCC. The preliminary separation based on pH adjustments before the methylene chloride extractions was no longer needed. Precision and accuracy data comparing Method 625 and modified Method 625 were published for over 80 analytes in 1983.[2] The analytes in this study are indicated by the entries "Ref. 2" in the last column of Table 4.1.

Status and Derivatives of Method 625. Method 625 was promulgated in wastewater regulations on 26 October 1984.[71] The method requires the pH adjustments before extraction and specifies two packed GC columns. However, the method does allow the use of capillary GC columns, but does not provide information to assist the analyst in the application of these columns. As of 1999, Method 625 had not been revised to modify the pH adjustments before extraction, provide information about FSCC, or incorporate other improvements in analytical techniques. The mass spectral performance criteria for DFTPP in Method 625 (Table 2.3) have not changed although improved criteria are used for other GC/MS methods for semivolatiles (see, e.g., Method 525.2). Therefore, technically, but not legally, Method 625 is obsolete. Users of Method 625 for regulatory purposes usually have approval from appropriate authorities to utilize some or all method improvements. Method 625 is historically important because of the innovations in its content and the precedents established for other analytical methods used for regulatory programs.

An analytical method very similar to Method 625 was used for many years in the USEPA's contract laboratory program, which analyzed ground water, surface water, soil, and sediment samples taken from abandoned hazardous waste sites. Another version of Method 625, Method 1625, was included in the 1984 wastewater regulations to describe isotope-dilution calibration for those samples where severe matrix effects cause significant deviations in extraction efficiencies and other problems (Chapter 2, "Quantitative Analysis").[71]

Method 8250 is a derivative of Method 625 that was published by the USEPA's Office of Solid Waste for the determination of semivolatiles in soils, solid wastes, and ground water.[72] This method is a departure from the integrated method style of the 600 series and does not include sampling techniques, extraction techniques, and other sample preparation procedures. These procedures are in separate documents referenced in Method 8250, which contains just the GC/MS and related procedures. For information on techniques used to extract semivolatiles from solid samples see the section of this chapter entitled "Semivolatile Organic Compounds in Air, Sediments, Soil, Solid Wastes, and Tissue".

Semivolatile Organic Compounds in Drinking Water

In 1986 the Congress passed and President Reagan signed amendments to the Safe Drinking Water Act which mandated the USEPA to establish national standards within 3 years for 83 drinking water contaminants that were named in the Act. The law also mandated national standards for 25 additional contaminants every 3 years beginning in 1990. Analytical methods were required to monitor drinking water for the semivolatile compounds that could be included in new regulations. This need presented an opportunity for the application of improved techniques that were not used in existing standard or USEPA analytical methods for organic compounds in water.

During the mid-1980s there was a growing interest in reducing the quantities of solvents used for the extraction of organic compounds from environmental samples. Solvents were increasingly expensive, evaporation of solvent extracts was often conducted in a fume hood and the vapors released into the atmosphere, and methylene chloride, benzene, and some other solvents were PPs. The USEPA designated methylene chloride one of 18 substances with a high priority for major reductions in the amounts released into the environment and it was placed on the reasonably anticipated to be a human carcinogen list.[4] Laboratory workers found that reagent blanks from Method 524.2 (VOCs) invariably contained methylene chloride when liquid–liquid extractions with this solvent were conducted in the same building.

A strategy for significantly reducing the amount of methylene chloride used in liquid–liquid extractions was to utilize the rapidly improving reverse-phase liquid chromatography (RPLC) packing materials to adsorb semivolatile organic compounds in water (Chapter 6, "Liquid Chromatography" for a brief description of RPLC materials).[73] The analytes are then eluted with just a few milliliters of methylene chloride and/or other solvents such as ethyl acetate, which is generally not useful for liquid–liquid extractions because of its water solubility. A few manufacturers were supplying RPLC packing materials in small plastic or glass cartridges during the mid-1980s. Drinking water is perhaps the perfect application for liquid–solid extraction because filtered water usually has very little particulate matter, which deposits in the solid-phase adsorbent and greatly reduces the flow rate of water sample through the device.

Method 525

Method 525 was developed by the USEPA to implement liquid–solid extraction with commercial RPLC cartridges in an integrated GC/MS method for semivolatile compounds in drinking water or other low-particulate waters.[74] The method specifies C_{18}–silica cartridges to partition a variety of semivolatiles from water, and internal standards for quantitative analysis are added to the water sample and extracted with the analytes. One liter of a typical low-particulate water passes through a cartridge containing about 1 g of adsorbent in about 2 h with a moderate vacuum assist. The adsorbed analytes and internal standards are eluted with a few milliliters of methylene chloride, and the eluate is concentrated and analyzed by GC/MS with continuous measurement of spectra. Recoveries above 85% have been obtained for most of the 42 analytes listed in Method 525. The method analytes are hexachlorobenzene, hexachlorocyclopentadiene, eight chlorinated biphenyl congeners, 10 chlorinated hydrocarbon pesticides, six phthalate ester plasticizers, 13 polycyclic aromatic hydrocarbons, and two triazine herbicides. The method detection limits for these analytes are usually $< 1\ \mu g/L$. However, most phenols are not retained on the C_{18}–silica although pentachlorophenol has been recovered, but with poor precision.

Method 525 was published by the USEPA in 1988 and it was approved in 1991 for monitoring eight semivolatile compounds which have maximum contaminant levels (MCLs) in drinking water in the United States.[75]

Method 525.1

Method 525.1 was published in 1991 and included the option of using 47 mm diameter poly(tetrafluoroethylene) (PTFE) disks impregnated with finely divided C_{18}–silica for liquid–solid extraction.[76] The disks are very efficient because about 500 mg of finely divided particles are evenly dispersed in the PTFE matrix, which is about 0.5 mm thick.[77] One liter of particulate-free water passes through a disk in 10 min or less with a moderate vacuum assist. Method 525.1 was approved in 1992 for monitoring seven additional semivolatile compounds that have MCLs in drinking water in the United States.[78]

Method 525.2

Method 525.2 is the most recent version in this series of methods and was published by the USEPA in 1995.[79] It has the option of cartridges or disks and specifies ethyl acetate and methylene chloride to elute a wide range of analytes. Method 525.2 has been single-laboratory validated for 118 analytes, mostly at a concentration of 0.5 μg/L, using the disk extraction and an ion-trap GC/MS system. Most recoveries are in the 80–120% range and method detection limits are generally below 1 μg/L. The 118 Method 525.2 analytes are included in Table 4.1 and Method 525.2 is shown as the GC/MS water method. Method 525.2 was approved in 1994 for monitoring 16 semivolatile compounds with MCLs in drinking water in the United States and six additional compounds that do not have MCLs.[80] The MCLs of the 16 compounds are listed in Table 4.1, and the additional compounds with monitoring requirements are designated by footnote "a" in the MCL column.

The cartridge or disk extraction technique was quickly adopted by many laboratories. Manufacturers competed to produce the most stable cartridges and disks with the lowest possible organic background and highest practical adsorption properties. The technique, unfortunately, was commercially advertised as *solid phase extraction* (SPE). This term is ambiguous because it does not identify both phases as is customary in analytical chemistry. Method 525 and subsequent versions use the term *liquid–solid extraction* (LSE) which is technically correct and precisely defines the technique. The LSE term is used by some investigators and is used in this book.

Modified GC/MS Standardization and Quality Control Criteria. An important issue in the development of the Method 525 series was the GC/MS standardization and quality control criteria. The provision used in USEPA Method 625 requires that the mass spectrometer be tuned to meet the key ion and relative abundance criteria established for decafluorotriphenylphosphine (DFTPP). The background and rationale for these criteria are described in the section of

Chapter 2 entitled "Standardization of Conventional GC/MS". The original DFTPP criteria were developed during the early 1970s when only about 50% of all quadrupole GC/MS systems gave a DFTPP $M^{+\cdot}$ ion greater than 40% relative abundance.

During the years since the publication of the first set of DFTPP criteria (Table 2.3) advances in mass spectrometer technology resulted in significant improvements in the performance of many instruments. In addition, several new types of mass spectrometer with greatly improved performance were introduced. The GC/MS systems based on these improved or new designs had significantly greater capabilities for transmitting and detecting ions with $m/z > 200$. One new design, the ion-trap mass spectrometer, was potentially very valuable because of its bench-top size, relatively low cost, and very high sensitivity. This sensitivity was clearly illustrated by the acquisition of the complete mass spectrum of DFTPP with very good signal/noise from just 100 pg injected into one of the first commercial GC/ion-trap instruments.[81] While the ion trap and some other improved types of instrument met the 4-bromofluorobenzene specifications in Table 3.2, they often did not meet the DFTPP specifications in Table 2.3 and used in several USEPA methods. These instruments typically gave a base peak $M^{+\cdot}$ ion at 442 Da while the DFTPP specifications called for a base peak at the m/z 198 fragment ion (Table 2.3).

It was incumbent on the USEPA to re-examine the DFTPP key ion and relative abundance criteria and make appropriate adjustments to take advantage of the capabilities of the improved and new mass spectrometer designs. This adjustment was made primarily by allowing either m/z 198 or 442 to be the base peak with the provision that the other key ions in the spectrum have appropriate relative abundances. Important criteria such as the acceptable relative abundance ranges of ions containing natural occurring isotopes were not changed. These modified DFTPP relative abundance criteria are incorporated into the Method 525 series and similar, or the same criteria are incorporated into some other USEPA methods for semivolatile compounds.

Alternative Extraction Methods for Semivolatiles in Water

Some other techniques are used to partition semivolatile compounds from water and concentrate them in a solvent for GC/MS analysis. The Grob closed loop stripping technique is used to extract and concentrate some nonpolar semivolatiles such as chlorinated hydrocarbons and alkylated benzenes. This technique is briefly described in the section of Chapter 3 entitled "Purge and Trap Technique". The solid-phase micro extraction technique has been demonstrated to extract and concentrate some semivolatile compounds. This technique is briefly described in the section of Chapter 3 entitled "Alternative Extraction Method for VOCs in Water".

In addition to C_{18}–silica, many other solid-phase adsorbents are used to partition compounds from water. These adsorbents include activated carbons of various types and synthetic organic polymers. Some of these adsorbents are listed in the section of Chapter 3 entitled "Sampling Techniques for VOCs in Air" and many have been evaluated for the extraction of a wide variety of compounds, including phenols, from water. The principles of the application of solid-phase adsorbents to the extraction of organic compounds from water have been reviewed.[82] In order to eliminate completely methylene chloride from Method 525.2, supercritical carbon dioxide has been used to elute semivolatile analytes from a solid phase adsorbent.[83,84]

Semivolatile Organic Compounds in Air, Sediments, Soil, Solid Wastes, and Tissue

It may seem inappropriate to consider semivolatile compounds in air in the same section as semivolatiles in solid phases such as soil and tissue. However, semivolatiles in air are often associated with particulate matter or aerosols, which are collected primarily on glass or quartz fiber filters. Some semivolatiles trapped on a filter are slowly vaporized during continuous air sampling and are collected on downstream adsorbents, for example, a polyurethane foam plug. The filters and adsorbents are extracted with organic solvents using techniques that are very similar to the techniques used to extract semivolatiles from soil and solid wastes. These extractions produce solutions of analytes that can be fractionated or immediately analyzed by GC/MS. Therefore, semivolatiles in air and in solid-phase matrices are determined using very similar techniques except for the initial collection of the analytes from an air sample.

The semivolatile organic compounds in Table 4.1, and many analogous compounds, are of considerable interest in air and solid-phase materials. Many of these compounds have low solubilities in water and they accumulate in urban dust, soil, sediment, and tissue (sediment is defined in the section of Chapter 3 entitled "Volatile Organic Compounds in Sediments, Soil, Solid Wastes, and Tissue"). These compounds are also widely dispersed in solid wastes whose disposal is controlled under Resource Conservation and Recovery Act regulations in the United States. The determination of semivolatile compounds in air and solid-phase matrices by GC/MS requires the extraction of the analytes into an organic solvent suitable for GC/MS.

Techniques for the extraction of semivolatiles from solid-phase samples include the time-honored Soxhlet extraction, which has been used for over 100 years, ultrasonic agitation and disintegration of solids dispersed in an organic solvent, supercritical fluid extraction with either pure or modified carbon dioxide, and a recently developed technique that uses both elevated temperature and pressure to increase the rate of dissolution of semivolatiles in an organic solvent. These techniques and some others have been widely studied and reviewed and several have been tested with a variety of solid phase matrices and documented as stand-alone extraction procedures.[72] A thorough review of extraction procedures is beyond the scope of this book, but some examples of applications of these techniques to several types of samples are given in this section and the last section of this chapter.

After a sample has been extracted and the semivolatiles concentrated in an organic solvent, a critical issue is whether a preliminary separation or fractionation of the substances in the extract is needed before GC/MS analysis. The decision to fractionate or not and the degree of fractionation depend on many factors including:

- The nature, organic content, and complexity of the sample.
- Whether the analytical strategy is broad spectrum or target analyte.
- The anticipated coeluting interferences, especially compounds that have ions with the same integer m/z values as target analytes.
- The selectivity of the extraction technique.
- The selectivity of the GC/MS technique.
- The laboratory's willingness to risk contamination of the injection port liner, GC column, and ion source that may require unscheduled system maintenance or replacement of some components.
- The detection limits required.

Fractionation techniques are often very effective because they largely remove and discard a significant portion of the components of the sample extract. Liquid–liquid extraction of an immiscible organic solvent sample extract with both strong aqueous acid and strong aqueous base solutions is a common technique. This removes basic and acidic components from the sample extract and may produce an extract containing mainly neutral compounds that is ready for injection into the GC/MS system. However, discarding the acidic and basic aqueous fractions, as is often done, also discards potentially important components of the sample that may be of great interest in a broad spectrum analytical approach. Therefore, depending on the general analytical strategy, it may be necessary to minimize or very carefully select the fractionation procedures to obviate the loss of potentially important sample components. Most environmental samples are probably analyzed using the target analyte strategy, and fractionation procedures are often optimized for recovery of the target analytes. Fractionation procedures are often called *clean-up* which is a term not used in this book and generally more appropriate for what a mechanic does to his hands after working on an old greasy automobile engine.

Extraction of slightly contaminated white sand from a pristine beach with a selective nonpolar solvent such as *n*-hexane or supercritical carbon dioxide should produce an extract that is ready for GC/MS. Unfortunately, this kind of clean extract is probably the exception and most solid-phase samples require some degree of fractionation to provide an appropriate extract or extracts for GC/MS. This is probably true even when very selective mass spectrometric techniques such as those described in Chapter 5 are used. Direct injection of complex extracts from very contaminated samples, for example, an organic sludge, into a GC/MS system is a sure method of causing severe background contamination, a significant reduction in the resolving power of the column, and/or a drastic reduction in the signal/noise of the entire system. An instrument used in this way will require more frequent maintenance and/or the replacement of injection port liners and expensive GC columns.

Effective liquid–liquid fractionation schemes have been developed to separate the components of a methylene chloride extract into stronger acid, weaker acid, basic, and neutral compound fractions.[85] The neutral components can be further fractionated into aldehydes, ketones, nonpolar compounds, and polar compounds. This type of broad spectrum fractionation is slow and costly and may not be used often. Open-column normal-phase liquid chromatography (LC) of solvent extracts on alumina, silica gel, florisil, and other polar solid phases are common techniques and are used to separate fractions containing various subgroups of analytes. The LC separations may be preceded by acid, base, or neutral water washes and other fractionation procedures. Gel permeation chromatography is often used to separate lipids and related groups of compounds from some target analyte groups.

Just as with the extraction techniques, a thorough review of fractionation procedures is beyond the scope of this book. The following sections contain brief descriptions of GC/MS methods and extraction, fractionation, and other techniques used for semivolatile analytes in air, wet and dry sediments, soils, fish tissue, and tree bark. Additional examples of these methods and techniques are contained in the last part of this chapter on chlorinated dibenzo-*p*-dioxins and dibenzofurans.

Semivolatiles in Air

Several USEPA methods in the toxic organic (TO) series are designed for the determination of semivolatiles in ambient air.[86] The target analytes in these methods are chlorinated hydrocarbon pesticides, herbicide carboxylic acid esters, nitrogen-based pesticides, phosphorus-based pesticides, polychlorinated biphenyls (PCBs), and polycyclic aromatic hydrocarbons (PAHs). These semivolatiles are often associated with particulates or aerosols, which are typically collected on glass or quartz fiber filters. However, some semivolatiles adsorbed on trapped particles have sufficient vapor pressures at ambient temperatures to be transported from the filter by the air flow during 4–24 h sampling periods. Downstream adsorbents are used to prevent losses of these analytes. Some semivolatile analytes, although not classified as VOCs (Chapter 3), have sufficient vapor pressures at some ambient temperatures to exist mainly in the vapor phase and they are also collected on the downstream adsorbent. Several of the TO series methods for semivolatiles are designed for multiple GC detectors or a mass spectrometer and are necessarily more complex and lengthy than methods that specify just GC/MS. Several of the TO methods also have target analytes that are not amenable to GC and require one of the techniques described in Chapter 6.

Method TO-4A.[86] The target analytes in Method TO-4A are indicated by the entry "TO-4A" in Table 4.1. Method TO-4A specifies a high flow rate air sampler capable of drawing ~8 standard (STD) cubic feet/min (0.225 STD m^3/min) through a quartz fiber filter followed by a glass cartridge containing a polyurethane foam (PUF) adsorbent. A sample volume of > 300 STD m^3

collected over 24 h is recommended. The quartz fiber filter and PUF are extracted together with 10% (v/v) diethyl ether in hexane in a Soxhlet apparatus. The method includes open-column LC fractionation procedures and GC/standard detector, GC/MS with SIM, and high-performance LC (ultraviolet detector) separation and measurement procedures. Some or all of the fractionation procedures may not be necessary with GC/MS, and the LC/MS techniques described in Chapter 6 are appropriate for the thermally sensitive analytes.

Method TO-10A.[86] The target analytes in Method TO-1OA are the same as those in Method TO-4A plus several additions. The significant difference between Method TO-4A and Method TO-10A is that the latter method specifies a low flow rate air sampler capable of drawing 1–5 STD L/min through an optional glass fiber filter followed by a glass cartridge containing either PUF alone or PUF and Tenax-TA™. This method uses a small portable sampler that is readily located where a high flow rate sampler may not be accommodated, for example, indoors, in a small space, or carried by a person. The detection limits are higher than those of Method TO-4A because of the much smaller volume of air sampled in 24 h. The adsorbents are extracted with 5% (v/v) diethyl ether in hexane in a Soxhlet apparatus, and the fractionation and extract analysis procedures are similar to those of Method TO-4A. This method is addressed in American Society for Testing and Materials (ASTM) standard D4861-94a.[87]

Method TO-13A.[86] The target analytes in Method TO-13A are the 19 polycyclic aromatic hydrocarbons (PAHs) indicated by the entry "TO-13A" in Table 4.1. The method specifies a high flow rate air sampler capable of drawing ~8 STD cubic feet/min (0.225 STD m^3/min) through a quartz fiber filter followed by a glass cartridge containing either PUF or a somewhat less preferred styrene–divinylbenzene copolymer (XAD-2). A sample volume of about 320 STD m^3 collected over 24 h is recommended. The quartz fiber filter and the adsorbent are extracted together in a Soxhlet apparatus with 10% (v/v) diethyl ether in hexane (PUF) or methylene chloride (XAD-2). The extract is concentrated by evaporation, and the concentrate is optionally fractionated on silica gel using open-column LC. The eluate containing the PAHs is concentrated and an aliquot is analyzed by GC/MS with SIM data acquisition. The GC/MS system must meet DFTPP criteria similar to those in Method 525.2 and PAHs labeled with deuterium are used as internal standards for quantitative analysis. This method is the basis for ASTM standard D6909-97.[88]

Methods TO-4A and TO-13A specify exactly the same sampling and extraction procedures when PUF is used as the downstream adsorbent. Therefore, all the TO-4A and TO-13A target analytes, plus many analogous nontarget analytes (Table 4.1), will be collected on the filters or adsorbents and concentrated in solvent extracts from the two methods. Selective detection of pesticides and PCBs in Method TO-4A and PAHs in Method TO-13A is accomplished by combinations of open-column LC fractionation, SIM data acquisition, and, in Method TO-4A, selective standard GC detectors. It would be reasonable to acquire a single sample extract and use this to determine the Method TO-4A and TO-13A analytes, using an unfractionated sample extract and a sensitive GC/MS instrument. Method TO-10A has the same target analytes as Method TO-4A and specifies similar sampling and extraction procedures, but employs a much smaller sample volume. Therefore, Method TO-10A has detection limits higher than those of Method TO-4A.

Dry Soil and Sediment

While dry soils, or soils having a low moisture content, may be sampled in the field, sediment is usually wet. Damp or wet soils and wet sediments are sometimes air-dried to give a solid that can be ground into uniform size particles with a mortar and pestle. The time required for drying can range from several days to a week and depends on the relative humidity, temperature, and air movement in the laboratory. The reason for drying a sample is that dry solids are easier to handle, weigh, and extract than wet solids. However, there is a risk of losses of some semivolatiles, for example, some chlorinated hydrocarbon pesticides, by vaporization and atmospheric transport, which is known to occur. Precautions must be taken to avoid contamination of the sample during the drying process. Laboratory workers and laboratory surfaces must be protected from contamination by dust and finely divided particles during sample handling. In a multilaboratory study of PCB contamination in the New Bedford, MA, harbor, air-dried sediments were extracted in a Soxhlet apparatus with 1:1 (v/v) hexane–acetone and fractionated by LC on florisil before GC/electron capture or GC/MS analysis of the PCB containing fraction.[89]

Wet Soil and Sediment

Direct analysis of wet soil (mud) and sediment has the advantages of saving the time required for drying, which can be several days to a week, avoiding losses of analytes that are vaporized during the drying process, and removing the risk of contamination of the sample during the drying process. However, wet solids are more difficult to handle than dry solids and an extraction with a water-miscible solvent is usually necessary before extraction of the analytes with a water-immiscible solvent. A second aliquot of the wet sample is usually weighed, dried, and weighed again to provide an estimate of the dry weight of the extracted sample and allow the analytical results to be reported on a dry weight basis.

This strategy was also used in the New Bedford Harbor PCB study.[89] Wet sediment was extracted by ultrasonic agitation and disintegration in acetone, the acetone extract was diluted with water, and the analytes were separated from the aqueous acetone solution by liquid–liquid extraction with hexane. The hexane extracts were fractionated by open-column LC on florisil, and the PCBs were determined by GC/MS and GC/electron capture. Another technique used in the New Bedford Harbor PCB study was the extraction of the wet sediment twice in a Soxhlet apparatus by using 2-propanol first and methylene chloride second. The extracts were combined, diluted with water, and the methylene chloride layer was separated and fractionated by LC on florisil before determination of the PCBs.

In another multilaboratory study of PCBs and chlorinated hydrocarbon pesticides in moist soil and wet sediments, two additional extraction techniques and two fractionation procedures were evaluated.[90] The samples were mixed with anhydrous sodium sulfate then extracted by ultrasonic agitation with a 1:1 (v/v) solution of methylene chloride and acetone. Samples were also mixed with methanol and extracted by shaking a closed sample container on a paint-can shaker. The methanol was decanted, the sediment was extracted with 9:1 (v/v) methylene chloride–methanol, and the extracts were combined. The extracts from both the ultrasonic agitation and paint-can shaker techniques were separately diluted with water and the methylene chloride layers were separated. Aliquots of the two extracts were fractionated by open-column LC on florisil and two additional aliquots were fractionated by gel permeation chromatography before GC/MS analysis. In this six-laboratory study, four different combinations of extraction and fractionation were evaluated with 14 chlorinated hydrocarbon pesticides in four types of samples and eight PCB isomer groups in five types of samples.

The conclusion from these two multilaboratory and multitechnique studies is that no combination of extraction technique and fractionation procedure is clearly best for all types of samples and analytes. Results depend on many factors including the texture and organic content of the soil or sediment, the analytes, the solvents, the extraction time, the ultrasonic power applied, and other variables. These techniques and others described in this chapter and in the scientific literature should be considered and judgment applied on the basis of the type of sample, the general analytical strategy, and the results of other studies.

Fish Tissue

The USEPA's laboratory in Duluth, MN, has been a leader for 30 years in the development of methods for the determination of semivolatiles, especially PCBs and chlorinated hydrocarbon pesticides, in fish tissue. For a national study of chemical residues in fish during the early 1990s, tissue samples were mixed with anhydrous sodium sulfate and extracted in a Soxhlet apparatus with 1:1 (v/v) hexane–methylene chloride.[91] After evaporation of the solvent, an aliquot of concentrated lipid was fractionated by gel permeation chromatography, and the fraction containing the analytes was further refined by open-column LC on silica gel before GC/MS analysis. This basic strategy has been well tested and the method and its performance are well documented.

Tree Bark

Chlorinated hydrocarbon pesticides have been extracted from tree bark with 1:1 hexane–acetone in a Soxhlet apparatus.[92] The acetone was evaporated and replaced with hexane, which was extracted multiple times with water to remove polar interferences. The hexane extract was fractionated by open-column LC on silica gel before GC/MS analysis using electron capture ionization (ECI), which is described in the section of Chapter 5 entitled "Chemical Ionization—Negative Ions". It is noteworthy that even with the selective ECI, both liquid–liquid extraction of the hexane extract with water and LC fractionation on silica gel were employed to isolate the analytes in a fraction with many potential interferences removed.

Supercritical Fluid Extraction

The interest in reducing the quantities of organic solvents used for the extraction of organic compounds from environmental samples was described at the beginning of the section of this chapter entitled "Semivolatile Organic Compounds in Drinking Water". This goal stimulated considerable research during the late 1980s and 1990s to find alternative extraction techniques. One technique investigated in detail is supercritical fluid extraction (SFE). Supercritical fluids are substances heated above their critical temperatures where they cannot be condensed regardless of pressure. The supercritical region exists over a range of pressures above the critical pressure, and supercritical fluids have physical properties similar to both liquids and gases.[93,94] The solvent strengths of supercritical fluids are comparable to those of liquid solvents and, like liquid solvents, depend on temperature. However, the solvent strengths of compressible supercritical fluids also depend on pressure, which is an important additional variable.

Many supercritical fluids have been used for SFE including low molecular weight hydrocarbons, fluorinated methanes, alcohols, ammonia, water, and nitrogen oxides. However, supercritical carbon dioxide is the dominant SFE solvent because of its low critical temperature (31 °C), relatively low critical pressure (72.9 atmos), and other properties. It is chemically unreactive, nonflammable, nontoxic, inexpensive, available in high purity, and environmentally inert. Most other supercritical fluids have undesirable properties including much higher critical temperatures and pressures, flammability, chemical reactivity, toxicity, and higher cost. However, supercritical CO_2 is rather nonpolar with a solvent strength, depending on the pressure, in the range between fluorinated aliphatic hydrocarbons and hexane.[93] This tends to limit SFE with pure CO_2 to less polar compounds such as some polycyclic aromatic hydrocarbons (PAHs), polychlorinated biphenyls (PCBs), chlorinated hydrocarbon pesticides, alcohols, and fatty acid esters of glycerol. Small portions of other solvents, for example, methanol, which are called modifiers, are often added to supercritical CO_2 to improve the solubility of more polar compounds.

In SFE, the solid sample is placed in a high-pressure cell and the supercritical fluid and modifier, if used, are introduced using high-pressure pumps. Most extractions employ a combination of a static process, where the fluid flow is stopped for some time to allow dissolution of the soluble compounds, and a dynamic process, where the fluid is continuously pumped through the sample.[95] The exit of the cell is connected to a restrictor, or a finely tapered tube, which maintains the required pressure until the fluid is decompressed into a small quantity of organic solvent. The CO_2 escapes into the atmosphere, and the analytes are dissolved in the organic solvent which may then be fractionated or immediately analyzed by GC/MS.

The SFE technique has been applied to the extraction of semivolatile compounds from many types of matrices including diesel particulate matter, sediment, soil, tissue, and urban dust.[94–98] Comparisons of SFE with Soxhlet, ultrasonic agitation, and other extraction techniques are also reported.[96,97] A significant limitation of SFE is the dependence of extraction efficiencies on the physical properties of the sample matrix.[95,98] While SFE is used by many investigators, more widespread applications may be limited by the cost of the required equipment, which is substantially greater than the cost of glassware and other materials used in traditional extractions.

Temperature and Pressure Assisted Solvent Extractions

The logical extension of SFE with organic solvent-modified CO_2 is the elimination of the CO_2 and the use of a pure solvent at an elevated temperature and pressure to extract semivolatile analytes from solid-phase samples. The sample and solvent are placed in a high-pressure cell which is heated in a GC oven. This technique has been evaluated with a variety of matrices including diesel particulate matter, oil shale, sediment, sewage sludge, soil, tissue, and urban dust containing aliphatic hydrocarbons, PAHs, PCBs, and chlorinated hydrocarbon pesticides.[99–101] The effects of temperature, pressure, solvent, sample matrix, and other variables were studied. This technique is considered equivalent to or better than Soxhlet and other established extraction procedures for temperature and pressure stable analytes. Like SFE, a limiting factor may be the cost of equipment required, but this is often offset by practical considerations including productivity requirements, solvent costs, the cost of cleaning equipment used in other techniques, and waste disposal costs. A variation on this approach is the application of microwave energy to provide the heat for the solvent extraction. This has been evaluated using sediment and soils containing PCBs.[102]

A related technique is the extraction of semivolatile analytes from solid-phase matrices with water at an elevated temperature followed by analysis of the water using one of the techniques described in this chapter.[103] This approach was evaluated with soil and urban dust containing PAHs and other semivolatile compounds. The sample and solvent are placed in a short section of stainless steel pipe capped at both ends, and the pipe is heated in a GC oven. The effects of heating time and temperature on recoveries were determined. The more water-soluble analytes were more efficiently recovered than some PAHs, which partitioned to the solid phase matrix when the extraction cell was cooled. The classical hot-water technique used to extract analytes from a solid-phase matrix is steam distillation.[104] In this technique the solid-phase sample is mixed with water in a distillation apparatus, and the distillate is collected and analyzed using one of the methods described in this chapter. All processes in which the sample is heated in the presence of water are limited to analytes that are thermally stable and resistant to hydrolysis.

Thermal Desorption

The USEPA's Office of Solid Waste has published Method 8275A, which describes the thermal desorption of semivolatiles from a soil, sludge, or solid waste sample directly into the injection port of the GC/MS instrument.[72] The sample is placed in a crucible in a closed system, and the compounds are vaporized by heating the crucible. Analytes are trapped in the first part of the GC column and the analysis is started by rapid temperature programming of the GC column oven. Method 8275A lists 41 analytes which are all thermally stable and generally unreactive PAHs, PCB congeners, and other similar compounds. The GC/MS used in Method 8275A must meet the DFTPP criteria contained in Method 8270C. Method 8275A may be of value for some types of samples but should be used with great caution to avoid overloading the GC/MS system with large quantities of analytes.

Method 8270C

Method 8270C is published by the USEPA's Office of Solid Waste for the determination of semivolatiles in extracts from solid waste, soils, air particulates, and water.[72] This method is not an integrated method in the style of the 500 and 600 series methods and does not include sampling techniques, extraction techniques, and other sample preparation procedures. These procedures are in separate documents referenced in Method 8270C. The method lists about 245 analytes including most of those in Table 4.1. Appropriate extraction procedures for specific compounds and classes of compounds are given in the method. Some of the listed analytes are marginal for GC/MS because of thermal instability, reactivity, and so on, and are better determined using the techniques described in Chapter 6.

Method 8270 provides the details of the GC/MS and quality control procedures, which are derived mostly from Methods 625 and the 525 series of methods. The method includes GC column performance criteria, and the mass spectrometer is required to meet the original DFTPP criteria in Table 2.3 or modified versions of these criteria. Limited performance data for some analytes extracted with specific procedures from several types of samples is included in the method.

Dibenzo-*p*-dioxins and Dibenzofurans

The chlorinated dibenzo-*p*-dioxins and dibenzofurans (bottom row of Chart 4.1) are an important class of semivolatiles that were described in the section of this chapter entitled "Classes of Semivolatile Organic Compounds". The evolution of mass spectrometric methods for this class of compounds began in the late 1960s and continued throughout the 1970s, 1980s, and 1990s. Until about the mid-1980s most environmental analyses were focused on the single target analyte 2,3,7,8-tetrachlorodibenzo-*p*-dioxin (2378-TCDD). In some of the early determinations of 2378-TCDD the sample was introduced with a direct insertion probe instead of by on-line GC, and high-resolution mass spectrometry (HRMS) was employed to provide selectivity for the analyte.[105] Enhanced analyte selectivity with HRMS is described in Chapter 5 using 2378-TCDD and some of its potential interferences to illustrate the extraordinary capabilities of this technique.

During the 1970s, packed-column GC provided a significantly better sample introduction system than the direct insertion probe, but HRMS continued to be used with the packed column for many determinations of 2378-TCDD. Selected ion monitoring (SIM) and HRMS provided high signal/noise and very low detection limits by only measuring ions within a narrow range of exact masses.[106] The exact and precise *m/z* measuring capability of HRMS also provided strong evidence to support the identification of 2378-TCDD. Significant improvements in sample preparation and preliminary fractionation procedures allowed reliable measurements of 2378-TCDD with packed-column GC and low-resolution MS. A method was validated for 2378-TCDD in fish tissue in the 10–100 ng/kg concentration range.[107] This approach was extended to the determination of tetra-, hexa-, hepta-, and octa-chlorodibenzo-*p*-dioxin congeners in industrial dust, fly ash, and sludge.[108] These analyses were made possible, in part, by using normal and reverse-phase high-performance liquid chromatography to separate extracts into fractions containing specific congeners (Chapter 6, "Condensed-Phase Separation Techniques"). The need for HRMS for 2378-TCDD measurements depends on several variables including the type of sample, the potential for coeluting interferences, the detection limits required, and the selectivity of the sample preparation procedures.[109]

High resolving power glass capillary GC columns made possible the separation of many tetra- through octa-chlorinated dibenzo-*p*-dioxin congeners.[42] The glass columns were largely replaced by fused silica capillary columns (FSCCs) that became commercially available from several suppliers during the early 1980s. Samples from some sources, for example, emissions from incinerators, were analyzed and found to contain not just 2378-TCDD but a variety of polychloro- and polybromo-dibenzo-*p*-dioxin and dibenzofuran congeners.[36,37] Analytical methods were greatly expanded to include the tetra- through octa-chlorodibenzo-*p*-dioxin and dibenzofuran congeners and some of the brominated compounds. Improvements in GC/MS instrument design and computer control systems allowed SIM, both at low and high mass spectrometer resolving powers, of groups of 12 or more ions and switching between multiple sets of ions as various congener groups eluted from the FSCC. Electron capture ionization was used for some analyses because of the very high sensitivity and high selectivity obtained with the more chlorinated congeners (Chapter 5, "Chemical Ionization—Negative Ions").[47] Pure authentic standards of many important congeners, including ^{13}C- and ^{37}Cl-labeled compounds, gradually became commercially available. These standards include the 17 tetra- through octa-chlorodibenzo-*p*-dioxin and dibenzofuran congeners with Cl in the 2, 3, 7, and 8 positions. The isotope-labeled congeners are widely used as internal standards and surrogate analytes in quantitative analysis and quality control.

This section contains summaries of some the analytical methods, especially those used in regulatory programs, developed during the period of roughly the late 1970s through the mid-1990s. These methods illustrate the strategies and techniques used and the concern for correct identifications of target analytes, quantitative accuracy, and analytical quality control. The identification criteria used in most methods is noteworthy and reflects the exclusive use of the SIM data acquisition strategy. Many analytical strategies were developed to utilize available equipment and skills or to control the costs of analyses. However, the unique potential health and other issues associated with this group of compounds often led to the utilization of very expensive HRMS.

Method 613

The first formal USEPA analytical method for a chlorinated dibenzo-*p*-dioxin or dibenzofuran was Method 613 for the single analyte 2378-TCDD. This method was included in the group of wastewater methods proposed on 3 December 1979.[70] The final version of Method 613 was published as part of USEPA wastewater regulations in 1984 and this summary is of that version.[71] Although Method 613 was probably used for only a short time with a small number of samples compared to later methods, it is historically important because it documented in the regulatory literature many features that were retained in the expanded methods described later in this section. Method 613 also affirmed to the regulated community and government regulatory officials that GC/MS, with both low-resolution MS and HRMS, was the reliable and preferred technique for the determination of controversial environmental pollutants such as 2378-TCDD. As of 1999, Method 613 is still published in its original 1984 version in the Code of Federal Regulations. However, other expanded and more precisely defined methods, which are discussed later in this section, are probably used for nearly all determinations of chlorinated dibenzo-*p*-dioxins and dibenzofurans in environmental samples.

Method 613 was designed for wastewater only, and an isotope-labeled internal standard, either the $^{37}Cl_4$-2378-TCDD or $^{13}C_{12}$-2378-TCDD, is added to the original sample before extraction of the analytes. This standard is used for quantitative analysis by isotope dilution calibration (Chapter 2, "Quantitative Analysis"). The method specifies a liquid–liquid extraction with methylene chloride and a change of solvent to hexane as the extract is concentrated by evaporation. The hexane extract is partitioned with aqueous sodium hydroxide and concentrated sulfuric acid to remove polar, acidic, and basic background compounds and potential interferences. The extract may optionally be fractionated by open-column LC on alumina and silica gel.

Both fused silica and glass capillary GC columns are specified with the requirement that 2378-TCDD must be resolved from the other 21 possible tetrachlorodibenzo-*p*-dioxin isomers (Table 4.2). This GC resolution could not be demonstrated until the late 1980s when appropriate mixtures of tetrachloro isomers became widely available. Electron ionization with either low-resolution MS or HRMS is specified and the instrument must be capable of SIM of at least four ions in a group. However, a resolving power requirement for the mass spectrometer is not specified. The ions monitored by SIM are *m*/*z* 320, 322, and 257 for 2378-TCDD (Figure 5.1), and either *m/z* 328 for $^{37}Cl_4$-2378-TCDD or *m/z* 332 for $^{13}C_{12}$-2378-TCDD. All these ions except *m/z* 257 are $M^{+\cdot}$ ions and exact masses are specified for SIM at high resolving power (Chapter 5, "High-Resolution Mass Spectrometry").

The identification criteria for 2378-TCDD in Method 613 require that the abundances of the four monitored ions maximize simultaneously at the retention time of 2378-TCDD in a calibration standard. The ratio of the abundances of *m/z* 320 and 322, which are the $M^{+\cdot}$ ions $^{35}Cl_4$-2378-TCDD and $^{35}Cl_3{}^{37}Cl$-2378-TCDD, must be within 10% of the value expected from the natural abundances of ^{35}Cl and ^{37}Cl. The signal/noise of the abundances of all ions must be at least 2.5, which implies that the ion at *m/z* 257, which is a fragment from the loss of COCl from the $M^{+\cdot}$, must be present. The stated MDL for 2378-TCDD in wastewater is 2 ng/L (Chapter 2, "Detection Limits"). The techniques and criteria in Method 613 are used, sometimes in expanded form, in most of the analytical methods developed for 2378-TCDD and other polychlorodibenzo-*p*-dioxin and dibenzofuran congeners.

The USEPA Contract Laboratory Program

During the early 1980s in the United States the USEPA initiated a large program, funded by a special tax which established a *Superfund*, to evaluate and, if necessary, remediate a large number of potentially hazardous chemical waste disposal sites. The USEPA implemented the contract laboratory program (CLP) to provide analyses of samples from these sites. The CLP specified an analytical method for the single target analyte 2378-TCDD, which utilized GC with low-resolution MS. This method was designed for concentrations of 2378-TCDD greater than 1 μg/kg, which was an action level in the *Superfund* program. The method was applied to tens of thousands of soil, sediment, solid waste, and other samples, and 2378-TCDD was found in some samples including sediments from the Love canal area near Niagara Falls, NY, and soils from the Times Beach area of eastern Missouri.

The method used for 2378-TCDD in the CLP program specifies the addition of both an internal standard, $^{13}C_{12}$-2378-TCDD, and a surrogate, $^{37}Cl_4$-2378-TCDD, to a small (10 g) sample of wet soil or sediment. Anhydrous sodium sulfate is mixed with the sample, which is allowed to dry before extraction with 1:7.5 (v/v) methanol–hexane solution in a closed container on a paint-can shaker. The filtered extract is concentrated by evaporation and fractionated by open column LC on basic and acidic silica gel, acidic alumina, and optionally on carbon. The concentrated fraction containing the analytes, internal standard, and surrogate is analyzed using a FSCC and low-resolution MS. The ions monitored with SIM are *m/z* 257, and 320, and 322 of 2378-TCDD, *m/z* 332 and 334 of the internal standard $^{13}C_{12}$-2378-TCDD, and *m/z* 328 of the surrogate $^{37}Cl_4$-2378-TCDD.

The analyte identification criteria used in the CLP are similar to those in Method 613 but slightly stronger. The 2378-TCDD in a sample had to elute from the GC column within 3 s of the internal standard, the relative abundance of *m/z* 257 had to be 20–45% of the base peak *m/z* 322, and the relative abundance of *m/z* 320 had to be 67–87% of the base peak. These strengthened criteria, compared to those of Method 613, are an indication of the trend of the 1980s toward methods that provide an unambiguous identification of 2378-TCDD in samples that could be used for evidence in regulatory actions and legal proceedings. The addition of two isotope-labeled analytes to the sample before extraction, an internal standard and a surrogate, is an indication of a similar concern for accurate quantitative analyses. The ratio of the amounts of the two labeled compounds recovered from the sample should be the same, within experimental error, as the ratio of the amounts added to the sample. This ratio is used to monitor method and laboratory performance in every sample.

High-Resolution Mass Spectrometry for 2378-TCDD

Although HRMS had been used with a direct insertion probe for some early 2378-TCDD measurements, the combination of GC and HRMS substantially improved the reliability of 2378-TCDD measurements in complex environmental samples. A strategy for the determination of 2378-TCDD in human milk, beef liver, fish, water, and sediment was developed in the late 1970s.[110] Most samples were heated with aqueous–alcoholic potassium hydroxide to hydrolyze the fats and oils before extraction with hexane. Extracts were fractionated by open-column LC on alumina and florisil, and the fraction containing 2378-TCDD was analyzed with a glass capillary GC column and high-resolution SIM. Water samples were extracted with methylene chloride, and the extract was processed as in Method 613.

The mass spectrometer was adjusted for a resolving power (10% valley) of 5000–9000 (Chapter 5, "High-Resolution Mass Spectrometry"). The ions monitored were: *m/z* 327.8847, which is the $^{37}Cl_4$-2378-TCDD internal standard $M^{+\cdot}$ ion; 319.8965, which is the $^{35}Cl_4$-2378-TCDD analyte $M^{+\cdot}$ ion; and 321.8936, which is the $^{35}Cl_3{}^{37}Cl$-2378-TCDD analyte $M^{+\cdot}$ ion. The number of ions used for SIM in this and other early to mid-1980s HRMS measurements was severely limited by the mass spectrometer technology of that period. Identification criteria were similar to those used in previous methods, but with the powerful additional requirement of ion abundances measured at the exact *m/z* values of the analyte and internal standard. This approach was later used for the determination of 2378-TCDD in human milk at the 0.1–10 parts per 10^{12} level.[111]

Significant improvements in sample preparation, GC, and HRMS techniques were apparent by the late 1980s.[46] In addition to the silica gel, florisil, and alumina LC fractionation procedures described previously, many sample preparation schemes incorporated a selective separation of planar chlorinated polycyclic aromatic compounds on a carbon-based adsorbent.[112] The separation of 2378-TCDD from the other tetrachlorodibenzo-*p*-dioxin isomers was regularly demonstrated.[46] Improved high-resolution mass spectrometers operating at resolving powers of 10,000 or more (10% valley) were capable of SIM of five or more ions at their exact *m/z* values. This allowed monitoring of the 2378-TCDD $M^{+\cdot}$ ions at 319.8965 and 321.8936, the $C_{11}H_4O^{35}Cl_2{}^{37}Cl$ fragment ion at 258.9298, and the $^{13}C_{12}$-2378-TCDD internal standard $M^{+\cdot}$ ions at 331.9368 and 333.9338.[46] Techniques were also developed to scan, during exact mass SIM, a narrow m/z range, for example, $\sim< 0.1$ Da, to locate the centroid of the peak and ensure that coeluting interferences with similar m/z were not present.[46,106]

Congeners with the 2378 Substitution Pattern

An important development of the 1980s was a consensus among toxicologists that 17 tetra-, penta-, hexa-, hepta-, and octa-chlorodibenzo-*p*-dioxin and dibenzofuran congeners substituted in the 2, 3, 7, and 8 positions are the most environmentally significant.[45] The chlorine substitution patterns of these 17 congeners are shown in Table 4.4, and the carbon position numbers of the parent compounds are shown in the bottom row of Chart 4.1. These 17 congeners are assigned toxicity weighting factors of 0.5–0.001 relative to 2378-TCDD, which is given a value of 1. The factors are used to calculate toxicity-weighted concentrations that are used in risk assessments and environmental regulations. The emphasis on this group of 17 polychlorodibenzo-*p*-dioxin (PCDD) and polychlorodibenzofuran (PCDF) congeners defined and somewhat simplified the analytical problem. Most analytical methods developed for regulatory monitoring in the United States specify the determination of the 17 specific congeners. Any of the other tetra- through hepta-chloro congeners identified in samples are usually determined as PCDD and PCDF isomer groups at each level of chlorination.

National Study of Chemical Residues in Fish

In 1986 the USEPA initiated a national study of chemical residues in fish.[113] This survey included the determination of PCDD and PCDF congeners in two species of fish from 388 major rivers, harbors, and lakes. The target analytes were the 17 environmentally significant PCDD and PCDF congeners (Table 4.4). Fish were either ground whole or skinned and filleted before grinding. The ground tissue was mixed with anhydrous sodium sulfate and fortified with 11 isotope-labeled tetra-, penta-, hexa-, hepta-, and octa-chlorodibenzo-*p*-dioxins and dibenzofurans and four unlabeled tetra- and penta-chloro congeners. The dry tissue was extracted with 1:1 (v/v) hexane–methylene chloride in a

Table 4.4 Chlorine substitution patterns in environmentally significant PCDDs and PCDFs

Level of chlorination	Dibenzo-*p*-dioxins	Dibenzofurans
Tetrachloro	2378	2378
Pentachloro	12378	12378, 23478
Hexachloro	123478, 123678, 123789	123478, 123678, 123789, 234678
Heptachloro	1234678	1234789, 1234678
Octachloro	12346789	12346789

Soxhlet apparatus, and the extract was concentrated by evaporation. The lipid extract was then fractionated on a column containing silica gel, potassium silicate, sodium sulfate, and celite/sulfuric acid, and the eluate was further fractionated on both florisil and carbon on silica gel.

The concentrated extract containing the target analytes was analyzed by GC/MS and SIM with a mass spectrometer resolving power of 5000 (10% valley). For both series of analytes, two exact *m/z* were monitored for each isomer group at each level of chlorination. The ions selected were generally the two most abundant from the $M^{+\cdot}$ ion clusters. Corresponding ions were monitored for the isotope-labeled internal standards. Isotope dilution calibration was used to calculate concentrations of analytes in samples when the isotope-labeled internal standard was the same isomer as the analyte. For some analytes the only available isotope-labeled internal standard was not the same isomer, but these quantitative analyses were not severely biased because similar isomers probably have similar extraction efficiencies (Chapter 2, "Quantitative Analysis"). This study employed an analytical strategy that was refined and somewhat expanded in subsequent USEPA methods for regulatory environmental monitoring that are described in this section.

Method 8280A

Method 8280A is published by the USEPA's Office of Solid Waste and specifies low-resolution MS for the determination of PCDD and PCDF congeners in water, soil, sediment, fly ash, and chemical wastes such as sludges, still bottoms, and fuel oil.[72] The target analytes are the 17 environmentally significant PCDD and PCDF congeners (Table 4.4) and the other tetra- through hepta-chloro congeners, which are determined as isomer groups at each level of chlorination. Unlike most other methods in the 8000 series, Method 8280A is an integrated analytical method that includes detailed procedures for sample extraction, extract fractionation, GC/MS, and quality control. Separate extraction procedures are described for the different types of samples. The method description is lengthy, very detailed, and includes most of the techniques and procedures used in Method 613, the CLP 2378-TCDD method, and the method used in the national study of chemical residues in fish. Detection limits are stated to be in the 10–50 ng/L range for water samples and the 1–50 μg/kg range for soil, fly ash, and chemical waste samples.

A mixture of five $^{13}C_{12}$-labeled PCDD and PCDF internal standards is added to the environmental sample. The mixture includes both 2378-TCDD and 2378-TCDF, a single hexaCDD, a single heptaCDF, and octaCDD. Solids and semisolids are extracted in a Soxhlet apparatus with toluene which, like benzene, is a good solvent for the analytes but, unlike benzene, is not a known human carcinogen. Water samples are partitioned with methylene chloride to extract the analytes. The concentrated extracts are treated with strong acid and base solutions and fractionated by open-column LC on silica gel, alumina, and carbon. Several additional isotope-labeled PCDD or PCDF congeners are added to the extracts before and after fractionation to determine recoveries of the internal standards in all samples and fortified analytes in control samples.

The final concentrated extract is analyzed by GC/MS; two GC columns are required to separate all of the 17 environmentally significant congeners. In order to ensure adequate detection of ions in the m/z range above 300, ion abundance criteria for the mass-scale calibration compound perfluorotri-n-butylamine (PFTBA) are specified. The abundances of the PFTBA ions at m/z 414 and 502 should be 30–50% of the abundance of the PFTBA ion at m/z 264. Four groups of 18 ions are specified for SIM with a 1 s cycle time per group and switching of groups as congeners elute from the GC column. Two $M^{+\cdot}$ ions and a $(M–COCl)^+$ fragment ion are monitored for each PCDD and PCDF isomer group at each level of chlorination (30 ions). Two $M^{+\cdot}$ ions are monitored for each isotope-labeled internal standard (10 ions) and several additional ions are monitored for other isotope-labeled quality control additives and potentially coeluting chlorinated diphenyl ethers.

Quality control criteria specified in Method 8280A for calibration solutions include a GC column resolving power test, ion abundance ratio limits for $M^{+\cdot}$ ions that are part of a cluster of ions caused by naturally occurring isotopes, signal/noise tests, and response factor precision tests. Response factors for quantitative analysis are calculated using abundances from the sum of the two $M^{+\cdot}$ ions monitored for analytes, internal standards, and other quality control materials. Analyte identification criteria include a GC retention time test, the presence of all monitored ions with a signal/noise of 2.5, maximizing of the abundances of the three analyte ions within 2 s, and isotope ratios for monitored $M^{+\cdot}$ ions within a specified range.

Isotope dilution calibration is used to calculate concentrations of the five analytes that have the same isomers present as isotope-labeled internal standards. For other analytes the isotope-labeled internal standard is not the same isomer, but these quantitative analyses are not severely biased because similar isomers probably have similar extraction efficiencies. This method could be expanded to include additional isotope-labeled internal standards for more or all the analytes. This potentially costly strategy is feasible, but, depending on the capabilities of the GC/MS system, the number of ions monitored will eventually negate the benefits of SIM.

Method 8290

Method 8290 is published by the USEPA's Office of Solid Waste and specifies HRMS for the determination of the 17 environmentally significant PCDD and PCDF congeners (Table 4.4) in paper pulp, fish tissue, human adipose tissue, and the types of samples addressed in Method 8280A.[72] A mass spectrometer resolving power of 10,000 (10% valley) is required with SIM data acquisition. Some directions in the method are specific for double-focusing magnetic deflection instruments and require modification for use with other types of high-resolution mass spectrometers.

Method 8290 is a fully integrated analytical method that includes detailed procedures for sample extraction, extract fractionation, GC/MS, and quality control. Separate extraction procedures are described for the different sample matrices and they are the same or similar to those described in Method 8280A and the method used in the national study of chemical residues in fish. The method description is lengthy, very detailed, and includes most of the techniques and procedures used in methods previously described in this section. The calibration ranges are 10–2000 pg/L for tetra- and penta-chlorodibenzo-*p*-dioxins and dibenzofurans in a 1 L water sample, and 1–200 ng/kg for a 10 g soil, sediment, fly ash, or tissue sample.

A mixture of nine $^{13}C_{12}$-labeled tetra-, penta-, hexa-, hepta-, and octa-chloro dibenzo-*p*-dioxin and dibenzofuran internal standards is added to each environmental sample before extraction. This mixture includes one PCDD isomer and one PCDF isomer at each level of chlorination except octaCDF. Fractionation procedures are provided to remove most coeluting interferences. The final concentrated extract is analyzed by GC/MS, and two GC columns are required to separate all of the 17 environmentally significant congeners. Five groups of 10 or fewer ions are specified for SIM with a 1 s cycle time per group and switching of groups as congeners elute from the GC column. Exact m/z values are specified for SIM and, unlike Method 8280A, the $(M{-}COCl)^+$ fragment ions are not monitored because more internal standards are used. Two $M^{+\cdot}$ ions are monitored for each PCDD and PCDF isomer group at each level of chlorination (20 ions) and two $M^{+\cdot}$ ions are monitored for each isotope-labeled internal standard (18 ions). An ion is monitored in four groups for potentially coeluting chlorinated diphenyl ethers, and a perfluorokerosene ion is used for spectrometer stability control (the lock mass) at high resolution. The quality control and quantitative analysis aspects of Method 8290 are the same as or similar to those of Method 8280A.

Method 23

Method 23 is published in federal regulations promulgating standards of performance for new stationary sources of air pollution.[114] Stationary sources include industrial boilers and furnaces and municipal solid-waste incinerators. No other types of samples are considered in this method. Method 23 specifies HRMS for the determination of total tetra-, penta-, hexa-, hepta- and octa-chlorinated dibenzo-*p*-dioxins and dibenzofurans. A mass spectrometer resolving power of 10,000 (10% valley) is required with SIM data acquisition. Method 23 is a fully integrated analytical method that contains detailed procedures for sampling, sample extraction, extract fractionation, GC/MS, and quality control. The technique for isokinetic sampling utilizes a sampling train that includes a glass fiber particulate filter and a styrene–divinylbenzene copolymer (XAD-2^{TM}) adsorbent in a glass cartridge.

A quality control provision of Method 23 is the addition of five isotope-labeled PCDD and PCDF congeners to the XAD-2 adsorbent before sampling. These surrogate analytes are not the same as the internal standards and are measured to determine losses of analytes during the sampling period. After sampling, a mixture of nine $^{13}C_{12}$-labeled tetra-, penta-, hexa-, hepta-, and octa-chlorodibenzo-*p*-dioxin and dibenzofuran internal standards is added to the combined glass fiber filter and XAD-2 adsorbent. The internal standards are the same as those used in Method 8290 except for one hexachlorodibenzofuran. The analytes, surrogates, and internal standards are extracted from the combined filter and adsorbent with toluene in a Soxhlet apparatus. The concentrated extract is fractionated by open-column LC on silica gel, alumina, and carbon and analyzed by GC/MS using an approach essentially the same as that of Method 8290. The concentrations of the PCDD and PCDF isomer groups are calculated by summing the individual isomer concentrations at each level of chlorination and a grand total PCDD + PCDF concentration is determined by summing all the isomer group concentrations.

Method TO-9A

Method TO-9A is a revised method which specifies HRMS for the determination of PCDDs, PCDFs, and some polybromo and bromopolychloro congeners in ambient air.[86] This method was used to determine the PCDDs and PCDFs in ambient air at McMurdo Station, Antarctica.[115] A mass spectrometer resolving power of 10,000 (10% valley) is required with SIM data acquisition. The target analytes are the 17 environmentally significant PCDD and PCDF congeners (Table 4.4) and the other tetra- through hepta-chlorodibenzo-*p*-dioxin and dibenzofuran congeners, which are determined as isomer groups at each level of chlorination. The tetra-, penta-, hexa-, and hepta-bromodibenzo-*p*-dioxins and dibenzofurans may also be determined along with the bromotrichloro and bromotetrachloro congeners. Method TO-9A is a fully integrated analytical method that includes detailed procedures for sampling, sample extraction, extract fractionation, GC/MS, and quality control. Detection limits are in the range 0.01–0.2 pg/m^3.

Method TO-9A uses the same sampling technique as Methods TO-4A and TO-13A, which are summarized in the section of this chapter entitled "Semivolatile Organic Compounds in Air, Sediments, Soil, Solid Wastes, and Tissue". The method specifies a high flow rate air sampler capable of drawing ~8 STD cubic feet/min (~0.225 STD m^3/min) through a quartz fiber filter followed by a glass cartridge containing polyurethane foam (PUF) adsorbent. A sample volume of about 325–400 STD m^3 collected over about 24 h is recommended. Just as in Method 23, the adsorbent is fortified before sampling with isotope-labeled surrogate analytes to determine losses of analytes during the sampling period. After sampling, a mixture of nine $^{13}C_{12}$-labeled tetra-, penta-, hexa-, hepta-, and octa-chlorodibenzo-*p*-dioxin and dibenzofuran internal standards is added to the combined glass fiber filter and PUF adsorbent. This mixture includes one PCDD isomer and one PCDF isomer at each level of chlorination except octachlorodibenzofuran. If brominated congeners are being determined, three isotope-labeled brominated internal standards are also added.

The quartz fiber filter and the adsorbent are extracted together in a Soxhlet apparatus with benzene or toluene. The particulate and adsorbed fractions are not measured separately because some analytes can vaporize from particles on the filter and are collected on the PUF. The extract is concentrated by evaporation and fractionated using the acid, base, and open-column LC techniques described in Method 8280A. The final concentrated extract is analyzed by GC/MS, and 12 groups of 4 to 16 exact *m/z* values are specified for SIM with a 1 s cycle time per group and switching of groups as congeners elute from the GC column. The PCDD and PCDF groups are measured with five sets of 8–12 ions and the polybromo- and bromopolychloro-dibenzo-*p*-dioxin and dibenzofuran congeners are measured with seven groups of 4–16 ions. The other aspects of the data acquisition, identification criteria, quantitative analysis, and quality control are similar to those of Method 8290.

Method 1613

Method 1613, revision B, is the USEPA mandated method for the determination of PCDDs and PCDFs in wastewater.[116] It is also the USEPA mandated method for the determination of 2378-TCDD in drinking water.[78] Method 1613 specifies HRMS for the determination of these analytes and includes procedures for other sample types including soil, sediment, sludge, and fish tissue. A mass spectrometer resolving power of 10,000 (10% valley) is required with SIM data acquisition. The target analytes are the 17 environmentally significant PCDD and PCDF congeners (Table 4.4) and the other tetra- through hepta-chlorodibenzo-*p*-dioxin and dibenzofuran congeners, which are determined as isomer groups at each level of chlorination. Method 1613 is a fully integrated analytical method that includes detailed procedures for sampling, sample extraction, extract fractionation, GC/MS, and quality control. Separate extraction procedures are described for the different sample matrices and they are the same or similar to those described in Method 8280A, Method 8290, and the method used in the national study of chemical residues in fish. A liquid–solid extraction procedure for water which is similar to that in Method 525.2 is also described. The method description is lengthy, very detailed, and includes most of the techniques and procedures used in methods previously described in this section. Detection limits are in the range 10–50 pg/L in water and 1–5 ng/kg in solids.

The samples are fortified with a mixture of 15 $^{13}C_{12}$-labeled tetra-, penta-, hexa-, hepta-, and octa-chlorodibenzo-*p*-dioxin and dibenzofuran internal standards. This mixture includes one or more PCDD isomers and one or more PCDF isomers at each level of chlorination except octachlorodibenzofuran. The extract is concentrated by evaporation and fractionated using the acid, base, and open-column LC techniques described in Method 8280A. The final concentrated extract is analyzed by GC/MS and five groups of 8 to 12 exact *m/z* values are specified for SIM with a 1 s cycle time per group and switching of groups as congeners elute from the GC column. Because more isotope-labeled internal standards are used than in the other methods described in this section, a few more analytes can be measured by the isotope dilution procedure. It is questionable whether the improved accuracy of these few measurements justifies the added expense of the additional standards. The other aspects of the data acquisition, identification criteria, quantitative analysis, and quality control are the same or similar to those of Method 8290 and the other HRMS methods.

Electron Capture Ionization

The USEPA has not published an analytical method that uses electron capture ionization (Chaper 5, "Chemical Ionization—Negative Ions"). However, this technique is employed by some investigators because of its high selectivity and sensitivity for many polychlorinated compounds.[47,117] Soil and sediment are mixed with anhydrous sodium sulfate, fortified with $^{13}C_{12}$-labeled chlorinated dibenzo-*p*-dioxin and dibenzofuran internal standards, and extracted with 2-propanol and dichloromethane in a Soxhlet apparatus. The extracts are concentrated by evaporation, and the solvent is changed to hexane before fractionation by open-column LC on silica gel and alumina. The fraction containing the analytes is analyzed by GC/MS, using electron capture ionization and SIM. Concentrations of analytes are determined as isomer groups at each level of chlorination. It is noteworthy that in spite of the selective ionization technique, fractionation of the sample extracts is used to minimize or eliminate interferences from coeluting chlorinated hydrocarbon pesticides, PCBs, and other electron capturing compounds.

*Summary of Dibenzo-*p*-dioxin and Dibenzofuran Analytical Methods*

Some may question why there are so many USEPA analytical methods for this series of important environmental contaminants. A reading of the preceeding summaries, or inspection of Table 4.3, reveals that each method has a somewhat different analytical approach or is primarily directed at a different type of environmental sample. For example, Methods 23 and TO-9A provide detailed sampling procedures for stationary source emissions and ambient air, respectively, but this kind of information is not needed in the methods for most types of condensed-phase samples. Where several different types of samples are considered in the same analytical method, for example, Methods 8280A, 8290, and 1613, the methods are lengthy and complex because separate extraction procedures must be described for most of the types of samples. The most duplication of content appears in the two HRMS methods 8290 and 1613, which were produced by different USEPA regulatory offices operating with different directives in different environmental legislation.

REFERENCES

1. Keith, L. H.; Telliard, W. A. *Environ. Sci. Technol.* **1979**, *13*, 416–423.
2. Eichelberger, J. W.; Kerns, E. H.; Olynyk, P.; Budde, W. L. *Anal. Chem.* **1983**, *55*, 1471–1479.

3. Code of Federal Regulations, Title 40, Part 141.
4. *Annual Report on Carcinogens*; 1998, URL http://ntp-server.niehs.nih.gov.
5. Hurst, R. E.; Settine, R. L.; Fish, F.; Roberts, E. C. *Anal. Chem.* **1981**, *53*, 2175–2179.
6. Fuhrer, U.; Ballschmiter, K. *Environ. Sci. Technol.* **1998**, *32*, 2208–2215.
7. Morris, P. J.; Quensen III, J. F.; Tiedje, J. M.; Boyd, S. A. *Environ. Sci. Technol.* **1993**, *27*, 1580–1586.
8. Carter, L. J. *Science* **1976**, *192*, 240–243.
9. Alford-Stevens, A. L.; Bellar, T. A.; Eichelberger, J. W.; Budde, W. L. *Anal. Chem.* **1986**, *58*, 2014–2022.
10. Rachdawong, P.; Christensen, E. R. *Environ. Sci. Technol.* **1997**, *31*, 2686–2691.
11. Cairns, T.; Siegmund, E. G. *Anal. Chem.* **1981**, *53*, 1183A–1193A.
12. Brown, M. P.; Werner, M. B.; Sloan, R. J.; Simpson, K. W. *Environ. Sci. Technol.* **1985**, *19*, 656–661.
13. Hillery, B. R.; Basu, I.; Sweet, C. W.; Hites, R. A. *Environ. Sci. Technol.* **1997**, *31*, 1811–1816.
14. Eichelberger, J. W.; Harris, L. E.; Budde, W. L. *Anal. Chem.* **1974**, *46*, 227–232.
15. Method 625, In *Guidelines Establishing Test Procedures for the Analysis of Pollutants Under the Clean Water Act; Final Rule and Interim Final Rule and Proposed Rule, Federal Register*; 26 October 1984, Vol. 49, 43234–43439; Title 40 Code of Federal Regulations Part 136, Appendix A.
16. Gebhart, J. E.; Hayes, T. L.; Alford-Stevens, A. L.; Budde, W. L. *Anal. Chem.* **1985**, *57*, 2458–2463.
17. Erickson, M. D.; Stanley, J. S.; Turman, J. K.; Going, J. E.; Redford, D. P.; Heggem, D. T. *Environ. Sci. Technol.* **1988**, *22*, 71–76.
18. Slivon, L. E.; Gebhart, J. E.; Hayes, T. L.; Alford-Stevens, A. L.; Budde, W. L. *Anal. Chem.* **1985**, *57*, 2464–2469.
19. Mullin, M. D.; Pochini, C. M.; McCrindle, S.; Romkes, M.; Safe, S. H.; Safe, L. M. *Environ. Sci. Technol.* **1984**, *18*, 468–476.
20. Frame, G. *Anal. Chem.* **1997**, *69*, 468A–475A.
21. de Voogt, P.; Haglund, P.; Reutergardh, L. B.; de Wit, C.; Waern, F. *Anal. Chem.* **1994**, *66*, 305A–311A.
22. Erickson, M. D. *Analytical Chemistry of PCBs*, 2nd ed.; CRC Lewis: Boca Raton, FL, 1997.
23. Carson, R. L. *Silent Spring*; Houghton Mifflin: New York, 1962.
24. Muir, D. C. G.; Norstrom, R. J.; Simon, M. *Environ. Sci. Technol.* **1988**, *22*, 1071–1079.
25. Simonich, S. L.; Hites, R. A. *Environ. Sci. Technol.* **1995**, *29*, 2905–2914.
26. Adams, W. J.; Kimerle, R. A.; Barnett Jr., J. W. *Environ. Sci. Technol.* **1992**, *26*, 1865–1875.
27. Wania, F.; Mackay, D. *Environ. Sci. Technol.* **1996**, *30*, 390A–396A.
28. Willett, K. L.; Ulrich, E. M.; Hites, R. A. *Environ. Sci. Technol.* **1998**, *32*, 2197–2207.
29. Dearth, M. A.; Hites, R. A. *Environ. Sci. Technol.* **1991**, *25*, 245–254.
30. Method 525.2, Revision 1.1, In *Methods for the Determination of Organic Compounds in Drinking Water*; Suppl. III, USEPA Report EPA/600/R-95/131, August 1995; URL http://www.epa.gov/nerlcwww/methmans.html.
31. Tomy, G. T.; Stern, G. A.; Muir, D. C. G.; Fisk, A. T.; Cymbalisty, C. D.; Westmore, J. B. *Anal. Chem.* **1997**, *69*, 2762–2771.
32. Private communication from Jack D. Henion, Cornell University, 16 June 1987.
33. Whiteside, T. *The Pendulum and the Toxic Cloud, The Course of Dioxin Contamination*; Yale University Press, New Haven, CT, 1979.

34. *Dioxin*; *Chem. Eng. News*, Special Issue, 6 June 1983.
35. Gough, M. *Dioxin, Agent Orange: The Facts*; Plenum Press, New York, 1986.
36. Karasek, F. W.; Hutzinger, O. *Anal. Chem.* **1986**, *58*, 633A–642A.
37. Thomas, V. M.; Spiro, T. G. *Environ. Sci. Technol.* **1996**, *30*, 82A–85A.
38. Kuehl, D. W.; Butterworth, B. C.; DeVita, W. M.; Sauer, C. P. *Biomed. Environ. Mass Spectrom.* **1987**, *14*, 443–447.
39. Kuehl, D. W.; Butterworth, B.; Marquis, P. J. *Chemosphere* **1994**, *29*, 523–535.
40. Huang, L. Q.; Tong, H.; Donnelly, J. R. *Anal. Chem.* **1992**, *64*, 1034–1040.
41. Tysklind, M.; Fangmark, I.; Marklund, S.; Lindskog, A.; Thaning, L.; Rappe, C. *Environ. Sci. Technol.* **1993**, *27*, 2190–2197.
42. Buser, H. R.; Rappe, C. *Anal. Chem.* **1984**, *56*, 442–448.
43. Waddell, D. S.; McKinnon, H. S.; Chittim, B. G.; Safe, S.; Boyd, R. K. *Biomed. Environ. Mass Spectrom.* **1987**, *14*, 457–464.
44. Swerev, M.: Ballschmiter, K. *Anal. Chem.* **1987**, *59*, 2536–2538.
45. Clement, R. E. *Anal. Chem.* 1991, *63*, 1130A–1139A.
46. Stanley, J. S.; Sack, T. M.; Tondeur, Y.; Beckert, W. F. *Biomed. Environ. Mass Spectrom.* **1988**, *17*, 27–35.
47. Koester, C. J.; Hites, R. A. *Environ. Sci. Technol.* **1992**, *26*, 1375–1382.
48. Buser, H. R. *Environ. Sci. Technol.* **1986**, *20*, 404–408.
49. Auger, P.; Malalyandl, M.; Wightman, R. H.; Bensimon, C.; Williams, D. T. *Environ. Sci. Technol.* **1993**, *27*, 1673–1680.
50. *NIST/EPA/NIH Mass Spectral Database*; National Institute for Standards and Technology: Gaithersburg, MD, 1992.
51. Jarnberg, U. G.; Asplund, L. T.; Egeback, A.-L.; Jansson, B.; Unger, M.; Wideqvist, U. *Environ. Sci. Technol.* **1999**, *33*, 1–6.
52. Paputa-Peck, M. C.; Marano, R. S.; Schuetzle, D.; Riley, T. L.; Hampton, C. V.; Prater, T. J.; Skewes, L. M.; Jensen, T. E.; Ruehle, P. H.; Bosch, L. C.; Duncan, W. P. *Anal. Chem.* **1983**, *55*, 1946–1954.
53. Philp, R. P.; Oung, J.-N. *Anal. Chem.* **1988**, *60*, 887A–896A.
54. Wang, Z.; Fingas, M.; Sergy, G. *Environ. Sci. Technol.* **1995**, *29*, 2622–2631.
55. Kvenvolden, K. A.; Hostettler, F. D.; Carlson, P. R.; Rapp, J. B.; Threlkeld, C. N.; Warden, A. *Environ. Sci. Technol.* **1995**, *29*, 2684–2694.
56. Boehm, P. D.; Page, D. S.; Gilfillan, E. S.; Bence, A. E.; Burns, W. A.; Mankiewicz, P. J. *Environ. Sci. Technol.* **1998**, *32*, 567–576.
57. Carpenter, R. A.; Hollowell, R. H.; Hill, K. M. *Anal. Chem.* **1997**, *69*, 3314–3320.
58. Wester, P. G.; De Boer, J.; Brinkman, U. A. T. *Environ. Sci. Technol.* **1996**, *30*, 473–480.
59. Glassmeyer, S. T.; De Vault, D. S.; Myers, T. R.; Hites, R. A. *Environ. Sci. Technol.* **1997**, *31*, 84–88.
60. Swackhamer, D. L.; Charles, M. J.; Hites, R. A. *Anal. Chem.* **1987**, *59*, 913–917.
61. Stern, G. A.; Muir, D. C. G.; Westmore, J. B.; Buchannon, W. D. *Biol. Mass Spectrom.* **1993**, *22*, 19–30.
62. Budde, W. L.; Eichelberger, J. W. *J. Chromatogr.* **1977**, *134*, 147–158.
63. Pereira, W. E.; Hostettler, F. D. *Environ. Sci. Technol.* **1993**, *27*, 1542–1552.
64. Kolpin, D. W.; Barbash, J. E.; Gilliom, R. J. *Environ. Sci. Technol.* **1998**, *32*, 558–566.
65. Keith, L. H., Ed. *Principles of Environmental Sampling*, 2nd ed.; American Chemical Society, Washington, DC, 1996.
66. Ballinger, D. G. *Environ. Sci. Technol.* **1979**, *13*, 1362–1366.
67. Poppiti, J. *Environ. Sci. Technol.* **1994**, *28*, 151A–152A.
68. Newman, A. *Anal. Chem.* **1996**, *68*, 733A–737A.

69. Kagel, R. O.; Stehl, R. H.; Crummett, W. B. *Anal. Chem.* **1979**, *51*, 233A.
70. *Guidelines Establishing Test Procedures for the Analysis of Pollutants; Proposed Regulations, Federal Register*; 3 December 1979, Vol. 44, 69464–69575.
71. *Guidelines Establishing Test Procedures for the Analysis of Pollutants Under the Clean Water Act; Final Rule and Interim Final Rule and Proposed Rule, Federal Register*; 26 October 1984, Vol. 49, 43234–43439; Title 40 Code of Federal Regulations Part 136, Table A.
72. *Test Methods for Evaluating Solid Waste, Physical/Chemical Methods*; USEPA Publication SW-846, 3rd Ed. and Updates I, II, IIA, IIB, and III; URL http://www.epa.gov.epaoswer/hazwaste/test/8xxx.htm.
73. Rostad, C. E.; Pereira, W. E.; Ratcliff, S. M. *Anal. Chem.* **1984**, *56*, 2856–2860.
74. Method 525, Revision 2.1, In *Methods for the Determination of Organic Compounds in Drinking Water*; USEPA Report EPA/600/4-88/039, December 1988; URL http://www.epa.gov/nerlcwww/methmans.html.
75. *National Primary Drinking Water Regulations*; *Final Rule Federal Register*; 30 January 1991, Vol. 56, 3526–3597; Title 40 Code of Federal Regulations, Part 141.
76. Method 525.1, Revision 2.2, In *Methods for the Determination of Organic Compounds in Drinking Water*, USEPA Report EPA/600/4-88/039, December 1988, Revised July 1991; URL http://www.epa.gov/nerlcwww/methmans.html.
77. Hagen, D. F.; Markell, C. G.; Schmitt, G. A.; Blevins, D. D. *Anal. Chim. Acta* **1990**, *236*, 157–164.
78. *National Primary Drinking Water Regulations; Synthetic Organic Chemicals and Inorganic Chemicals*; Final Rule, *Federal Register* 17 July 1992, Vol. 57, 31776–31849; Title 40 Code of Federal Regulations Part 141.
79. Method 525.2, Revision 1.1, In *Methods for the Determination of Organic Compounds in Drinking Water*, Suppl. III, USEPA Report EPA/600/R-95/131, August 1995; URL http://www.epa.gov/nerlcwww/methmans.html.
80. *National Primary and Secondary Drinking Water Regulations*: *Analytical Methods for Regulated Drinking Water Contaminants*; Final Rule, *Federal Register* 5 December 1994, Vol. 59, 62456–62471; Title 40 Code of Federal Regulations, Part 141.
81. Eichelberger, J. W.; Budde, W. L. *Biomed. Environ. Mass Spectrom.* **1987**, *14*, 357–362.
82. Hennion, M.-C.; Pichon, V. *Environ. Sci. Technol.* **1994**, *28*, 576A–583A.
83. Ho, J. S.; Budde, W. L. *Anal. Chem.* **1994**, *66*, 3716–3722.
84. Ho, J. S.; Tang, P. H.; Eichelberger, J. W.; Budde, W. L. *J. Chromatogr. Sci.* **1995**, *33*, 1–8.
85. Colgrove, S. G.; Svec, H. J. *Anal. Chem.* **1981**, *53*, 1737–1742.
86. *Compendium of Methods for the Determination of Toxic Organic Compounds in Ambient Air*, 2nd ed.; USEPA Report EPA/625/R-96/010b, January 1999.
87. *Standard D4861-94a, Standard Practice for Sampling and Selection of Analytical Techniques for Pesticides and Polychlorinated Biphenyls in Air, Annual Book of ASTM Standards*; Vol. 11.03, American Society for Testing and Materials: W. Conshohocken, PA, 1998.
88. *Standard D6909-97, Standard Test Method for the Determination of Gaseous and Particulate Polycyclic Aromatic Hydrocarbons in Ambient Air (Collection on Sorbent-Backed Filters with Gas Chromatographic/Mass Spectrometric Analysis), Annual Book of ASTM Standards*; Vol. 11.03, American Society for Testing and Materials: W. Conshohocken, PA, 1998.
89. Alford-Stevens, A. L.; Budde, W. L.; Bellar, T. A. *Anal. Chem.* **1985**, *57*, 2452–2457.

90. Alford-Stevens, A. L.; Eichelberger, J. W.; Budde, W. L. *Environ. Sci. Technol.* **1988**, *22*, 304–312.
91. Marquis, P. J.; Hanson, R. L.; Larsen, M. L.; DeVita, W. M.; Butterworth, B. C.; Kuehl, D. W. *Chemosphere* **1994**, 29, 509–521.
92. Simonich, S. L.; Hites, R. A. *Environ. Sci. Technol.* **1997**, *31*, 999–1003.
93. Smith, R. D.; Wright, B. W.; Yonker, C. R. *Anal. Chem.* **1988**, *60*, 1323A–1336A.
94. Hawthorne, S. B. *Anal. Chem.* **1990**, *62*, 633A–642A.
95. Taylor, L. T. *Anal. Chem.* **1995**, *67*, 364A–370A.
96. Maio, G.; von Holst, C.; Wenclawiak, B. W.; Darskus, R. *Anal. Chem.* **1997**, *69*, 601–606.
97. Heemken, O. P.; Theobald, N.; Wenclawiak, B. W. *Anal. Chem.* **1997**, *69*, 2171–2180.
98. Benner, B. A. *Anal. Chem.* **1998**, *70*, 4594–4601.
99. Richter, B. E.; Jones, B. A.; Ezzell, J. L.; Porter, N. L.; Avdalovic, N.; Pohl, C. *Anal. Chem.* **1996**, *68*, 1033–1039.
100. Fisher, J. A.; Scarlett, M. J.; Stott, A. D. *Environ. Sci. Technol.* **1997**, *31*, 1120–1127.
101. Schantz, M. M.; Nichols, J. J.; Wise, S. A. *Anal. Chem.* **1997**, *69*, 4210–4219.
102. Lopez-Avila, V.; Benedicto, J.; Charan, C.; Young, R.; Beckert, W. F. *Environ. Sci. Technol.* **1995**, *29*, 2709–2712.
103. Hageman, K. J.; Mazeas, L.; Grabanski, C. B.; Miller, D. J.; Hawthorne, S. B. *Anal. Chem.* **1996**, *68*, 3892–3898.
104. Veith, G. D.; Kiwus, L. M. *Bull. Environ. Contam. Toxicol.* **1977**, *17*, 631–636.
105. Baughman, R.; Meselson, M. *Environ. Health Perspect.* **1973**, *5*, 27.
106. Gross, M. L.; Sun, T.; Lyon, P. A.; Wojinski, S. F.; Hilker, D. R.; Dupuy, A. E.; Heath, R. G. *Anal. Chem.* **1981**, *53*, 1902–1906.
107. Lamparski, L. L.; Nestrick, T. J.; Stehl, R. H. *Anal. Chem.* **1979**, *51*, 1453–1458.
108. Lamparski, L. L.; Nestrick, T. J. *Anal. Chem.* **1980**, *52*, 2045–2054.
109. Hummel, R. A.; Shadoff, L. A. *Anal. Chem.* **1980**, *52*, 191–192.
110. Harless, R. L.; Oswald, E. O.; Wilkinson, M. K.; Dupuy, A. E.; McDaniel, D. D.; Tai, H. *Anal. Chem.* **1980**, *52*, 1239–1245.
111. Heath, R. G.; Harless, R. L.; Gross, M. L.; Lyon, P. A.; Dupuy, A. E.; McDaniel, D. D. *Anal. Chem.* **1986**, *58*, 463–468.
112. Smith, L. M.; Stalling, D. L.; Johnson, J. L. *Anal. Chem.* **1984**, *56*, 1830–1842.
113. Marquis, P. J.; Hackett, M.; Holland, L. G.; Larsen, M. L.; Butterworth, B. C.; Kuehl, D. W. *Chemosphere* **1994**, *29*, 495–508.
114. *Standards of Performance for New Stationary Sources*; Title 40 Code of Federal Regulations, Part 60, Appendix A.
115. Lugar, R. M.; Harless, R. L.; Dupuy, A. E.; McDaniel, D. D. *Environ. Sci. Technol.* **1996**, *30*, 555–561.
116. *Guidelines Establishing Test Procedures for the Analysis of Pollutants; EPA Method 1613, Federal Register*; 15 September 1997, Vol. 62, 48394–48442; Title 40 Code of Federal Regulations, Part 136, Appendix A.
117. Brzuzy, L. P.; Hites, R. A. *Environ. Sci. Technol.* **1995**, *29*, 2090–2098.

5

Strategies for Enhanced Analyte Selectivity, Ion Composition, Lower Detection Limits, Molecular Weight Determination, and Structure

Chapters 3 and 4 are concerned with organic compounds of environmental interest that are amenable to separation by gas chromatography (GC). The analytical methods described in these chapters specify, with a few exceptions, open tubular (capillary) GC columns, nominal 70-eV electron ionization (EI), and low-resolution mass spectrometry (MS). This combination of GC and MS is defined as conventional GC/MS and it is the *killer application* of MS in environmental analysis (Chapter 2, "Conventional Gas Chromatography/Mass Spectrometry").

Conventional GC/MS, however, does have limitations (Chapter 2, "Limitations of Mass Spectrometry"). The purpose of this chapter is to introduce and describe several mass spectrometric strategies and techniques that can provide valuable supplemental information and often overcome most of the limitations of conventional GC/MS. Several of these strategies were briefly mentioned or described in Chapters 2–4. The price which must be paid for these additional capabilities varies widely and depends on the problem to be solved, the particular technique selected, and the type of instrumentation required. Some instrumentation, for example, some high-resolution or tandem mass spectrometers, can cost more than 10 times the cost of a basic conventional GC/MS system.

Techniques to address some of the limitations of conventional GC/MS were described in Chapter 2 or by reference to other chapters in this book, for example, Chapter 6 for compounds not amenable to GC. Limitations of conventional GC/MS not considered or addressed in Chapter 2 include insufficient selectivity for target analytes in very complex samples, minimal inferential information about ion composition, detection limits that are sometimes too high, missing $M^{+\cdot}$ ion and therefore molecular weight information, and inadequate information to deduce molecular structure. These limitations are described in more detail, and the techniques considered to address these limitations are identified in this section:

- A major advantage of conventional GC/MS is that EI is a universal ionization technique. Therefore, every atom or molecule that has sufficient vapor pressure and thermal stability, from helium to very complex organic or inorganic molecules, can be converted into an ion in the gas phase by EI. This universal ionization characteristic is very important because it allows the determination of a wide variety of organic and inorganic substances under similar conditions and with similar sensitivities. The disadvantage of the universal EI technique is limited selectivity, which is often needed to detect target analytes at low concentrations in very complex samples. Selectivity is the ability to find clear signals from target analytes or unknowns in the presence of numerous interferences, coeluting substances, and general chemical background. General chemical background, which is often called *chemical noise,* is produced by GC septum bleed, GC column bleed, and especially by residues from previous samples. Examples of complex samples, where selectivity may be important, are soil extracts from chemical waste sites, tissue extracts, and very contaminated industrial wastewater. Even with high-resolution GC separation conditions, coelution of several components is common with these types of samples.
- Knowledge of the composition of ions is required to interpret mass spectra that may be needed to identify unknown substances. The composition of some small *m/z* ions observed by conventional GC/MS can be inferred from general knowledge of $M^{+\cdot}$ ion fragmentation mechanisms. For example, the presence of an ion at *m/z* 77 usually is a strong indication of a stable phenyl ion ($C_6H_5^+$). This conclusion is usually supported by the presence of a lower abundance ion at *m/z* 51, which is produced by the expulsion of a neutral C_2H_2 from the phenyl ion. However, most ions observed with low-resolution MS, especially those with *m/z* $>\sim100$, cannot be assigned compositions without knowing the identity of the compound and the details of the mechanisms of fragmentation of the $M^{+\cdot}$ and the larger fragment ions.
- While detection limits are generally very good with conventional GC/MS, there are some analytes, for example, 2,3,7,8-tetrachlorodibenzo-*p*-dioxin and related compounds, that are of considerable interest at detection limits below those attainable with conventional GC/MS and selected ion monitoring (SIM). This problem is usually called inadequate sensitivity, which may be caused by several factors including the inherently low ionization efficiency of EI.
- One of the limitations of conventional GC/MS mentioned in Chapter 2 is excessive fragmentation of the $M^{+\cdot}$ ions of some compounds.

When this occurs the abundance of the $M^{+\cdot}$ may be too low to observe and valuable information about the molecular weight of the substance is not available. The absence of a $M^{+\cdot}$ is not uncommon in conventional GC/MS.

- Nontarget analytes are often found in samples and sometimes these cannot be identified using the techniques described in Chapter 2 ("Identification Criteria"). Sometimes these substances are obscure or their structures have not been reported in the chemical literature, and conventional GC/MS does not generate sufficient information to determine their structures.

Four strategies are considered in this chapter to address one or more of the limitations of conventional GC/MS. These strategies are also applicable to most of the techniques for compounds not amenable to GC, which are described in Chapter 6.

One strategy for increased analyte selectivity, lower detection limits, and molecular weight determination is the selection of a chemical or related ionization technique in place of EI. Several of these techniques are described and they sometimes produce abundant ions from target analytes as a function of composition and structure and few, if any, ions from interferences and chemical background. With a selective ionization technique it is often possible to determine one member of a group of coeluting substances and minimize or eliminate altogether the signals from the other compounds. These ionization techniques are also sometimes effective in lowering detection limits and often provide information to aid in molecular weight determination.

High-resolution mass spectrometry (HRMS) and tandem MS can provide increased analyte selectivity, information about ion composition, lower detection limits, and information to aid structure determinations. These techniques are especially helpful with samples that contain analytes at concentrations similar to or much less than the concentrations of coeluting substances and/or general chemical background. Selectivity is enhanced with these techniques by separating and rejecting from detection ions that are interferences. In this process some analyte ions may also be rejected and there is usually a loss in analyte signal as selectivity is increased. The loss in signal is often compensated by a significant reduction in chemical noise, and the signal/noise for the analyte may actually be increased substantially. Exact mass measurements with HRMS can provide information about ion composition, which is crucial in determining structure. Tandem MS can provide details of ion fragmentations which support interpretation of spectra and structure determination.

Chemical derivatization is a strategy to increase selectivity, lower detection limits, aid in molecular weight determination, and support structure determination. Derivatization is perhaps more frequently considered as a technique to increase the vapor pressure and thermal stability of compounds not amenable to GC and it is considered for that purpose in Chapter 6. However, by appropriate selection of the derivative, the other objectives may also be achieved.

Chemical and Related Ionization Techniques

In EI, gas-phase $M^{+\cdot}$ ions are formed by the interaction of gas-phase molecules with nominal 70-eV electrons in a high vacuum, typically 10^{-6} Torr or less. Under these high vacuum conditions in magnetic deflection, linear quadrupole, and time-of-flight spectrometers (ion-beam type instruments), the $M^{+\cdot}$ ions rarely interact with one another or with neutral molecules or fragments that are present in very low concentrations in the ion source. In beam type instruments, $M^{+\cdot}$ ions are continuously formed in the ion source as long as sample is present. Depending on their internal energies, some $M^{+\cdot}$ ions may spontaneously and rapidly decompose, which is called a unimolecular reaction, to give product ions that may fragment further. In magnetic deflection and quadrupole instruments this mixture of ions is continuously extracted from the ion source and focused into an ion beam that is separated according to m/z values to give the conventional EI mass spectrum. Time-of-flight instruments extract ions in batches from the ion source and separate these ions according to their m/z values.

Some unimolecular decompositions occur while the ions are in transit from the ion source to the separation element and the detector. Ions that decompose in this way are called metastable ions, and the products of decomposition of metastable ions can be measured under some conditions and provide valuable information about fragmentation mechanisms.[1–3] Under similar conditions of temperature, pressure, and so on, $M^{+\cdot}$ decompositions in a high-vacuum ion source are reasonably reproducible, and similar EI mass spectra can be produced with different types of instruments in different laboratories.

High Pressure Ion Sources

In the 1950s, experiments were conducted with ion sources constructed so that a higher gas pressure, typically 0.2–2 Torr, could be maintained in the source while a high vacuum was maintained in the other parts of the mass spectrometer.[4] When methane was introduced at the higher pressures and bombarded with nominal 70-eV electrons, the most abundant ions initially formed from methane (Reactions 5.1 and 5.2) were found to react rapidly with abundant neutral methane molecules to give product ions (Reactions 5.3 and 5.4) which did not react further with methane.[4–6] Reactions 5.3 and 5.4 are called ion–molecule reactions.

$$CH_4 + e^- \longrightarrow CH_4^{+\cdot} + 2e^- \qquad \text{(Rx 5.1)}$$

$$CH_4^{+\cdot} \longrightarrow CH_3^+ + H^\cdot \qquad \text{(Rx 5.2)}$$

$$CH_4^{+\cdot} + CH_4 \longrightarrow CH_5^+ + {}^\cdot CH_3 \qquad \text{(Rx 5.3)}$$

$$CH_3^+ + CH_4 \longrightarrow C_2H_5^+ + H_2 \qquad \text{(Rx 5.4)}$$

In the mid-1960s it was discovered that if low concentrations of analyte molecules were introduced into an ion source containing ionized methane in the 0.2–2 Torr range, the most abundant CH_5^+ and $C_2H_5^+$ ions would react rapidly with the analyte molecules to form several kinds of analyte ions that could be separated and detected by MS.[7] This new ionization technique was called chemical ionization (CI) and the CH_5^+ and $C_2H_5^+$ ions were called reagent ions. Because of the much higher concentration of methane, EI of analyte molecules was not observed. The CI mass spectra were different from EI spectra and were often simpler and easier to interpret, but they provided information complementary to EI spectra.

Subsequently, experiments were conducted at atmospheric pressure, using much more energetic electrons to ionize the reagent gases. Under these conditions the frequency of collisions among molecules and ions increases enormously, the rates of ion–molecule reactions that produce chemical ionization are greatly increased, and the system rapidly attains a thermodynamic equilibrium of reactants and products.[8] Ions are drawn from this atmospheric plasma into the high vacuum of a mass spectrometer through a pinhole orifice, and atmospheric pressure ionization (API) spectra are measured that have many of the characteristics of CI spectra.

Finally, in the 1980s it was discovered that CI could be accomplished under normal conditions with ion-storage type mass spectrometers such as the Fourier transform mass spectrometer[9] and the Paul ion trap.[10] With these ion-storage instruments, reagent ions are formed at much lower pressures of reagent gas, that is, $\sim 10^{-6}$ Torr, held in the trap, and are allowed to react for periods in the range 1–1000 ms with sample molecules. The CI spectra measured with ion-storage spectrometers are often very similar to those obtained under classical conditions, although some differences such as fewer adduct ions have been observed.[10]

During the last 30 years the CI and API spectra of tens of thousands of compounds have been measured and these techniques are routinely used for chemical analysis and to provide molecular weight information. Following the CI discovery, many reagent gases besides methane were investigated and numerous ion–molecule reactions, analogous to Reactions 5.3 and 5.4, were studied in standard elevated pressure and atmospheric pressure ion sources and ion-storage spectrometers. The formation of negative ions under CI conditions was investigated and several negative ion techniques were found to be of great value in environmental and other studies. Several reviews and books which describe CI processes and spectra have been published and these should be consulted for additional background, history, theory, and applications of CI.[1,3,5,6,8,11] Most GC/MS and other mass spectrometer system manufacturers offer some form of CI as a standard capability or an optional accessory.

There are many applications of CI and API, but this chapter is concerned with applications in GC/MS. Other actual or potential applications of CI and API are described in Chapters 6 and 8. Three important GC/MS applications are:

- Molecular weights of analytes cannot always be determined with EI because $M^{+\cdot}$ ions are not always observed. This is a significant problem with nontarget analytes whose structures are unknown. However, CI is usually a lower energy or softer ionization process and ions are frequently observed with m/z values close to the $M^{+\cdot}$ ion. These are often called pseudo $M^{+\cdot}$ ions and they allow determination of molecular weight.
- Enhanced analyte selectivity can be achieved because some analytes are much more efficiently ionized than others by some reagent ions or ionizing conditions. This selectivity may preclude the need for SIM data acquisition, or provide an additional boost to the selectivity of SIM.
- Increased analyte sensitivity may also be observed for some substances because of higher ionization efficiencies with CI than EI and reduced fragmentation with CI.

Chemical Ionization—Positive Ions

The first part of this section is concerned with standard CI in a beam type mass spectrometer with an ion source designed to maintain a 0.2–2 Torr pressure of reagent gas. The actual optimum ion source pressure for CI in beam instruments varies but is usually in this range. In standard CI the reagent gas is ionized with nominal 70-eV electrons continuously emitted from a filament, and the ions are separated by m/z and detected in the beam instrument. Many of the characteristics of this form of CI are also observed in API and in CI at normal pressures in ion-storage mass spectrometers. Some special features of API are discussed near the end of the section.

When analyte molecules (M) are introduced at low concentrations into a 0.2–2 Torr pressure of ionized methane, they react rapidly with the predominate CH_5^+ and $C_2H_5^+$ reagent ions to form several kinds of analyte ions that can be separated and detected by MS. The main types of reactions observed are:

$$CH_5^+ + M \longrightarrow MH^+ + CH_4 \qquad \text{(Rx 5.5)}$$

$$C_2H_5^+ + MH \longrightarrow (M-1)^+ + C_2H_6 \qquad \text{(Rx 5.6)}$$

$$C_2H_5^+ + M \longrightarrow MC_2H_5^+ \qquad \text{(Rx 5.7)}$$

Reaction 5.5 is the transfer of a proton from a reagent ion to the molecule M to form an even-electron $(M + 1)^+$ ion and methane. Reaction 5.6 is the abstraction of a hydride ion from a neutral molecule by a reagent ion to form an even-electron $(M - 1)^+$ ion and ethane. Reaction 5.7 is the formation of an adduct ion by the combination of a reagent ion and a neutral molecule. These product ions are routinely observed in CI spectra, and the MH^+ and $(M - 1)^+$ ions are often called pseudo $M^{+\cdot}$ ions. The CI reactions are less energetic than those of conventional 70-eV EI, and compounds which do not produce stable $M^{+\cdot}$ ions by EI often form stable MH^+, $(M - 1)^+$, and adduct ions with CI.

Therefore, CI is a valuable tool for determining molecular weight when $M^{+\cdot}$ ions are not observed in EI spectra, but with CI the $M^{+\cdot}$ ions are not directly observed.

In addition to methane, the most common reagent gases used in CI are isobutane, ammonia, water, and mixtures of these. As with methane, EI at 70-eV or higher is used to create reagent ions in the reagent gas at about 0.2–2 Torr in the ion source. Isobutane produces almost exclusively the *tert*-butyl ion $C_4H_9^+$, ammonia the ammonium ion and solvated ammonium ions, and water generates solvated protons. These reagent ions react with analyte molecules to give product ions analogous to those shown in Reactions 5.5–5.7. However, the product ions produced and their relative abundances depend on several factors, which are discussed in the following paragraphs.

The proton affinities (PAs) of the conjugate base of the reagent ion and of the analyte determine whether a proton transfer reaction will occur (Reaction 5.5). Proton affinities are thermodynamic quantities expressed in kcal/mol that measure enthalpy changes on protonation.[1] Proton affinities of the conjugate bases of several important reagent ions and PA ranges for representative analytes and some classes of compounds are listed in Table 5.1.[12] In general, proton transfers will occur from reagent ions whose conjugate bases have smaller PAs to analytes with larger PAs. In Table 5.1, proton transfers will occur from protonated substances whose conjugate bases are positioned higher in the table to unprotonated substances positioned lower in the table.

Substances with larger PAs are typically compounds containing atoms or functional groups with unshared electron pairs or unsaturation that can readily accept a proton. A large difference between the PA of the conjugate base of the reagent ion and the PA of the analyte molecule gives a strongly exothermic proton transfer reaction, and these molecules often produce MH^+ ions. However, the MH^+ ion abundance will be reduced by fragmentation of MH^+, which is caused by transfer of excess energy to MH^+ in the exothermic

Table 5.1 Proton affinity ranges for representative analytes and proton affinities of the conjugate bases of several important reagents ions

Proton affinity (kcal/mol)	Representative analytes and conjugate bases
129.9	Methane (conjugate base of CH_5^+)
130–165	Fluoromethanes except CF_4, ethane, propane, CO_2, N_2O, NO_2, SO_3, HCl
162.6	Ethene (conjugate base of $C_2H_5^+$)
165.2	Water (conjugate base of H_3O^+)
165–175	Chloroethane, bromoethane, formaldehyde, HCN, trifluoroacetic acid
175–200	Alcohols, aldehydes, nitriles, benzene, toluene, propene, chlorobenzene
191.7	Isobutene (conjugate base of *tert*-$C_4H_9^+$)
204.1	Ammonia (conjugate base of NH_4^+)
200–225	Ketones, ethers, esters, alkylated benzenes, dienes
225–250	Amines, amine oxides, amides, *N*-heterocyclics, phosphines, other bases

Source: Data from Ref. 12.

reaction. For example, the CH_5^+ ion transfers protons to many organic molecules with PAs in the 185–235 kcal/mol range, but the resulting MH^+ ions fragment somewhat because of the exothermic reaction.

On the other hand, the conjugate base of the *tert*-$C_4H_9^+$ ion is isobutene, which has a PA of 191.7. This is roughly in the middle of the range of PAs of many organic molecules. Therefore, proton transfers in isobutane CI are less exothermic and little fragmentation of the MH^+ ions is observed. Because ammonia has a high PA, proton transfers from ammonium ions to many organic compounds are precluded. Therefore, ammonia CI is selective for the more basic nitrogen and phosphorus compounds with PAs larger than that of ammonia. An application of this selectivity is the identification of amines and other nitrogen-containing compounds in a complex fraction of a coal-derived mineral oil with ammonia GC/CI/MS.[13] These identifications would have been much more difficult or precluded with conventional GC/MS.

The hydride affinities (HAs) of the reagent ion and the $(M-1)^+$ ion determine whether a hydride transfer reaction will occur (Reaction 5.6). The HAs of CH_5^+ and $C_2H_5^+$ are both 272 kcal/mol, and the HA of the *tert*-$C_4H_9^+$ ion is 232 kcal/mol.[5] Therefore, the two reagent ions from methane will give more exothermic hydride ion transfers from analyte molecules than will the reagent ion from isobutane. Also, the excess energy from the hydride transfers to the methane reagent ions will cause more fragmentation of the $(M-1)^+$ ions than hydride transfers from analyte molecules to the *tert*-$C_4H_9^+$ ion. Generally, hydride transfers will occur from analytes which do not have electron-withdrawing atoms or functional groups, for example, aliphatic and alicyclic hydrocarbons, to reagent ions with high HAs.

Adduct ions (Reaction 5.7) are observed in CI spectra and their presence can reduce the abundances of other ions in the CI spectrum. The adduct ions are sometimes considered a nuisance because they can give a false indication of the molecular weight. Adduct ions can be recognized by checking the m/z differences between the highest m/z ions in the spectrum. These differences, for example, m/z 17 (CH_5^+), 18 (NH_4^+), 29 ($C_2H_5^+$), and 57 ($C_4H_9^+$) are characteristic of the reagent ions and the reagent gas and can serve as a useful confirmation of the m/z of the $M^{+\cdot}$.

By selection of an appropriate reagent gas, it is possible to achieve considerable analytical selectivity for a specific analyte or a group of similar analytes, enhance the formation of particular ions, and obtain information about the molecular weights of the substances. Many reagent gases besides those mentioned in this section have been investigated for selectivity and other special purposes. However, selectivity can be impacted by other variables that are discussed in the next section.

Other Variables That Affect CI Spectra. The temperature of the ion source and of the reagent gas can have a significant effect on CI spectra, which are often strongly temperature dependent.[14] Increased temperatures have the effect of reducing the abundances of adduct ions and increasing fragmentation, especially with isobutane. Similarly, the pressure of the reagent gas in the ion source can have a significant effect on the observed CI spectra.[5,15] During CI with elevated pressures of hydrocarbon gases, decomposition products accumulate on the ion source walls, ion focusing lenses, ceramic insulators, slits, quadrupole rods, and other components, and these carbonaceous deposits eventually degrade spectrometer performance, especially m/z resolution and ion transmission. The degree of cleanliness of the ion source is a variable that can affect the quality and character of the observed CI spectra. The maintenance required to clean ion source parts to achieve acceptable CI performance is a definite disadvantage although it is alleviated somewhat with instrument systems that have easily interchangeable ion volumes and other ion source parts.

Finally, the experimental conditions used for CI and the spectra produced can depend on the design of the mass spectrometer. Nearly all of the early research in CI was conducted with magnetic deflection or quadrupole ion-beam type mass spectrometers with ion sources designed to maintain the 0.2–2 Torr pressures needed to produce the reagent ions. As previously discussed, ion-storage type mass spectrometers provided a new dimension to CI by allowing reagent ions and sample molecules to react for relatively long times at normal pressures.[9,10]

There is little incentive to develop databases of positive-ion CI mass spectra analogous to the large collections of nominal 70-eV EI mass spectra. This is probably because many CI spectra contain few, if any, fragment ions, but there are many experimental variables, including reagent gas, temperature, pressure, cleanliness of the ion source, and spectrometer design, which complicate the collection of spectra. Information about CI spectra of individual compounds and classes of compounds is available in the original research literature and some books.[1,3,11] Unfortunately, in some reports, experimental conditions are poorly or incompletely documented and reported CI spectra may be difficult to reproduce with different types or even the same model of mass spectrometer.

Although it is clearly possible to predict the selectivity of various reagent ions for specific compound structures, and to anticipate the kinds of ions that will be formed, it is usually necessary for investigators to determine by experimentation whether CI offers advantages of sensitivity compared to EI. Sensitivity can vary widely and depends on the ionization efficiency of the reagent ion with the particular analyte, coeluting substances, background contamination, the degree of fragmentation of various ions, the design and condition of the mass spectrometer, and other factors. For specific substances sensitivity with CI can be much greater, about the same, or less than with conventional EI.

Charge Exchange Ionization. In addition to hydrogen-containing reagent gases, aprotic gases such as He, Ar, N_2, and CO can be used to induce chemical ionization. Electron ionization of nitrogen at 0.2–2 Torr gives the reagent ion $N_2^{+\cdot}$ (Reaction 5.8), which reacts rapidly with low concentrations of analyte molecules to give $M^{+\cdot}$ ions (Reaction 5.9). The energy released in the recombination of the $N_2^{+\cdot}$ ion with an electron is higher than the ionization potentials of most molecules, and therefore $M^{+\cdot}$ is usually formed with sufficient internal energy to cause some fragmentation similar to that observed in conventional 70-eV EI.[6,16] Reagent ions from He, Ar, CO, and other aprotic gases also ionize analyte molecules, but each has a different recombination energy and deposits more or less energy in the $M^{+\cdot}$ ion. Milder, lower energy charge exchange with less fragmentation of the $M^{+\cdot}$ occurs with the reagent ion $NO^{+\cdot}$, which is formed by EI of nitric oxide. With this reagent ion other reactions also occur including adduct formation, hydride abstraction, and hydroxide abstraction.[6,17]

$$N_2 + e^- \longrightarrow N_2^{+\cdot} + 2e^- \qquad \text{(Rx 5.8)}$$

$$M + N_2^{+\cdot} \longrightarrow M^{+\cdot} + N_2 \qquad \text{(Rx 5.9)}$$

Charge exchange ionization at 0.2–2 Torr with filament electrons has been sparingly used in environmental MS. Simultaneous charge exchange ionization and methane CI of polycyclic aromatic hydrocarbons (PAHs) has been investigated using helium as GC carrier gas and a mixed reagent gas consisting of 5–25% methane in argon.[18] Charge exchange with $Ar^{+\cdot}$ gave the $M^{+\cdot}$ while protonation with the methane reagent ions gave the MH^+ ions. The $MH^+/M^{+\cdot}$ abundance ratio depends on the ion source temperature, ion source pressure, reagent gas composition, cleanliness of the ion source, and the specific PAH. By controlling the four operating parameters the $MH^+/M^{+\cdot}$ abundance ratio can be used to distinguish PAH isomers and calibrate and measure the relative amounts of coeluting chrysene (1,2-benzophenanthrene) and triphenylene (9,10-benzophenanthrene) isomers (Chart 5.1).

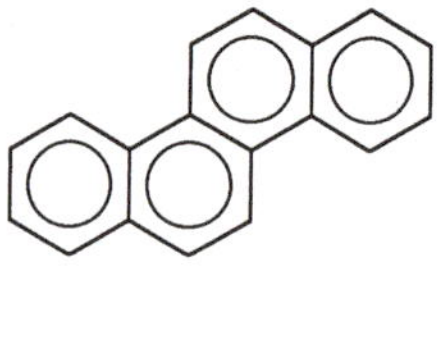

Chrysene 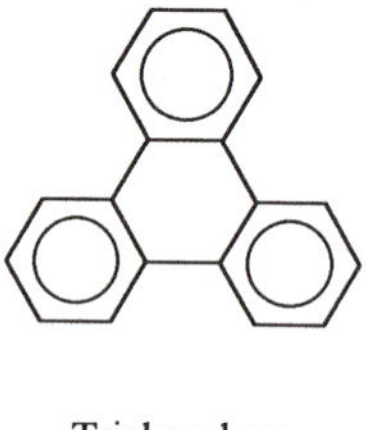 Triphenylene

Chart 5.1

It is noteworthy that 10 years later the same principal investigator selected the same GC column and conventional GC/MS to measure nearly the same group of PAHs in vegetation samples.[19] However, with EI GC/MS the coeluting chrysene and triphenylene isomers were measured together. This decision to use conventional GC/MS instead of the more informative charge exchange/CI method is an example of the realities and practical trade-offs that are often required in a MS laboratory. Most laboratories have at least one, and often several, conventional GC/MS instruments calibrated, tuned, performance-checked, and ready to use. Unless a laboratory routinely uses positive-ion CI, an instrument prepared for CI may not be available. The reconfiguration, tuning, and recalibration of an instrument for CI may require a day or more and was not judged cost-beneficial in the analyses of PAHs in vegetation.[19] Charge exchange ionization is an important ionization process in atmospheric pressure ionization (API), which is discussed in the next section.

Atmospheric Pressure Ionization. The logical extension of CI in an ion source designed to maintain 0.2–2 Torr is ionization at atmospheric pressure. At the much higher pressure, the frequency of collisions among molecules and ions increases enormously, the rates of ion–molecule reactions that cause CI are greatly increased, and the system rapidly attains a thermodynamic equilibrium of reactants and products. Because of the great excess of reagent ions over sample molecules, and the rapid attainment of thermodynamic equilibrium, analyte ionization efficiency is very high and API is potentially much more sensitive than other methods of ionization. This method of ionization also has other advantages, which are presented in subsequent sections of this chapter. However, it is not feasible to generate high-energy electrons at atmospheric pressure with a conventional electrically heated metal filament.

Atmospheric pressure chemical ionization/mass spectrometry (APCI/MS) was first accomplished using very high energy (0.067 MeV) electrons emitted from the radioactive isotope ^{63}Ni to ionize nitrogen gas.[20] The ions formed, including $N_4^{+\cdot}$ and H_3O^+ (from traces of water in the nitrogen), reacted with very small quantities of vaporized analyte molecules by charge exchange and protonation to form $M^{+\cdot}$ and MH^+ ions that were separated and measured in a quadrupole mass spectrometer. The ions were drawn into the mass spectrometer from the API source through a pinhole orifice. Later, a corona discharge needle at 800–1000 V was employed to generate ionizing electrons, and both techniques were found to produce identical APCI mass spectra.[21] Other typical CI reactions, including hydride abstraction and adduct formation, were also observed under API conditions. A short review of the early research, instrumentation, and applications of API MS is available.[8]

The combination of GC and APCI has not been widely used or developed commercially. This is probably because EI and the standard 0.2–2 Torr CI techniques are adequate for almost all GC/MS applications, and APCI requires a significantly different ion source design that cannot be easily incorporated into existing mass spectrometers. However, APCI is widely used for the determination of compounds not amenable to GC (Chapter 6) and for field and continuous measurements (Chapter 8).

Chemical Ionization—Negative Ions

Negative ions can also be produced by CI. Obviously, the separation and detection of negative ions requires a mass spectrometer that is designed for this purpose. General purpose commercial instrument systems manufactured since the mid-1980s usually have this capability, although sometimes it is an optional accessory that may not be available on the lower-cost bench-top instruments. Negative-ion MS has been known since the earliest days of MS, but the discovery of practical techniques to generate abundant negative ions in the gas phase did not occur until the early 1970s. Analytes with several or more strong electron-withdrawing groups, for example, NO_2 , Cl, and F, or with a delocalized π-electron system that can support the negative charge, are most likely to form negative ions in high abundances.

Under conventional high vacuum conditions, that is, $\sim 10^{-6}$ Torr or less, and with nominal 70-eV EI, most organic compounds produce few negative ions compared to the production of positive ions. If the analyte molecule contains electronegative elements or groups, reasonable abundances of low *m/z* fragment anions such as Cl^-, F^-, $NO_2{}^-$, CN^-, and OH^- are observed. The reaction that produces these anions is ion-pair formation (Reaction 5.10), which gives an equal number of positive and negative ions.[22] Higher *m/z* negative ions and molecular anions (M^-) are observed with only a few classes of compounds such as highly fluorinated compounds, polynitro compounds, and some boron, phosphorus, sulfur, and metal-containing compounds. Because of the general absence of M^- ions and higher *m/z* anions which reflect composition and structure, negative-ion MS under conventional EI conditions is rarely used.

$$\text{M–X} + e^-(>\sim 20\,\text{eV}) \longrightarrow \text{M}^+ + \text{X}^- + e^- \qquad \text{(Rx 5.10)}$$

Electron Capture Ionization (ECI). If the ionizing electron energy is reduced to the range 0–15 eV, two additional negative ion producing reactions are observed (Reactions 5.11 and 5.12). Reaction 5.11 is called resonance capture, resonance electron capture, or simply electron capture, and Reaction 5.12 is called dissociative resonance capture.[22,23] Electron capture generally requires very low energy electrons (~0–1 eV), which are called thermal electrons. Reaction 5.11 is sometimes described as a *soft* ionization because at temperatures of about 100 °C or below fragmentation of $MX^{-\cdot}$ is often minimal. Reaction 5.12 is described as a *hard* dissociative electron attachment.[24] Reaction 5.11 or 5.12 or both may be observed depending on the composition and structure of the analyte, the energy distribution of the electrons in the ion source, the temperature, and other factors. Experimental conditions are often selected to minimize or avoid Reaction 5.12 because it usually produces only low *m/z* fragment anions, such as Cl^-, which do not provide much information about the analyte molecule. However, conditions conducive to Reaction 5.12 can be used to an advantage. The chromatography peaks in a complex sample that contain chlorinated compounds can be determined by SIM of Cl^- ions at *m/z* 35 and 37 with an ion source temperature of 250 °C.[24]

$$\text{M–X} + e^-(\sim 0\text{–}1\,\text{eV}) \longrightarrow \text{MX}^{-\cdot} \qquad \text{(Rx 5.11)}$$

$$\text{M–X} + e^-(\sim 0\text{–}15\,\text{eV}) \longrightarrow \text{M}^{\cdot} + \text{X}^- \qquad \text{(Rx 5.12)}$$

The abundant thermal electrons needed for electron capture cannot be produced with a conventional, electrically heated metal filament in a high vacuum. The ECI process was made readily accessible by the important discovery that elevated gas pressures in an ion source led to the formation of secondary electrons with the low energies required for electron capture.[25,26] Large populations of secondary electrons are produced during the ionization of some reagent gases at 0.2–2 Torr by high-energy electrons (Reaction 5.1).[22] The distribution of energies of the secondary electrons includes thermal energy levels that are produced by the primary ionization process itself and by multiple collisions of electrons with neutral reagent gas molecules. Therefore, ECI is almost always observed with an elevated pressure CI ion source or at atmospheric pressure.[8,23] With ion-storage type mass spectrometers, electron capture can be accomplished at normal lower pressures.[27] Although some important progress has been made in recent years to develop instrumentation to produce thermal electrons without a moderating reagent gas, these instruments are not yet commercially available.[28–30]

Electron capture ionization (Reaction 5.11) is the most important negative-ion technique used in chemical analysis.[23] High purity methane or isobutane are the standard reagent gases used to generate the plasma of thermal electrons needed for electron capture. This ionization technique has the advantages of simplicity, high sensitivity for some analytes, high selectivity, and often little or no fragmentation of the $M^{-\cdot}$ ion, which allows determination of molecular weight.

The reproducibility of electron capture negative-ion mass spectra within the same instrument and among different types of instruments has been studied in considerable detail.[31–36] Among the many instrumental variables that affect the ions formed and their abundances are the mass spectrometer design, the distribution of secondary electron energies, the filament emission current, the temperature of the plasma in the ion source, the charge density of the plasma in the ion source, the ion source pressure, the bulk reagent gas composition, the cleanliness of the ion source, trace impurities in the reagent gas, and even the sample concentration. Generally, $M^{-\cdot}$ ions are subject to greater fragmentation at temperatures above about 100 °C. A high purity reagent gas is essential because trace quantities of oxygen, water, or other substances in the reagent gas can form reagent ions (Reaction 5.15) that react with the analyte to reduce the abundance of $M^{-\cdot}$ ions and increase the abundances of other ions (Reaction 5.16). The concentration dependence of the abundances of some ions and other troublesome variables have inhibited the use of ECI in some studies.

In a survey of the residues of 21 pesticides and representative polychlorinated biphenyls in fish, conventional GC/MS was used instead of the more selective and often more sensitive ECI, in part, because the former technique has well established quantitative capabilities as described in Chapters 2 and 4.[37] On the other hand, ECI was selected by other investigators for the determination of similar analytes in sediment samples.[38] Classes of compounds that are particularly amenable to ECI include a variety of brominated compounds, polychlorinated aromatic hydrocarbons, polychlorinated hydrocarbon pesticides, polynitro compounds, and phosphorus-containing pesticides.[23] Some target analytes that are determined using ECI are noted in Chapter 4. A compilation of electron capture negative-ion mass spectra of environmental contaminants and related compounds has been published.[39]

Ion–Molecule Reactions. Besides electron capture and dissociative resonance capture, negative ions can also be produced by ion–molecule reactions in an ion source operating at 0.2–2 Torr,[40] at atmospheric pressure,[8] or at lower pressures in an ion-storage mass spectrometer.[27] A variety of reagent gases and mixtures of gases have been used to generate negative ions, and a number of ion–molecule reactions have been discovered.[8,11,40]

Electron ionization of methylene chloride at 1 Torr produces abundant chloride ions (Reaction 5.13) which attach to various analytes including carboxylic acids, amines, phenols, and amides to produce MCl^- ions (Reaction 5.14).[26]

$$CH_2Cl_2 + e^- \longrightarrow CH_2Cl^{\cdot} + Cl^- \qquad \text{(Rx 5.13)}$$

$$M + Cl^- \longrightarrow MCl^- \qquad \text{(Rx 5.14)}$$

The chloride attachment reaction is rather mild with little energy released, which is in contrast to the often more exothermic protonation reaction (Reaction 5.5). Therefore, fragmentation of MCl^- is minimal and some electrophilic analytes are detected with very high sensitivity and selectivity. With relatively large quantities, that is, 10–100 μg, of highly chlorinated analytes, for example, the polychlorinated diphenylethane type pesticides, the methylene chloride can be omitted as a reagent gas. The Cl^- reagent ions are generated by a dissociative capture reaction of the analyte (Reaction 5.12) with methane or isobutane serving as the reagent gas to supply the secondary electrons.[41] This process is called self-chemical ionization (self-CI) because the reagent ions are fragments from the decomposition of analyte ions. An alternative mechanism in which the chloride ion is transferred from the neutral analyte molecule to the molecular anion has been suggested.[33] For general purpose use, especially trace analyses at the low nanogram level and below, a mixture of 93% methane or isobutane, 5% methylene chloride or methyl chloride, and 2% oxygen has been recommended as the reagent gas for chloride attachment CI.[40]

Oxygen diluted with nitrogen or other gases absorbs electrons to form the superoxide reagent ion (Reaction 5.15).[42] The superoxide ion can attach to electrophilic molecules (Reaction 5.16) to form reactive intermediates[42] or observable adducts.[43] The use of oxygen in higher concentrations is constrained by its high oxidizing power which significantly shortens filament lifetime when filaments are used to generate ionizing electrons. However, discharge-type ionization devices allow the use of oxygen and other oxidizing agents.[43] Radical anions formed by electron capture (Reaction 5.11) also react with oxygen to form oxygenated species that are diagnostically useful (Reaction 5.17 where X = Cl, for example):[42]

$$O_2 + e^- \longrightarrow O_2^{-\cdot} \qquad \text{(Rx 5.15)}$$

$$M + O_2^{-\cdot} \longrightarrow MO_2^{-\cdot} \qquad \text{(Rx 5.16)}$$

$$M{-}X^{-\cdot} + O_2 \longrightarrow MO^- + XO^{\cdot} \qquad \text{(Rx 5.17)}$$

Hydroxide reagent ion is formed from several combinations of reagent gases including 5% water in methane (Reaction 5.18).[40] and mixtures of methane and nitrous oxide (Reactions 5.19 and 5.20).[44] Electron ionization of the methane provides the secondary electrons shown in Reactions 5.18 and 5.19. Hydroxide ion is a mild but not very selective reagent that is an effective deprotonating agent and produces abundant M-1 negative ions with many analyte molecules (Reaction 5.21):

$$H_2O + e^- \longrightarrow H^{\cdot} + OH^- \qquad \text{(Rx 5.18)}$$

$$N_2O + e^- \longrightarrow N_2 + O^{-\cdot} \qquad \text{(Rx 5.19)}$$

$$MH + O^{-\cdot} \longrightarrow M^{\cdot} + OH^- \qquad \text{(Rx 5.20)}$$

$$MH + OH^- \longrightarrow (M-1)^- + H_2O \qquad \text{(Rx 5.21)}$$

Other ion–molecule reactions are known that produce negative ions, and the reader is referred to other reviews and books for more details and information about these.[1, 3, 8, 11, 40] In environmental MS, the chloride attachment reaction has been used as a sensitive and selective method of screening complex matrices for the presence of electrophilic analytes.[40] The ion–molecule reactions described in this section have been applied to important environmental analytes including halogenated hydrocarbons[45] and polycyclic aromatic hydrocarbons.[46] Nevertheless, in spite of the high sensitivity and selectivity for some compounds, negative ions from ion–molecule reactions have not been widely used in environmental analyses. This may be due, in part, for the need to prepare, or have systems available to handle, mixed reagent gases or the need for quantitative analysis in many environmental studies. Quantitative analyses should be reliable with negative-ion CI, but development of this approach may be inhibited by the need to control many variables in the instrument system.

Nomenclature. Electron capture ionization (ECI) has been historically considered part of the general area of negative-ion chemical ionization (NICI),[47] or as it is frequently called in the scientific literature, negative chemical ionization (NCI).[40] Logically, the term *chemical ionization* should only apply to the ionization of analytes by ion-molecule reactions, and there is a recent trend to distinguish ECI from NICI because electron capture is not an ion–molecule reaction.[23] However, because the thermal secondary electrons required for electron capture are frequently generated using standard CI conditions, and anions formed by electron capture often react with reagent gases in ion–molecule reactions, it is likely that for many years ECI will be considered by many investigators to be a part of NICI. In this book ECI is distinguished from NICI in support of the use of logical nomenclature whenever possible. The term NCI is not used because it implies the reverse reaction of chemical ionization to reform the neutral analyte.

Atmospheric Pressure Ionization (API). Negative ions are also produced with an API ion source.[8] Oxygen, nitrogen, water, and other reagent gases are used with electrons emitted from a corona discharge, or the radioactive isotope ^{63}Ni used in earlier research, to generate thermal electrons and negatively charged reagent ions. Analyte molecules capture these electrons, and the analyte molecular anions and analyte neutral molecules react with the atmospheric gases and negative reagent ions, respectively, as shown in Reactions 5.11–5.21. However, as with positive ions, the frequency of collisions between molecules and ions is greatly increased at atmospheric pressure, the rates of ion–molecule reactions are enhanced, and a state of thermodynamic equilibrium is rapidly attained. Again, because of the great excess of reagent ions over sample molecules, and the rapid attainment of thermodynamic equilibrium, analyte ionization efficiency is very high and API is potentially much more sensitive than other methods of ionization. Although some interesting GC/API/MS results have been reported,[48] the most significant applications of negative-ion API are with condensed-phase sample introduction systems used to determine compounds not amenable to GC (Chapter 6).

High-Resolution Mass Spectrometry

Resolution is the degree of separation between adjacent ions that have different m/z values, usually 1 Da or less. Resolving power is the ability of a mass spectrometer to distinguish between adjacent ions that have different m/z values. High-resolution mass spectrometry (HRMS) with double-focusing instruments consisting of electrostatic and magnetic sectors has been available for many years and was described in detail in John Beynon's classic 1960 book.[49] In 1999 the electrostatic and magnetic sector technology was probably mature as indicated by the general peaking of new instrument developments during the late 1980s.[50] However, the availability of powerful, low-cost computer systems, high-field cryogenic superconducting magnets, and several new ion-focusing techniques have led to the development and commercialization of the Fourier transform mass spectrometer, the reflection time-of-flight mass spectrometer, and the ion-trap mass spectrometer. These instruments are capable of high resolving power and have opened a new era of HRMS.[50] Substantial power, versatility, and ease of use has already been demonstrated for some of these instruments, and the prospects for further significant developments in HRMS, including lower costs, are good.

Detailed information about HRMS is beyond the scope of this book, and readers are referred to other books and reviews for details of the construction, principles of operation, and performance of various types of instruments.[1,49–54] In this section the concepts of theoretical resolving power, resolution, and actual mass spectrometer resolving power are defined and described. The definitions of resolution and resolving power are those recommended by the American Society for Mass Spectrometry.[55,56] Also considered in this section are the capabilities of HRMS to provide high analyte selectivity, very low detection limits, and information about ion composition, which is valuable for molecular structure determination. The definitions of resolution and resolving power and the capabilities of HRMS are illustrated using 2,3,7,8-tetrachlorodibenzo-*p*-dioxin (Figure 5.1). Analytical methods for the determination of the polychlorodibenzo-*p*-dioxins and dibenzofurans with HRMS and SIM are described in Chapter 4. In this section the terms *mass* and *masses* are used frequently for convenience, but *m/z* is implied when the reference is to ions measured in a mass spectrometer.

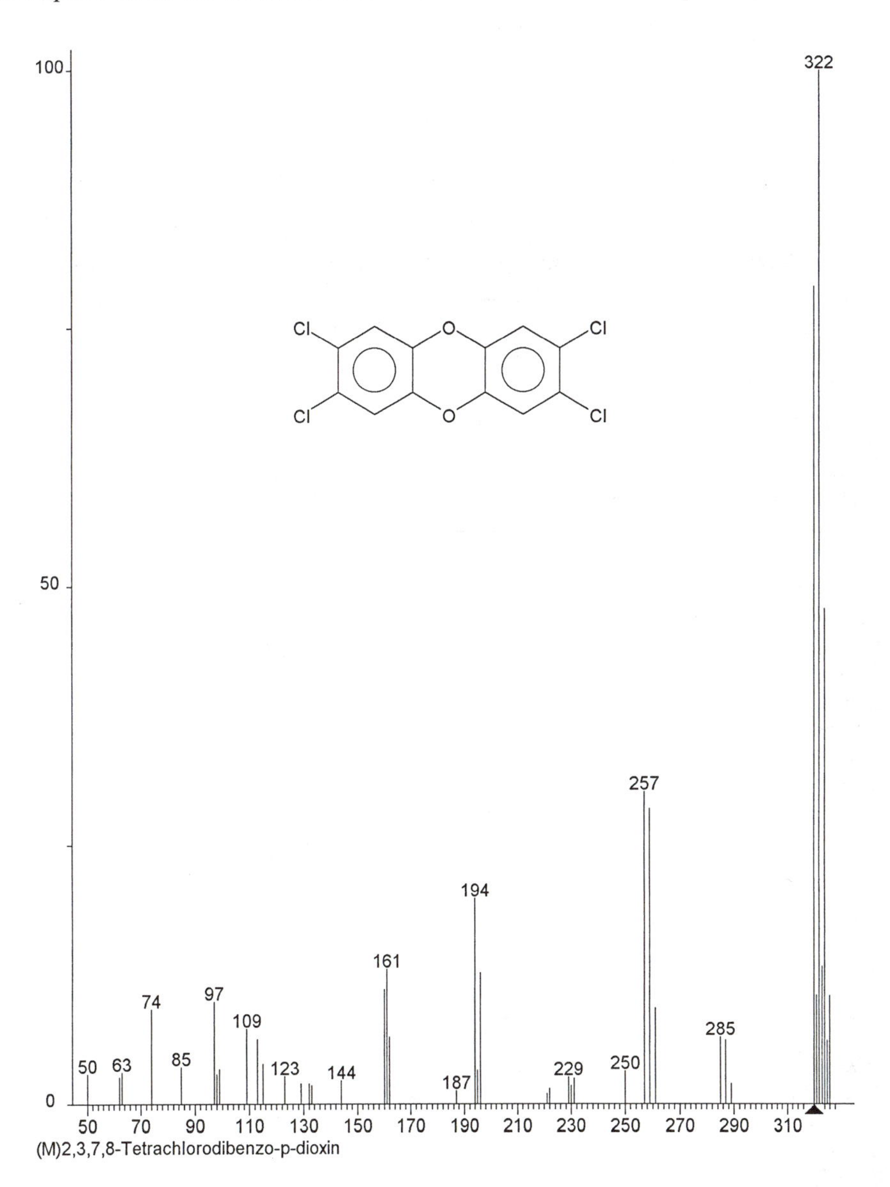

Figure 5.1 The EI mass spectrum and structure of 2,3,7,8-tetrachlorodibenzo-*p*-dioxin. (Reproduced with permission from Ref. 71. Copyright 1992 U.S. Department of Commerce on behalf of the United States.)

Theoretical Resolving Power

The theoretical resolving power (TRP) necessary to separate two ions differing in m/z is defined as $m/\Delta m$ where m is the mean of the two masses and Δm is the difference between the two masses. In order to separate an ion of m/z 600 from an ion of m/z 601, a TRP of about 600 is required and this is commonly called low-resolution MS. In order to separate the $CO^{+\cdot}$ ion from the $N_2^{+\cdot}$ ion, a TRP of about 2500 is required and this is often called medium-resolution MS. The exact masses used in this and other HRMS calculations are listed in Table 5.2.

$$CO^{+\cdot} = 12 + 15.99491 = 27.99491$$
$$N_2^{+\cdot} = (2 \times 14.00307) = 28.00614$$
$$TRP = m\,(28)/\Delta m\,(0.01123) = 2493$$

In order to separate the $M^{+\cdot}$ ion of 2,3,7,8-tetrachlorodibenzo-*p*-dioxin (2378-TCDD) at m/z 320 containing only ^{35}Cl (Figure 5.1) from the $M^{+\cdot}$ ion of 1,1-bis(4-chlorophenyl)-2,2-dichloroethene (4,4′-DDE) at m/z 320 containing two ^{35}Cl and two ^{37}Cl (Figure 5.2), a TRP of about 9000 is required:

Tetrachlorodibenzo-*p*-dioxin ($C_{12}H_4{}^{35}Cl_4O_2$):

Table 5.2 Monoisotopic integer masses, exact masses, and natural abundances of nuclides commonly found in organic compounds

Nuclide	Symbol	Monoisotopic integer mass (Da)	Exact mass (Da)	Natural abundance (%)
Hydrogen	H	1	1.007825	99.985
Deuterium	D	2	2. 01410	0.015
Carbon	C	12 (by definition)	12 (by definition)	98.90
Carbon	C	13	13.003355	1.10
Nitrogen	N	14	14.003074	99.64
Nitrogen	N	15	15.000109	0.36
Oxygen	O	16	15.994915	99.76
Oxygen	O	17	16.999131	0.04
Oxygen	O	18	17.999160	0.20
Fluorine	F	19	18.998403	100
Silicon	Si	28	27.976927	92.23
Silicon	Si	29	28.976495	4.67
Silicon	Si	30	29.973771	3.10
Phosphorus	P	31	30.973762	100
Sulfur	S	32	31.972071	95.02
Sulfur	S	33	32.971458	0.75
Sulfur	S	34	33.967867	4.21
Sulfur	S	36	35.967081	0.02
Chlorine	Cl	35	34.968852	75.77
Chlorine	Cl	37	36.965903	24.23
Bromine	Br	79	78.918336	50.69
Bromine	Br	81	80.916289	49.31
Iodine	I	127	126.904473	100

Source: Data from Ref. 57.

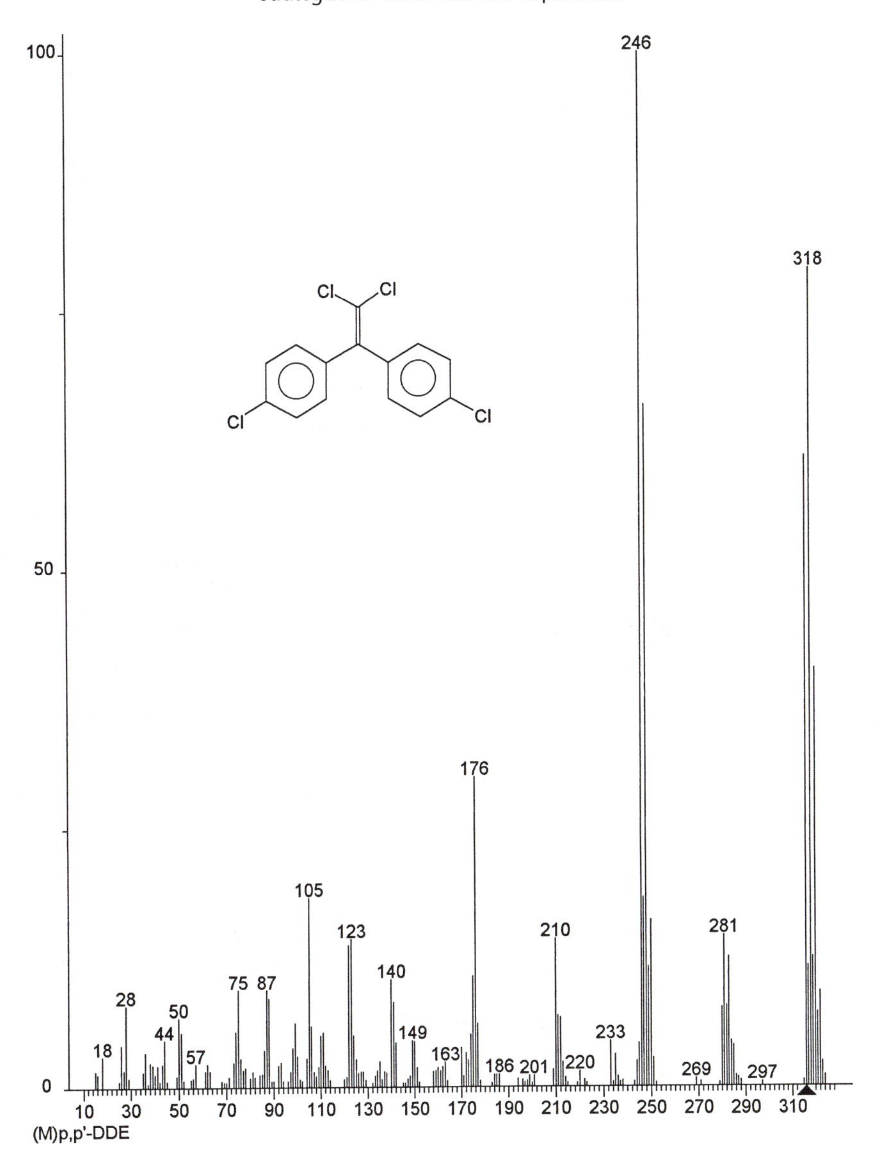

Figure 5.2 The EI mass spectrum and structure of 4,4′-DDE. (Reproduced with permission from Ref. 71. Copyright 1992 U.S. Department of Commerce on behalf of the United States.)

(12 × 12.00000) + (4 × 1.00782) + (4 × 34.96885) + (2 × 15.99491)
= 319.8965
1,1-Bis(chlorophenyl)-2,2-dichloroethene ($C_{14}H_8{}^{35}Cl_2{}^{37}Cl_2$):
(14 × 12.00000) + (8 × 1.00782) + (2 × 34.96885) + (2 × 36.96590)
= 319.93206
TRP = m (320)/Δm (0.03556) = 8999

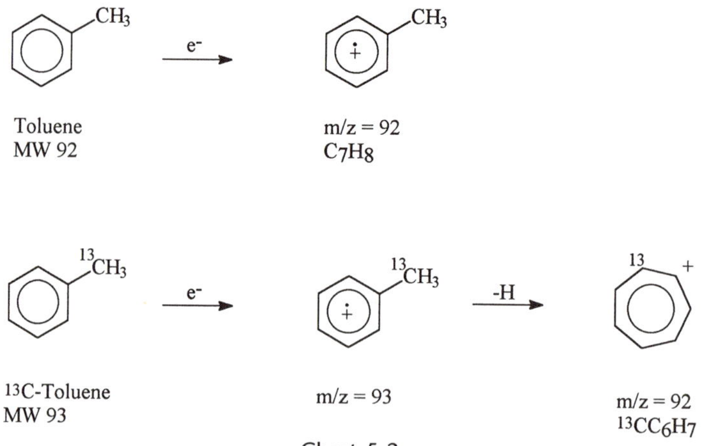

Toluene
MW 92

m/z = 92
C_7H_8

13C-Toluene
MW 93

m/z = 93

m/z = 92
$^{13}CC_6H_7$

Chart 5.2

Similar calculations could be made for an endless number of compositions, some of which require a TRP of 100,000 or more. If the nuclides that 2378-TCDD and 4,4′-DDE have in common, that is, $C_{12}H_4{}^{35}Cl_2$, are subtracted from the respective molecular compositions, the masses of the residuals $^{35}Cl_2O_2$ and $C_2H_4{}^{37}Cl_2$ are revealed as the cause of the Δm (0.03556). This pair of compositions, which have an integer mass of 102, is called a doublet. Any two ion compositions that differ by this doublet composition will have the same Δm (0.03556). If two ions containing this doublet had an integer mass of 600, the TRP required would be about 17,000.

Many common doublets have been identified and cataloged.[49] One example is the ^{12}CH–^{13}C doublet which has a Δm of 0.004468 Da. This doublet is fairly common because fragment ions produced by the loss of a single hydrogen atom will also have a ^{13}C-containing ion with the same integer mass as the ion that lost the hydrogen. For example, toluene (Chart 5.2) forms an abundant $M^{+\cdot}$ ion, but also has an abundant $(M\text{–}H)^+$ ion whose ^{13}C ion has the same integer mass as the $M^{+\cdot}$ ion. A TRP of 20,590 is required to resolve the ^{12}CH–^{13}C doublet in the $M^{+\cdot}$ ion of toluene. The ^{12}CH–^{13}C doublet will be present in many spectra, but is rarely resolved because of the high TRP required (Table 5.3).

Table 5.3 Theoretical resolving power required to separate the CH–^{13}C doublet

Integer mass	Approximate TRP
100	22,400
200	44,500
300	67,000
400	89,500
500	112,000

Resolution

The TRP is an estimate of the resolving power necessary to separate a doublet, but an incomplete measure because it does not specify the degree of separation between the ions, which have slightly different masses. Resolution may be defined as the m/z difference, Δm_x, that exists between two adjacent peaks that are of equal height and shape with a specified amount of overlap indicated by x. The overlap may be measured by the height of the valley between the peaks as a percentage of the height of the peaks, using the same baseline for each measurement. Figure 5.3 is an illustration of two adjacent peaks that are of equal height and shape. The height of the valley is 3 units and the height of the peaks is 30 units, which gives a 10% valley between the peaks. The valley height may be any percentage of the peak heights, but for purposes of comparison the 1, 10, and 50% valley standards are typically used. Since the x in Δm_x designates the amount of overlap, the 10% valley resolution in Figure 5.3 would be indicated as $\Delta m_{10\%\text{V}}$. If the peaks in Figure 5.3 represented the 2378-TCDD and 4,4′-DDE ions discussed earlier in this section, the resolution would be given as $\Delta m_{10\%\text{V}} = 0.03556$ Da. If the peaks in Figure 5.3 differed by 1 Da and were resolved at the baseline or at least 10% valley, this would be commonly called low or unit resolution. A high percentage of all mass spectrometers normally operate at unit resolution and produce low-resolution mass spectra.

While the $\Delta m_{10\%\text{V}}$ and the other standards of resolution are elegant and precise, they are not generally practical measures of resolution in the spectrum of an analyte because it is extremely rare to observe two peaks of exactly the same

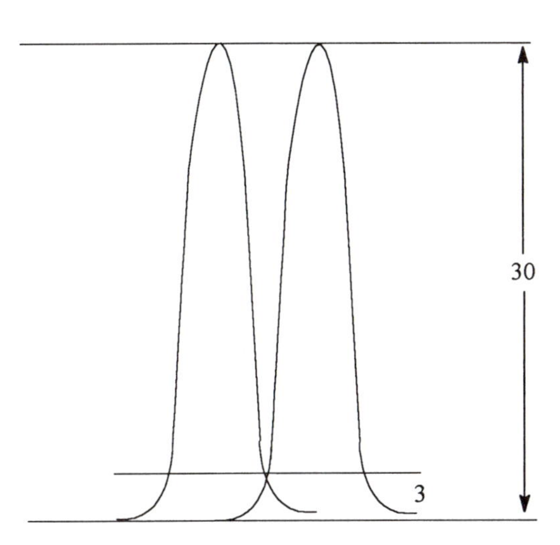

Figure 5.3 An illustration of two adjacent mass peaks that are of equal height and shape with a 10% valley between the peaks.

height and shape resolved to some standard degree such as a 10% valley. When adjusting the resolving power of a spectrometer, as discussed in the next paragraph, it is possible to introduce doublets of the same height and shape into the instrument and this is routinely done, but for most other applications and real-world samples some other estimate of resolution based on a single peak is needed. Figure 5.4 shows a single peak illustrating two commonly used measures of resolution. The peak width (in Da) at 5% of the peak height (PW5%PH) is approximately equal to the 10% valley standard resolution, and the peak width (in Da) at half peak height (PWHPH) is a frequently used estimate of resolution. These estimates are designated as $\Delta m_{PW5\%PH}$ and Δm_{PWHPH}, respectively, and the resolutions displayed in Figure 5.4 are $\Delta m_{PW5\%PH} = 0.008$ Da and $\Delta m_{PWHPH} = 0.004$ Da.

Mass Spectrometer Resolving Power

The actual mass spectrometer resolving power, that is, the ability of a mass spectrometer to distinguish between ions differing slightly in m/z is defined as $m/\Delta m_x$ where m and Δm_x are as defined in the previous sections. The process used to adjust or determine a mass spectrometer's resolving power depends on the type of instrument in use. The details of these procedures are beyond the scope of this book, but several techniques used with double-focusing electrostatic and magnetic sector instruments are briefly described because this type of instrument is used for many environmental and other applications. The resol-

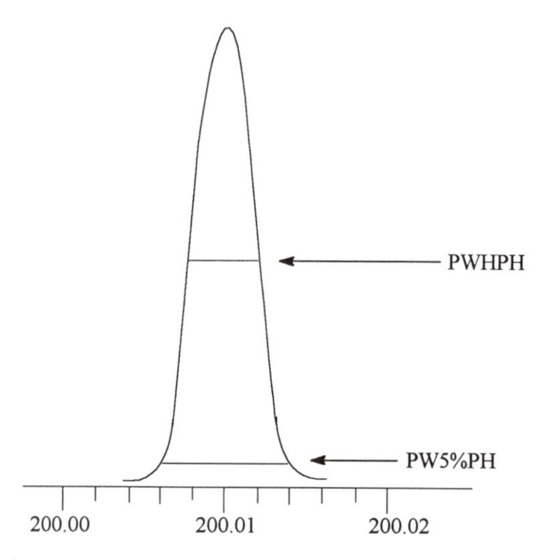

Figure 5.4 A single mass peak illustrating two commonly used measures of resolution. The peak width at 5% of the peak height (PW5%PH) is approximately equivalent to the 10% valley definition of resolution, and the peak width at half peak height (PWHPH) is a frequently used estimate of resolution.

ving power of other types of high-resolution mass spectrometers, such as the Fourier transform (FT) and ion trap (IT) instruments, depends on operating parameters including the length of time that a transient is recorded (FT), the analyzer pressure (FT), the scan speed (IT), and others, but the basic definitions of resolution and resolving power are the same. The PWHPH is often used to estimate the resolving power of these instruments.[50, 52] Environmental and other applications using the newer types of high-resolution mass spectrometers are rare, but should increase in the future.

The resolving power of a double-focusing electrostatic and magnetic sector instrument can be adjusted by introducing a doublet into the ion source from a heated reservoir inlet system. The inlet maintains a constant pressure of the doublet in the ion source over a period of several minutes or more. For example, a mixture of benzene and pyridine provides a doublet at *m/z* 79 caused by a single ^{13}C in the $M^{+\cdot}$ of benzene and the $M^{+\cdot}$ of pyridine (^{13}CH–N doublet). The composition of the mixture in the reservoir can be adjusted so that the doublet ions are of equal abundance. The doublet ion peak profiles are displayed on the monitor of the spectrometer data system, and the spectrometer ion source potentials, slit widths, and other operating parameters are adjusted until the doublet peaks have the same shape and the desired resolution, for example,, a 10% valley. The TRP required to separate the benzene–pyridine ^{13}CH–N doublet is 9746. After tuning the spectrometer as described above, the resolution, $\Delta m_{10\%V}$, will be 0.00811 and the spectrometer resolving power, $m/\Delta m_{10\%V}$, will be set at 9746. If the peaks in Figure 5.3 represented the 2378-TCDD and 4,4′-DDE ions discussed earlier, the resolution, $\Delta m_{10\%V}$, would be 0.03556 Da and the mass spectrometer resolving power, $m/\Delta m_{10\%V}$, would be about 9000.

The procedure described in the previous paragraph, while not uncommon, requires the availability of appropriate doublets, a suitable inlet system, and a method of displaying the resolved doublet in real time while spectrometer tuning adjustments are made. For convenience, especially when such doublets and/or inlet systems are not available, a single mass peak from a mass calibrant ion is displayed on the data system and the PW5%PH or the PWHPH is used to measure and adjust the spectrometer's resolving power. The resolutions displayed in Figure 5.4, $\Delta m_{PW5\%PH} = 0.008$ Da and $\Delta m_{PWHPH} = 0.004$ Da, correspond to the spectrometer resolving powers $m/\Delta m_{PW5\%PH} = 25{,}000$ and $m/\Delta m_{PWHPH} = 50{,}000$, respectively, at mass 200. Top of the line vintage 1980s double-focusing electrostatic and magnetic sector instruments can achieve resolving powers of 75,000–150,000.[50] More recently, reflection time-of-flight mass spectrometers have been reported to have resolving powers of 10,000–15,000.[58] A resolving power of 10,000 has been reported for commercial bench-top ion-trap GC/MS systems operating with a nonstandard slow scan speed.[59] Fourier transform spectrometers can routinely achieve resolving powers of 200,000[60] and a resolving power of 1. 5×10^6 has been reported.[52]

The adjustment of a double-focusing electrostatic and magnetic sector mass spectrometer's resolving power is done at constant magnetic field strength. This is sometimes called the static resolving power. The singlet or doublet ion peak profiles are scanned and displayed by slight changes in the ion source accelerating potential and/or the electrostatic sector potential. Once the adjustment of static resolving power is complete, it should remain constant during subsequent data acquisition provided that conditions are the same or similar, that is, constant magnetic field strength with focusing of selected ions by modest changes ($\sim$10%) in the accelerating and electrostatic potentials. The data acquisition technique used for the adjustment of static resolving power is very similar to the standard technique for SIM with a double-focusing sector instrument. Therefore, the double-focusing sector mass spectrometer is readily tuned to a specific resolving power, and the same conditions are used for SIM data acquisition.

If the spectrometer is subsequently operated by scanning the magnetic field to repetitively observe complete mass spectra, the resolving power may not be stable, and will very likely be less than it was during the tuning operation. Many environmental and other applications of HRMS use SIM data acquisition because it provides the stability necessary to maintain resolving power, which is necessary for enhanced selectivity. This technique also provides the signal/noise needed for typical low concentration measurements and the capability for exact mass measurements that are discussed in subsequent sections.

Enhanced Analyte Selectivity

The HRMS–SIM technique is used to provide high selectivity for target analytes. This technique is especially valuable when analytes are present in concentrations similar to or below concentrations of coeluting compounds and/or chemical noise with the same integer but different exact masses. The polychlorodibenzo-*p*-dioxins and dibenzofurans, for example, 2378-TCDD (Figure 5.1), are found in very low concentrations in many types of environmental samples that potentially contain coeluting interferences. Chart 5.3 shows the structures of polychloro $M^{+\cdot}$ and fragment ions that are potential interferences in the determination of 2378-TCDD or any tetrachlorodibenzo-*p*-dioxin. The $M^{+\cdot}$ ions at m/z 320 and 322 are typically used to monitor 2378-TCDD and other tetrachlorodibenzo-*p*-dioxins (Chart 5.3). Potential interferences which have $M^{+\cdot}$ or fragment ions at m/z 320 and 322 include the 24 possible heptachlorobiphenyl isomers, 4,4′-DDE, and numerous isomers of methoxytetrachlorobiphenyl and tetrachlorophenyl benzyl ether (Chart 5.3).

Contemporary analytical methods for 2378-TCDD and related compounds, which are summarized in Chapter 4, specify a variety of post-extraction fractionation procedures to separate interferences from the target analytes. However, these procedures are not always 100% efficient with regard to every possible interference, and fractionation procedures may not be available for some inter-

	SIM Ions (m/z)	Theoretical Resolving Power Required
$M^{+\cdot}$ ion of 2,3,7,8-tetrachlorodibenzo-*p*-dioxin (2378-TCDD)	319.8965 321.8936	
Fragment ions from the loss of Cl_2 from the $M^{+\cdot}$ ions of any of 24 possible heptachlorobiphenyl isomers	321.8677	12,440
$M^{+\cdot}$ ion of 1,1-bis(4-chlorophenyl)-2,2-dichloroethene (4,4'-DDE)	319.9321 321.9292	9,000 9,000
$M^{+\cdot}$ ions of many possible isomers of methoxytetrachlorobiphenyl	319.9329	8,800
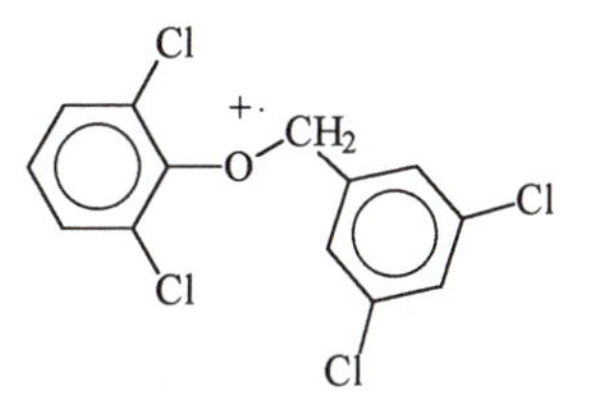 $M^{+\cdot}$ ions of many possible isomers of tetrachlorophenyl benzyl ether	319.9329	8,800

Chart 5.3

ferences. Furthermore, it is desirable to minimize the number of fractionation procedures because inadvertent losses of analytes can occur during sample extract manipulations and these procedures increase the time and cost of analyses. Therefore, HRMS–SIM analytical methods are often used to supplement fractionation and these typically specify a mass spectrometer resolving power of 10,000, which is adequate to separate most interferences (Chart 5.3). The heptachlorobiphenyls produce fragment ions which require a TRP of 12,440 (Chart 5.3), but these are well-known interferences and fractionation procedures are designed and tested to ensure separation of them from the target analytes. The USEPA regulatory analytical methods that use HRMS–SIM for this series of compounds are summarized in Chapter 4.

Sometimes the nature of the interference or the knowledge that an interference even exists is not known until HRMS measurements are made. For example, a HRMS–SIM technique was used to distinguish and measure 1,2,3,7,8-pentachlorodibenzo-*p*-dioxin (PCDD) at 550 parts per 10^{15} in an extract of human milk despite a coeluting hexachlorobiphenyl.[61] The PCDD has an ion used in the analysis at *m/z* 357 and the hexachlorobiphenyl also has an ion at *m/z* 357; these differ in mass by 0.00725 Da. The presence of the hexachlorobiphenyl was not known until the HRMS measurement was made. Another example is the recognition of the pesticide mirex (Chart 4.1) coeluting with bis(3,4-dichlorophenyl)diazene oxide in an extract of a sample from a waste site.[62] Both compounds have ions at *m/z* 335 that differ in mass by 0.11618 Da.

Signal/Noise and Sensitivity

The SIM data acquisition technique used with quadrupole and magnetic sector instruments increases the signal/noise for an analyte compared to the acquisition of complete mass spectra. This increase in signal/noise occurs because the one or few ions monitored for each analyte are focused on the detector a very high percentage of the time and the signal is integrated over this period. Any random noise present in the signal is similarly integrated, but since the random noise can be both positive and negative, the net effect is an increase in signal/noise. With complete spectrum acquisition, a range of ions of different *m/z* are monitored and the analyte's specific ions are focused on the detector for only a small percentage of the time.

The sensitivity of HRMS–SIM with a double-focusing sector mass spectrometer will always be less than the sensitivity of SIM at unit resolution. This must be so because high resolving power in a double-focusing sector instrument is achieved by literally discarding a substantial percentage of analyte ions that are not narrowly focused. With a high resolving power, fewer ions pass through the focusing slits to the detector, and the sensitivity is reduced compared to low resolution. The loss of sensitivity with HRMS is compensated by a general reduction in interferences from coeluting compounds and chemical noise with the same integer but different exact masses. Therefore, the signal/noise is usually increased with HRMS. Nevertheless, resolving powers greater than about 20,000 are not generally feasible for low nanogram–picogram measurements with double-focusing sector instruments because of the loss of sensitivity.[61]

Exact Mass Measurements and Ion Composition

If HRMS can provide high target analyte selectivity by detecting ions within a very narrow m/z range, then this technique can also be used to measure the exact m/z of ions of unknown composition. Exact m/z values are used to calculate possible elemental compositions of ions using the principle that for any given exact m/z only a limited number of combinations of the elements will have masses that correspond to the measured exact m/z. The determination of the elemental compositions of one or more ions in a mass spectrum can provide very valuable information needed to identify unknown substances and deduce molecular structures. However, this elemental composition technique is highly dependent on several factors including the accuracy and precision of the exact m/z measurements and the number and kind of elements that are considered in the calculation of possible elemental compositions.

Table 5.4 shows an unknown ion of m/z 236.1095 Da measured with three 99% confidence intervals of accuracy and precision. In order to limit the possible elemental compositions to a reasonable number, It is assumed that the ion contains only ^{12}C, ^{1}H, a maximum of three ^{14}N, and a maximum of five ^{16}O. These are reasonable assumptions that are usually made based on the history of the sample, the sample preparation, inlet system, and other information (Chapter 2, "Identification Criteria"). The ion compositions whose masses are within the three measurement confidence intervals were calculated using a simple computer program that also considered basic rules of chemical bonding. The exact masses used in the calculation are shown in Table 5.2 and the number of ion compositions found for each measurement are shown in Table 5.4.

The measurement of m/z to within 0.1 Da is not very helpful because, even with the constraints, 37 compositions are found which have exact masses within this confidence interval. The more accurate and precise measurement to within 0.01 Da or 42 parts per 10^6 (ppm) reduces the number of compositions to a manageable five, which are shown in Table 5.5. In some closed systems, for example, a manufacturing plant using a limited number of starting materials, this level of confidence may be adequate to select the correct composition. If this ion is known to be a $M^{+\cdot}$ ion, three compositions can be discarded because they

Table 5.4 The number of compositions corresponding to m/z 236.1095 as a function of the measurement accuracy and precision and composition constraints

Measured m/z	99% Confidence interval	Maximum number of		Number of compositions
		nitrogens	oxygens	
236.1095	+ or −0.1	3	5	37
236.1095	+ or −0.01	3	5	5
236.1095	+ or −0.002	3	5	1

Table 5.5 Calculated exact masses and compositions that are within the range 236.0995–236.1195 Da

Exact mass	Composition
236.11877	$C_{15}H_{14}N_3$
236.10754	$C_{16}H_{14}NO$
236.11609	$C_{12}H_{16}N_2O_3$
236.10352	$C_{11}H_{14}N_3O_3$
236.10486	$C_{13}H_{16}O_4$

contain an odd number of nitrogens. For an environmental or some other type of sample that could contain almost anything, more information is needed to make the correct selection. If the measurement is made with an accuracy and precision to within 0.002 Da (8.5 ppm), only one of the compositions in Table 5.5 has an exact mass within the range 236.1075–236.1115 Da. This level of confidence in exact mass measurements is well within the capabilities of several types of mass spectrometer.[50] However, the accuracy and precision required to limit the possible compositions to one or a few increases dramatically as the *m/z* increases and the constraints on the number and kind of elements that may be present are removed.

High-resolution mass spectrometry is not required for exact *m/z* measurements. There are reports in the literature demonstrating exact measurements at low resolution using a single-focusing magnetic deflection mass spectrometer or a quadrupole mass spectrometer.[50] However, low resolution measurements are best limited to $M^{+\cdot}$ ions for several reasons. Fragment ions are sometimes formed with some kinetic energy which, in a magnetic deflection mass spectrometer, has the effect of shifting the peak centroid to a higher *m/z*. In addition, fragment ions with the same integer *m/z*, but different compositions and exact *m/z*, are often formed from the same $M^{+\cdot}$ or other ion and this causes broadening of the ion peak and shifting the peak centroid, which can give an erroneous exact *m/z* measurement. Ions containing naturally occurring isotopes also cause complications in exact *m/z* measurements of fragment ions (see Table 5.3). In environmental and some other types of samples, coelution of compounds with the same integer $M^{+\cdot}$ ions is a real possibility, for example, the compounds in Chart 5.3 with a $M^{+\cdot}$ ion at *m/z* 320. This can cause distortion and broadening of the $M^{+\cdot}$ ion peak and an erroneous exact mass measurement. Therefore, exact mass measurements of $M^{+\cdot}$ ions at low resolution in complex samples is also risky.

The vast majority of exact *m/z* measurements are made with high resolving power, usually greater than about 8000. However, high resolving power does not guarantee *m/z* measurement accuracy or precision. Other factors affect the quality of exact *m/z* measurements with all types of mass spectrometer.[50] For example, measurements made with a double-focusing sector mass spectrometer in the HRMS–SIM mode at constant magnetic field strength, as described previously, can provide very accurate and precise *m/z* measurements.[61,62] However, scanning of the magnetic field in the continuous measurement of spectra (CMS) mode of data acquisition with the same type of instrument produces m/z measurements with lower accuracy and precision.

Tandem Mass Spectrometry

Tandem MS is the linking together in space or time of two independently operating m/z analyzers. Ions are formed in an ion source, separated in the first analyzer, undergo ion–molecule reactions in a collision cell, and the ionic products of these reactions are measured in the second analyzer. This versatile technique, which is often described as mass spectrometry/mass spectrometry (MS/MS), has some important analytical applications including high analyte selectivity and very low detection limits. Tandem MS can also provide information about ion fragmentation mechanisms, which is valuable for the identification of unknowns and molecular structure determination. This technique is also useful for identifying certain classes of analytes in complex mixtures and chromatographic effluents.

Tandem MS was first explored with double-focusing instruments consisting of magnetic and electrostatic sectors, but was made more practical by connecting together multiple quadrupole analyzers.[63–66] These types of instruments are called tandem-in-space mass spectrometers because the individual analyzers are literally separated in space. Subsequently, it was recognized that some analogous information could be obtained with a single ion-storage type analyzer, such as a Fourier transform mass spectrometer or an ion trap, by a sequence of timed ion storage, isolation, reaction, and measurement events.[67,68] Tandem MS in an ion-storage type instrument is called tandem-in-time, which is the logical equivalent of tandem-in-space MS.[69]

Figure 5.5 is a diagram of a tandem-in-space instrument system. Samples are introduced and ions are formed using any of a variety of inlet systems and ionization techniques that are described in Chapters 2 and 5–8. All the ions are injected into the first analyzer, which is operated in either the SIM mode or in the CMS mode. The second analyzer also is operated in either the SIM or the CMS mode. Table 5.6 gives the names of the four tandem MS data acquisition strategies and the data acquisition techniques used in the individual analyzers. The strategies in which the first analyzer is operated in the SIM mode are discussed first, and the other strategies are described later in this section.

Collision Induced Dissociation

When the first analyzer is operated in the SIM mode, ions of a specific m/z are selected, for example, the $M^{+\cdot}$ ion in Reaction 5.22, and focused into the collision cell (Figure 5.5). The m/z of the selected ion can be changed as a function of time as different substances elute from a chromatograph, and the selected ion can be a fragment ion. The collision cell contains a neutral target gas, for example, argon, and collisions of the selected ion with the target gas atoms cause fragmentations of the selected ion, to give product ions (Reaction 5.22).

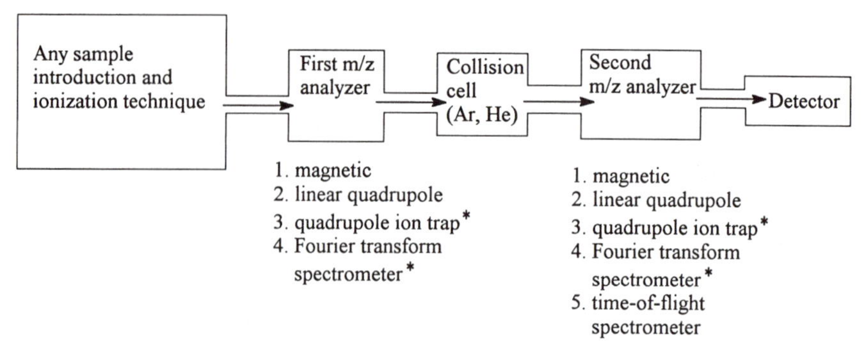

Figure 5.5 A diagram of a tandem-in-space mass spectrometer system.

This process is called collision induced dissociation (CID) or collision activated dissociation (CAD). If the second analyzer is used in the CMS mode to analyze the mixture of product ions, this is called a product ion scan or sometimes a daughter ion scan (Table 5.6). The resulting mass spectrum displays the m/z and abundances of the ions formed from the $M^{+\cdot}$ or fragment ion by CID. The CID spectra of $M^{+\cdot}$ ions are generally very similar to conventional EI spectra and they can be processed and interpreted using most of the strategies presented in Chapter 2. Collision induced dissociation can also occur in some other parts of a mass spectrometer, for example, some inlet systems, and these processes are described in subsequent chapters as appropriate.

$$M^{+\cdot} + Ar \longrightarrow A^+ + B^+ + C^+ \qquad \text{(Rx 5.22)}$$

The first analyzer of a tandem-in-space instrument is most often a quadrupole although other types of mass spectrometer, including double-focusing sector instruments, are used. The collision cell is often a quadruple analyzer operating in a nonanalyzer mode to gather and focus into the second analyzer the ions formed by CID. If the first analyzer, the collision cell, and the second analyzer are all quadrupoles, the whole system is called a triple quadrupole spectrometer.[63] Various combinations of analyzers have been used in tandem MS, and the history, principles of operation, and some applications are found in several early and more recent reviews.[63–66]

Table 5.6 Data acquisition strategies used in tandem mass spectrometry

Tandem data acquisition strategy	First analyzer	Second analyzer
Product ion scan	SIM	CMS
Selected reaction monitoring (SRM)	SIM	SIM
Precursor ion scan	CMS	SIM
Neutral loss scan	CMS	CMS

Direct Mixture Analysis

During the early years of development of tandem MS it was quickly recognized that the first analyzer of a tandem mass spectrometer was acting as a very rapid and specific separation device, analogous to a very fast and efficient gas chromatograph, and the second analyzer was capturing the information from the separated ion's CID spectrum.[63,64] It was suggested that preliminary separations by GC or other techniques could be eliminated and mixtures analyzed directly and more rapidly by MS/MS. While some demonstrations of this capability for environmental samples showed promise,[70] economic and other factors inhibited the widespread use of tandem MS for direct analyses of most environmental and other samples.

During the 1980s the bench-top GC/MS system was much lower in price than a tandem instrument, and the performance of basic GC/MS systems improved dramatically with the introduction of bonded-phase fused silica capillary GC columns. The demand for parts per 10^9 level detection capabilities and reasonably accurate and precise quantitative analyses of very complex environmental samples also strongly favored the conventional GC/MS approach. However, when samples are introduced into the mass spectrometer using one of the desorption ionization or related probe techniques described in Chapter 6, MS/MS provides a separation capability for those techniques that are not well suited to on-line chromatographic separations. Direct mixture analysis with tandem MS also has significant advantages over conventional GC/MS for continuous measurements in the field or in a processing plant. Some of these applications have been developed and are discussed in Chapter 8.

Ion-Storage Spectrometers

A product ion scan is conducted in an ion-storage mass spectrometer by forming ions just outside the trap, or in an external ion source, storing in the trap only those ions with a specific m/z, causing CID by energizing the stored ions in the presence of a target gas, and measuring the abundances of the fragments as a function of m/z.[67–69] This sequence of events separated in time is referred to as tandem-in-time MS, which is the logical equivalent of tandem-in-space. An ion-storage spectrometer allows multiple stages of selected ion storage and fragmentation by repeating the processes as many times as needed to obtain the desired information (Reactions 5.23 and 5.24):

$$B^+ + He \longrightarrow D^+ + E^+ \qquad \text{(Rx 5.23)}$$

$$D^+ + He \longrightarrow F^+ + G^+ \qquad \text{(Rx 5.24)}$$

In this general example, the primary fragment ion from Reaction 5.22, B^+, is stored in the trap, and the other product ions are ejected. The B^+ ion is then energized to cause CID (Reaction 5.23) and the second-generation fragment ion D^+ is selectively stored. Finally, the process is repeated (Reaction 5.24) and the third-generation product ions F^+ and G^+ are measured in the product ion scan.

This multistage process is abbreviated MS^n where $n = 1, 2, 3$, and so on, and indicates the number of selective storage and fragmentation events before the product ion scan. The sequence in Reactions 5.22–5.24 would be designated MS^4. While in principle this sequence of fragmentations could be accomplished with a tandem-in-space instrument, it would require four separate analyzers, three collision cells, a large room, and considerable financial resources.

Identification of Unknown Compounds and Molecular Structure

A standard strategy to identify unknown compounds and assign molecular structure is the interpretation of EI mass spectra in terms of a series of fragmentation reactions of the $M^{+\cdot}$ and fragment ions.[1–3] If the compositions and structural features of most fragment ions can be assigned, and pathways of fragmentation determined, it is possible to combine this information and develop a hypothesis for the composition and structure of the intact molecule. Unfortunately, this process is frequently very difficult because, other than losses of small atoms or groups of atoms from the $M^{+\cdot}$, the sequence of fragmentation reactions of various ions is not usually obvious and is often very complex.

Several techniques are used to determine the products of fragmentation reactions of various ions in a mass spectrum. Some fragmentation mechanisms have been elucidated by labeling certain atoms of model compounds with isotopes, for example, deuterium, which causes m/z shifts in some fragment ions.[1–3] This technique has limited value with unknown substances, but can be used if the unknown has exchangeable hydrogens that are readily replaced by deuteriums. The study of metastable ion decompositions in magnetic deflection instruments is a historically important technique to determine ion fragmentation processes, but this technique is not available with most types of mass spectrometer.[2] While CI, ECI, and the other ionization techniques described in this chapter and Chapter 6 are very valuable because they provide molecular weight information, they usually do not provide the fragment ions needed to identify unknowns and determine molecular structure.

Tandem MS provides a convenient and efficient strategy to determine fragmentation mechanisms of model compounds and unknown substances. The product ion scan is particularly useful because a single $M^{+\cdot}$ or fragment ion can be isolated in the collision cell, and the products of the decomposition of that ion can be determined in the product ion scan. In an ion-storage mass spectrometer the analysis of a sequence of ion fragmentation reactions is facilitated by the MS^n technique. The CID spectra of most $M^{+\cdot}$ ions are very similar to EI spectra and, therefore, the fragmentation pathways determined by CID are directly applicable to the interpretation of EI mass spectra. For example, a fragment ion of m/z 43 in the spectrum of an unknown could be either $H_3C{-}C{=}O^+$ or $C_3H_7^+$.[63] Knowledge of the composition of m/z 43 is important information for the identification of an unknown. When the unknown ion of m/z 43 is submitted to CID in a tandem instrument, a very low abundance ion is observed at m/z 27 which is assigned the composition $C_2H_3^+$ (Chart 5.4).

However, when a *m/z* 43 fragment ion from an aliphatic hydrocarbon, which is known to be $C_3H_7^+$, is submitted to CID, an approximately 40% abundant *m/z* 27 is measured in the product ion scan (Chart 5.4). The $C_2H_3^+$ ion is readily formed by losses of a methyl radical and a hydrogen atom from the $C_3H_7^+$ ion. If the unknown ion was $C_3H_7^+$, a relatively abundant *m/z* 27 ion would be expected. Since the abundance of *m/z* 27 is very low in the CID spectrum of the unknown ion, it is very likely to be $H_3C-C=O^+$ which can form a $C_2H_3^+$ only by the unlikely breaking of a carbon–oxygen multiple bond.

Unknown ion $\xrightarrow{\text{CID}}$ $C_2H_3^+$

m/z = 43 m/z = 27 of very low abundance

$C_3H_7^+$ $\xrightarrow{\text{CID}}$ $C_2H_3^+$

m/z = 43 from an aliphatic hydrocarbon m/z = 27 (40% relative abundance)

Chart 5.4

Selected Reaction Monitoring

For maximum selectivity and signal/noise the second analyzer of a tandem-in-space instrument is also operated in the SIM mode, and the abundance of one of the fragment ions, for example, B^+ in Reaction 5.22, is monitored during the elution of the molecule M. This technique is called selected reaction monitoring (SRM), which measures the specific transformation $M^{+\cdot} \rightarrow B^+$ and provides a highly selective measure of the original molecule M (Table 5.6). Only those molecules M that give the ion $M^{+\cdot}$, which fragments by CID to give B^+, will be measured by this technique. The advantage of SRM is tremendous selectivity for a specific substance and the exclusion of many potential interferences including chemical noise and coeluting substances with the same integer masses. Although coeluting compounds may produce multiple $M^{+\cdot}$ or fragment ions with the same integer masses as the target analyte $M^{+\cdot}$, and these ions will be injected into the collision cell, it is unlikely that substances other than the target analyte will produce the B^+ fragment ion by CID. Selected reaction monitoring is the most widely used MS/MS technique for quantitative analysis of very complex samples. With ion-storage type spectrometers, the storage and isolation of $M^{+\cdot}$, CID, and measurement of B^+ are separated in time within the single analyzer. However, B^+ and all other product ions are measured together because there is no advantage with SIM in an ion-storage mass spectrometer.

An indication of the selectivity of SRM can be obtained by examining the spectra in the 1992 release of the NIST/EPA/NIH mass spectral database which contains 62,235 EI spectra.[71] Consider the determination of 2378-TCDD by SRM. The reaction monitored is the CID of the *m/z* 320 $M^{+\cdot}$ ion to give the *m/z* 257 ion by loss of CO and a ^{35}Cl atom. Figure 5.1 shows the EI spectrum of 2378-TCDD, which contains these ions. The abundant *m/z* 320 $M^{+\cdot}$ ion is monitored in the first analyzer or isolated and stored in an ion-storage type instrument, and fragmented by CID; the abundance of the *m/z* 257 ion is monitored in the second analyzer or measured in the ion-storage mass spectrometer. Of the 62,235 EI spectra in the database, only 689 spectra have a *m/z* 320 ion between 5 and 100% relative abundance (RA). More importantly, only 89 spectra have both *m/z* 320 and 257 ions between 5 and 100% RA. Most of these are compounds that are unlikely to be in an environmental sample or elute from a high-resolution gas chromatograph in the same retention time window (RTW) as 2378-TCDD.

The major interferences in the determination of 2378-TCDD by SRM are the other 21 possible TCDD isomers whose *m/z* 320 ions will also enter the collision cell and undergo CID to give *m/z* 257 ions. Just as with HRMS–SIM, these isomers must not be present or must be separated on the GC column in order to achieve the selective measurement of 2378-TCDD. Other potential interferences in the SRM determination of 2378-TCDD are the hexachlorodibenzo-*p*-dioxins (10 isomers possible), which give low abundance ($\sim < 10\%$), *m/z* 320 and 257 ions. These compounds are not likely to elute from the column in the same RTW as the tetrachloro isomers. The *m/z* 320 ion in the spectrum of coeluting 4,4′-DDE (Figure 5.2) would also enter the collision cell, but does not fragment to give a *m/z* 257 ion. The SRM technique has been evaluated for the determination of 2378-TCDD and related compounds with several types of tandem mass spectrometer systems. These include a double-focusing analyzer consisting of electrostatic and magnetic sectors, a hybrid double-focusing quadrupole instrument, a triple quadrupole, and an ion-trap spectrometer.[59, 72–74]

Other Tandem Data Acquisition Strategies

When the first analyzer of a tandem-in-space instrument is operated in the CMS mode, two additional selectivity enhancing data acquisition strategies are possible that cannot be directly duplicated with an ion-storage type mass spectrometer. If the second analyzer is operated in the SIM mode, for example, monitoring *m/z* 257 ions, then only ions that leave the collision cell with *m/z* 257 are detected. This acquisition strategy is called a precursor ion or parent ion scan because the chromatogram would consist of a series of peaks corresponding to molecules that give precursor ions which undergo CID to produce *m/z* 257 ions that are detected after the second analyzer (Table 5.6). These molecules can have any molecular weight $\geqslant 257$, the only requirement being that either the $M^{+\cdot}$ ion has a *m/z* of 257 and passes through the collision cell without CID, or the $M^{+\cdot}$ or a fragment ion undergoes CID to give *m/z* 257 ions.

The 22 TCDD isomers would be detected with a precursor ion strategy because their *m/z* 320 and 322 $M^{+\cdot}$ ions fragment by loss of CO and either ^{35}Cl or ^{37}Cl, respectively, to give *m/z* 257 ions (Figure 5.1). All other possible losses of combinations of CO and ^{35}Cl or ^{37}Cl from the TCDD $M^{+\cdot}$ ion cluster (*m/z* 320–328) give ions in the range *m/z* 259–263 and these ions would not be detected in the precursor ion scan. The potential TCDD interference 4,4′-DDE would not be detected because it does not have a *m/z* 257 ion in its spectrum (Figure 5.2). Of the 62,235 EI spectra in the 1992 release of the NIST/EPA/NIH mass spectral database,[71] only 2380 spectra have an *m/z* 257 ion between 5 and 100% RA. Since most of these would be very unlikely to be in an environmental sample and elute from a GC column in the RTW of the 22 TCDD isomers, the precursor ion scan provides a chromatogram with high selectivity for the class of TCDDs, but it is not nearly as selective as SRM. The precursor ion scan strategy can be applied to many classes of analytes that give a common ion on CID.

A second and related selective scan strategy for tandem-in-space instruments is the neutral loss scan (Table 5.6). In this acquisition strategy, both analyzers are operated in the CMS mode with a constant time delay between the scans of the two analyzers. This delay or offset corresponds to a specific mass difference, for example, *m/z* 63, which is indicative of a neutral loss in the collision cell. Using the TCDD example again, as *m/z* 320 ions (Figure 5.1) from the first analyzer enter the collision cell the second analyzer is set to measure *m/z* 257 ions, which correspond to the loss of a neutral CO and a ^{35}Cl atom, that is, mass 63, from *m/z* 320 ions. As *m/z* 322 ions from the first analyzer enter the collision cell the second analyzer is set to measure *m/z* 259 ions, which again corresponds to the loss of a neutral CO and a ^{35}Cl atom from *m/z* 322 ions. This process continues as ions are repetitively scanned in the first analyzer, and ions with their masses reduced by *m/z* 63 are measured in the second analyzer. The chromatogram produced by this strategy consists of a series of peaks that correspond to molecules which give $M^{+\cdot}$ or fragment ions that undergo CID with loss of neutrals with a mass of 63 Da. These molecules could be of any molecular weight, the only requirement being that the $M^{+\cdot}$ or a fragment ion undergoes CID to lose a mass 63 neutral.

The 22 TCDD isomers would be detected with a neutral loss strategy because most of their $M^{+\cdot}$ ions (*m/z* 320, 322, 324, and 326) fragment by loss of mass 63 to give *m/z* 257, 259, 261, and 263 ions, respectively (Figure 5.1), although *m/z* 263 is too weak to observe. A potential advantage of the neutral loss scan, compared to the precursor ion scan, is an increased total signal for a given component which has multiple $M^{+\cdot}$ ions caused by the distribution of Cl isotopes. Each of these $M^{+\cdot}$ ions can lose the same neutral, 63 in this example, to give multiple measured product ions. The potential TCDD interference, 4,4′-DDE, would not be detected with this strategy because there is no apparent neutral loss of mass 63 in its spectrum (Figure 5.2).

Of the 62,235 EI spectra in the 1992 release of the NIST/EPA/NIH mass spectral database,[71] 3676 spectra show a loss of mass 63 from the $M^{+\cdot}$ ion to give a fragment ion with at least 5% of the abundance of the $M^{+\cdot}$ ion. Many compounds in this group are counted two or more times because of naturally occurring isotopes, so the actual number of compounds that show a loss of 63 from the $M^{+\cdot}$ ion may be much smaller than 3676. However, many fragment ions may also lose mass 63 neutrals, which will increase the number of possible compounds. This analysis suggests that the neutral loss scan is less selective than the precursor ion scan, but it is a valuable technique for locating chromatographic peaks from certain classes of compounds.

Chemical Derivatization

Chemical derivatization is the conversion of a compound or a group of structurally related compounds into another compound or class of compounds by reaction with a chemical reagent or reagents. Derivatization is an historically important strategy to increase analyte vapor pressure and thermal stability and facilitate GC. The use of derivatives for that purpose is discussed in Chapter 6. In addition, by appropriate selection of the derivative, analyte selectivity and sensitivity can be enhanced and derivatives can facilitate the determination of molecular weight and molecular structure. Atoms or groups of atoms can be added to the analyte, which allows the derivative to be readily determined whereas the analyte cannot be distinguished from chemical noise or coeluting substances. Several specific benefits provided by derivatization are listed and some of these have been applied to environmental analyses:

- The derivative has a chromatographic retention time that is displaced from interferences and chemical noise.
- The derivative is more efficiently ionized than the original analyte, coeluting substances, or chemical noise.
- The $M^{+\cdot}$ or $M^{-\cdot}$ ions of the derivative are stable and observed, but the corresponding ions of the original analyte are not observed.
- The mass spectrum of the derivative is characteristic of molecular structure whereas the spectra of the original analyte and coeluting structural isomers are indistinguishable.
- The spectrum of the higher molecular weight derivative has higher m/z ions that are displaced from the lower m/z chemical noise that obscures the ions of the original analyte.
- The added atoms of the derivative provide a characteristic isotope distribution abundance pattern that aids in the recognition of the analyte.

It is beyond the scope of this book to describe the many types of derivatives that have been used with MS to increase analyte selectivity or sensitivity, or aid in molecular weight or structure determination. Several examples of derivatives that are useful for environmental and other analyses are described and additional information on chemical derivatization is available in the chemical literature.[75–77]

Acetylation of aromatic amines has been used to distinguish these weak bases from isomeric alkylated heterocyclic nitrogen compounds.[78] These two classes of compounds are found together in the same types of environmental samples, for example, wastewater from a synthetic fuels plant. The compounds 2-methylaniline (Figure 5.6) and 2-ethylpyridine (Figure 5.7) are structural isomers that have very similar EI mass spectra and similar GC retention characteristics.[71,78] Because there are other structural isomers with similar EI spectra and

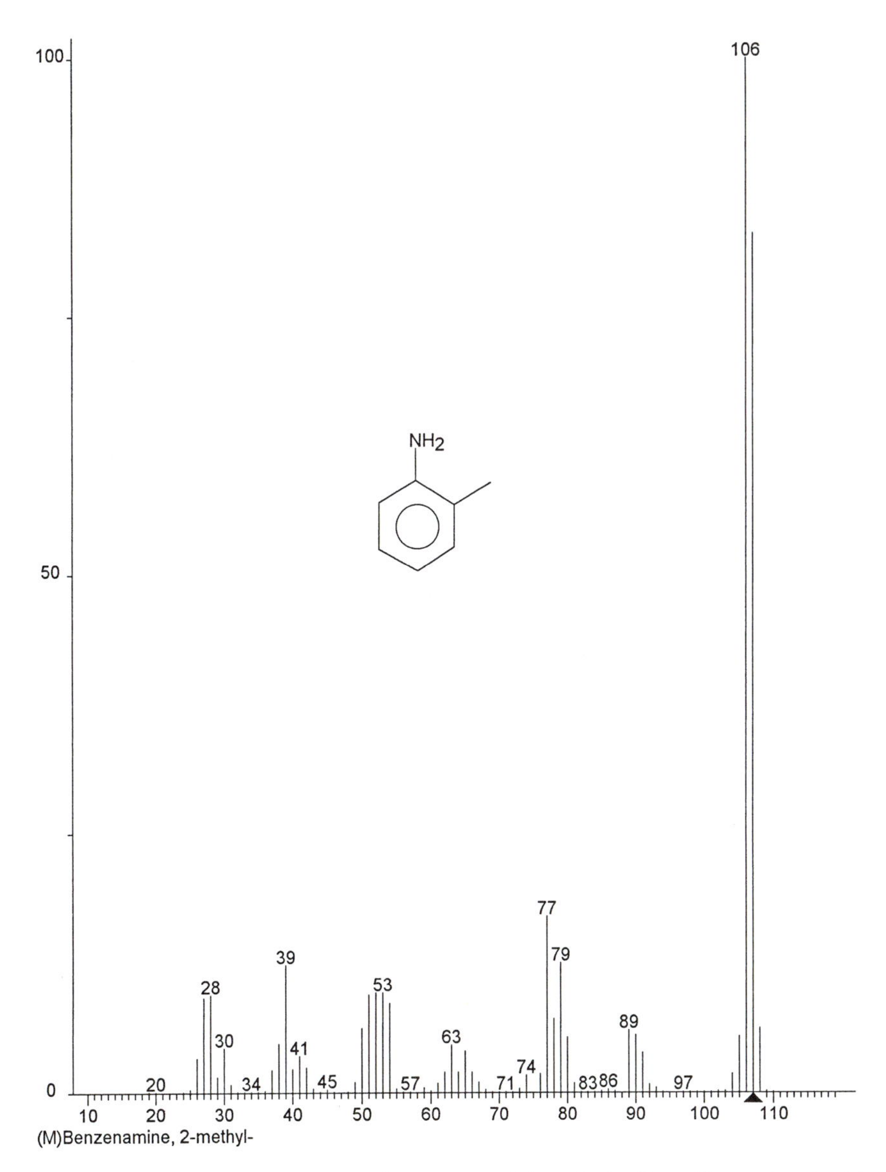

Figure 5.6 The EI mass spectrum and structure of 2-methylaniline. (Reproduced with permission from Ref. 71. Copyright 1992 U.S. Department of Commerce on behalf of the United States.)

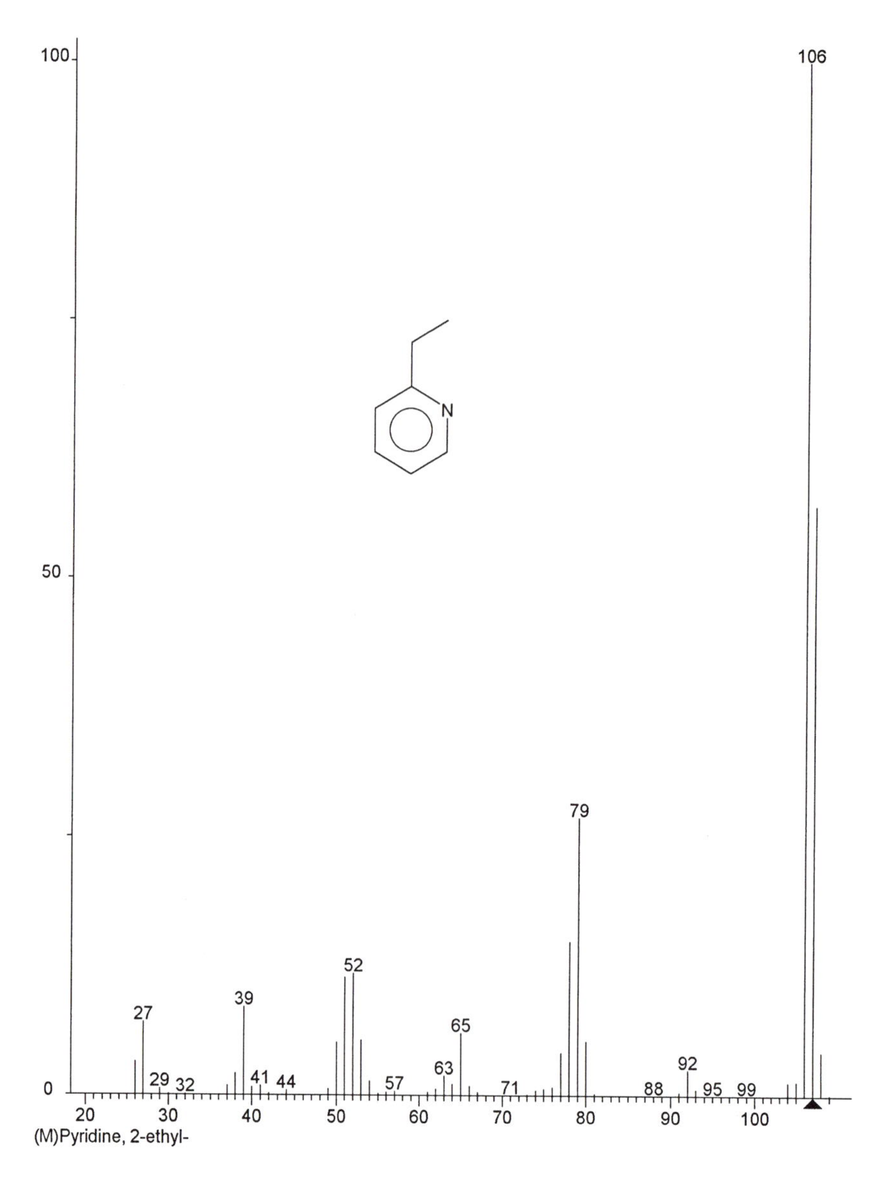

Figure 5.7 The EI mass spectrum and structure of 2-ethylpyridine. (Reproduced with permission from Ref. 71. Copyright 1992 U.S. Department of Commerce on behalf of the United States.)

GC retention characteristics, positive identification of specific compounds by conventional GC/MS is difficult. Small differences in RAs, for example, the *m/z* 77–79 ions in Figures 5.6 and 5.7, cannot be used to distinguish between these compounds because these differences could be caused by coeluting substances or general chemical noise.

When a sample or sample extract containing these compounds is treated with acetic anhydride under appropriate conditions, the anilines are acetylated to form N-substituted acetamides (Figure 5.8[71]), but the pyridines are unaffected.[78] The acetylated anilines have GC retention characteristics different from those of the alkylated pyridines and $M^{+\cdot}$ ions that are 42 Da higher in *m/z*, which allows the two groups of compounds to be readily distinguished. The fluorinated acetylating reagents perfluoroacetic anhydride and perfluoropropionic anhydride produce analogous derivatives.[79] The fluorine atoms add significant characteristic mass to the derivative but do not adversely affect GC retention behavior. In addition, the perfluoro groups provide high sensitivity and selectivity with electron capture ionization (ECI).

Pentafluorobenzyl bromide is a useful derivatizing agent for phenols, thiols, carboxylic acids, and other weak acids (Chart 5.5). The fluorinated ethers, thioethers, and esters form negative ions by ECI, which provides additional selectivity and sensitivity for these classes of compounds. Alkylated phenols in a Lake Erie sediment were determined by using this technique, which gave detection limits about 100 times lower than those of conventional GC/MS.[80] The $M^{-\cdot}$ anions initially formed by ECI fragment by loss of pentafluorobenzyl radicals and give anions characteristic of the original phenols (Chart 5.5).

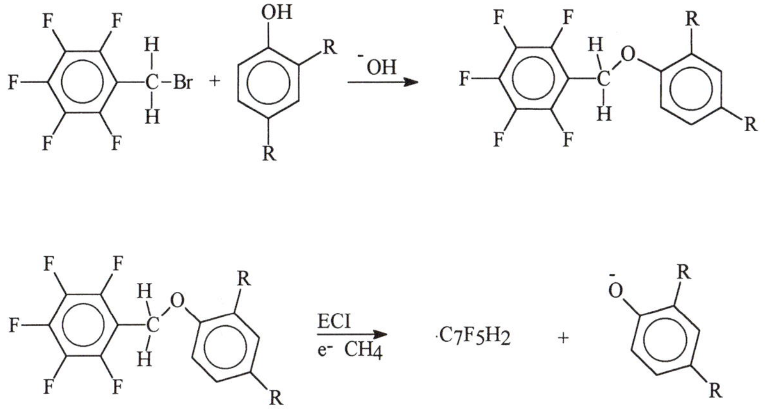

Chart 5.5

Aliphatic aldehydes in the C_1–C_{14} range, for example, *n*-hexanal, are products of the ozone disinfection of water containing natural organic material from decaying vegetation and other sources.[81] Low molecular weight aldehydes are also products of the photochemical oxidation of hydrocarbons in air, and analytical methods for these analytes are described in Chapter 3 (under "Volatile Organic Compounds in Air"). Aldehydes are chemically reactive, somewhat

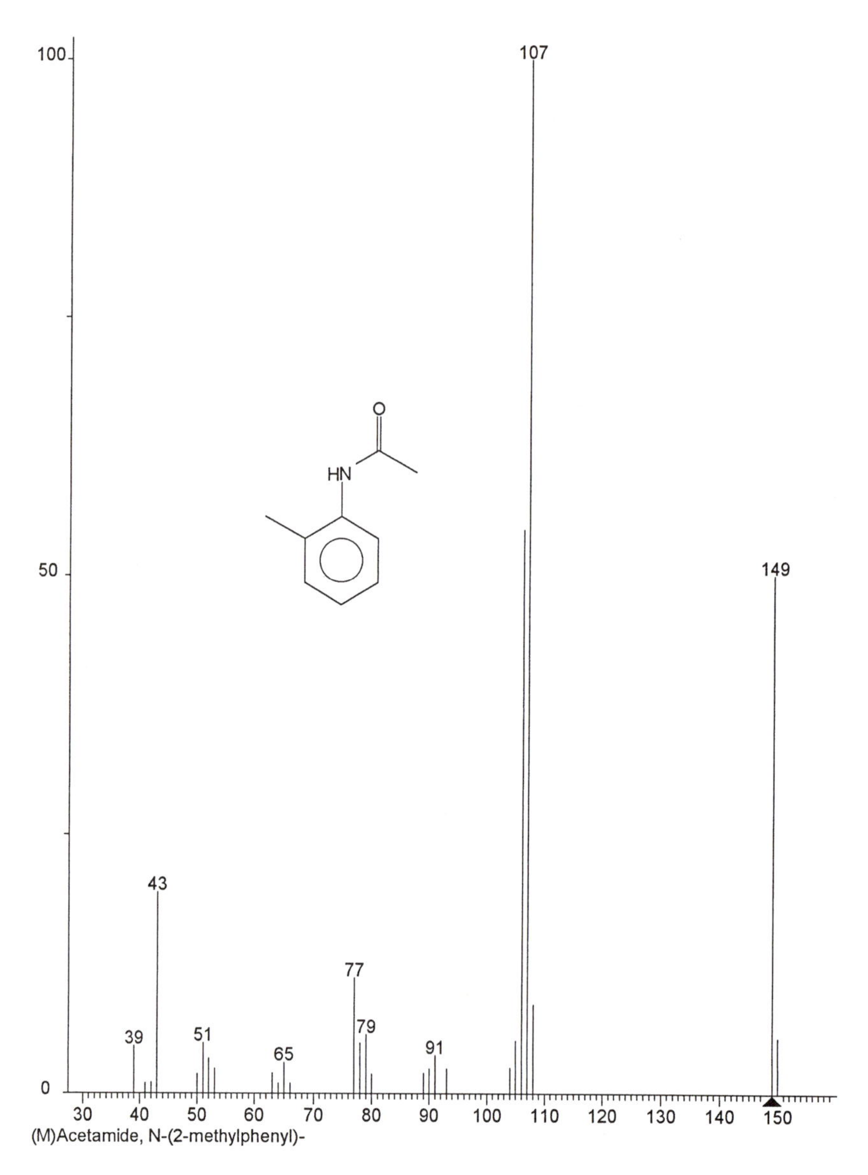

Figure 5.8 The EI mass spectrum and structure of *N*-(2-methylphenyl)acetamide. (Reproduced with permission from Ref. 71. Copyright 1992 U.S. Department of Commerce on behalf of the United States.)

thermally unstable, and generally not amenable to separation by GC. When EI mass spectra of aliphatic aldehydes are measured by using a direct insertion probe, the spectra are generally devoid of $M^{+\cdot}$ ions or they are present in very low RAs. The EI spectrum of *n*-hexanal in Figure 5.9 does not have a $M^{+\cdot}$ ion at *m*/*z* 100 and contains mostly low *m*/*z* fragment ions.[71] The identification of aldehydes in environmental and other samples is inhibited by the absence of $M^{+\cdot}$ ions in their EI spectra.

The 2,4-dinitrophenylhydrazone is a standard derivative of aldehydes and ketones and is prepared by their reaction with 2,4-dinitrophenylhydrazine. This derivative is used to determine photochemical oxidation products in air and was discussed in Chapter 3 (under "Volatile Organic Compounds in Air"). However, 2,4-dinitrophenylhydrazones are not amenable to separation by GC, but they are readily separated by using one of the condensed-phase separation techniques described in Chapter 6. A significant advantage of these derivatives is that, in general, 2,4-dinitrophenylhydrazones give abundant $M^{+\cdot}$ ions in their EI spectra.[82] The structure and the EI spectrum of the 2,4-dinitrophenylhydrazone of *n*-hexanal are shown in Figure 5.10; the spectrum has a 25% $M^{+\cdot}$ ion at *m*/*z* 280.[71] This spectrum was probably measured by using a direct insertion probe, but spectra containing molecular weight information should be obtainable with most of the mass spectrometric techniques described in Chapter 6.

If an EI spectrum of the 2,4-dinitrophenylhydrazone of an unknown aldehyde or ketone can be obtained, for example, by using the particle–beam interface described in Chapter 6, and the $M^{+\cdot}$ ion identified in the spectrum, the molecular weight of the unknown can be readily determined. The mass attributable to the 2,4-dinitrophenylhydrazine residue, 196, is subtracted from the *m*/*z* of the hydrazone $M^{+\cdot}$ ion and 16 is added to account for the missing oxygen. The result, using the hydrazone depicted in Figure 5.10 as an example, is 100, which is the *m*/*z* of the missing $M^{+\cdot}$ of *n*-hexanal in Figure 5.9. The $M^{+\cdot}$ of a hydrazone of an aldehyde or ketone with a straight-chain alkyl group containing four or more carbons can usually be verified using a fragment ion in its EI spectrum. In these molecules there will be an odd-electron even-mass ion at some lower *m*/*z* which corresponds to the loss of an olefin in a McLafferty rearrangement.[82] This ion is at *m*/*z* 224 in the spectrum of the 2,4-dinitrophenylhydrazone derivative of *n*-hexanal and corresponds to the loss of C_4H_8 (Figure 5.10).

An alternative derivative for aldehydes and ketones that is amenable to separation by GC was described in Chapter 3 (under "Volatile Organic Compounds in Air"). The reagent *O*-(2,3,4,5,6-pentafluorobenzyl)hydroxylamine reacts with aldehydes and ketones in the aqueous phase to produce oximes (Chart 3.2).[81] However, the oximes of most aldehydes and some ketones do not contain $M^{+\cdot}$ ions or have very low $M^{+\cdot}$ ion abundances in their EI spectra. The dimethylhydrazones of some aldehydes and ketones are reported to be amenable to separation by GC, and give abundant $M^{+\cdot}$ ions, but these derivatives are not widely used for the determination of aldehydes and ketones in environmental and other samples.[83]

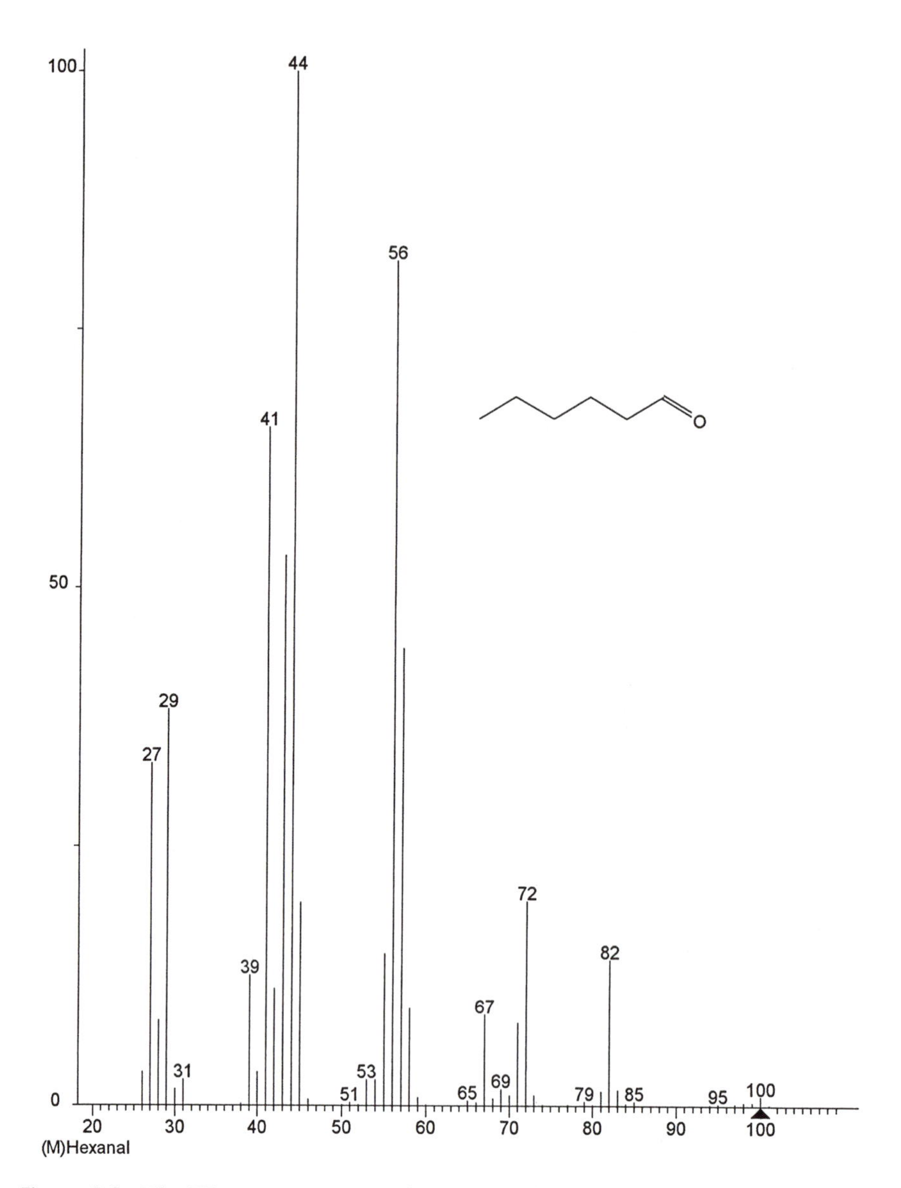

Figure 5.9 The EI mass spectrum and structure of *n*-hexanal. (Reproduced with permission from Ref. 71. Copyright 1992 U.S. Department of Commerce on behalf of the United States.)

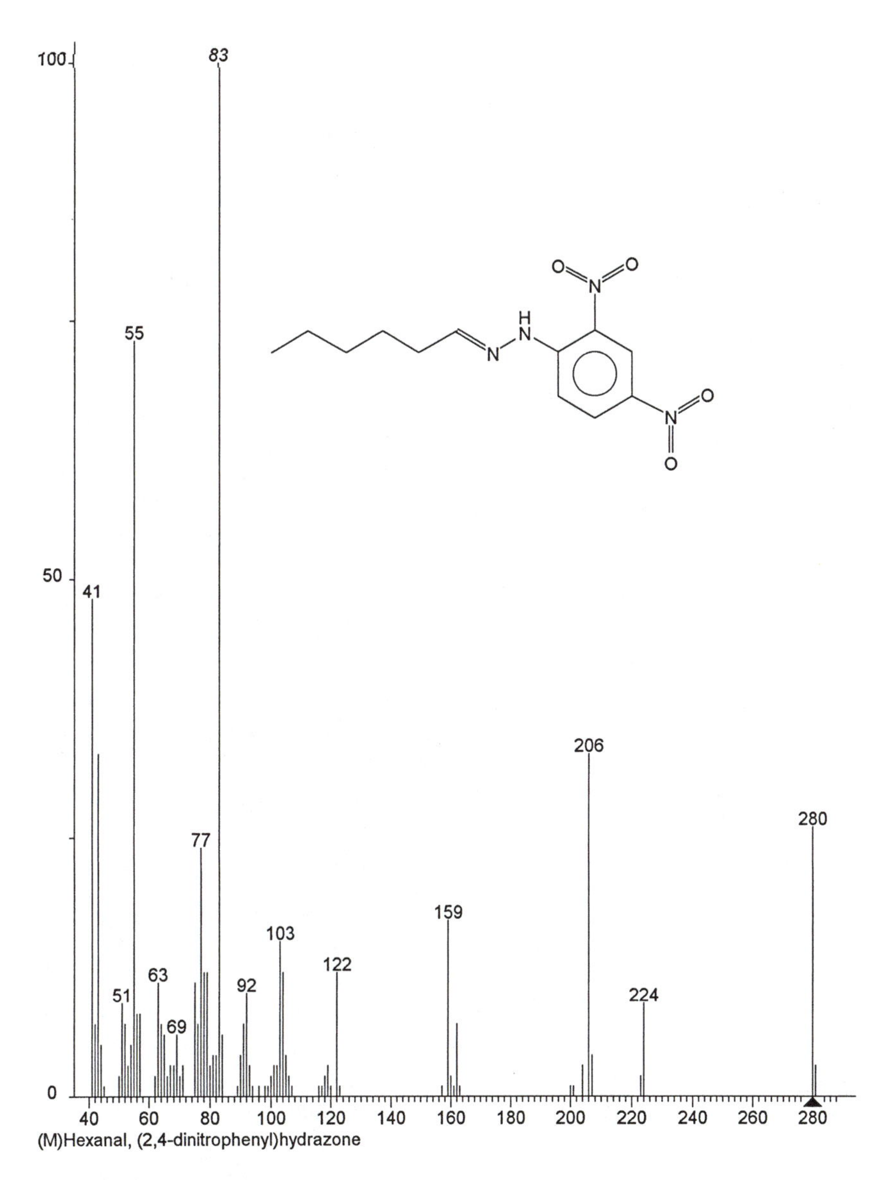

Figure 5.10 The EI mass spectrum and structure of the 2,4-dinitrophenylhydrazone of *n*-hexanal. (Reproduced with permission from Ref. 71. Copyright 1992 U.S. Department of Commerce on behalf of the United States.)

Disadvantages of Derivatization

Chemical derivatization also has disadvantages, which have tended to increase research to find alternative techniques of increasing selectivity and sensitivity, for example, the ionization and other techniques discussed in previous sections of this chapter. These negative aspects have reduced the use of derivatization in some monitoring studies even when alternative techniques were not available:

- Derivatization requires the analyte to have a chemically reactive functional group, and many compounds are not chemically reactive or do not have an easily derivatized functional group.
- Derivatization requires additional time, reagents, glassware, and sometimes the isolation of the analytes in a special solvent. These factors add to the cost of the analysis.
- Competing reactions can occur with very reactive derivatizing reagents, and these can reduce yields of the desired derivatives and form other products which add to the chemical noise and complexity of the sample.
- Reaction conditions required for derivatization, for example, a very low or high pH and/or heat, may alter the composition and structures of other analytes in the mixture.
- Conversion yields may vary widely among a series of structurally related compounds and will rarely be 100% for all members of a series. Yields less than 100% reduce sensitivity.
- The variability caused by competing reactions and variable yields adds to the total variability of quantitative analyses.

Summary of Strategies

The strategies described in this chapter are valuable and provide several approaches to enhanced selectivity for specific analytes, and information about ion composition, lower detection limits, molecular weight determination, and structure. Most large industrial, government, and university laboratories that support chemical analyses should have available most or all of the techniques described in this chapter. Smaller laboratories may have only one or a few techniques available. Principal investigators, designers of environmental monitoring studies, and research personnel should be aware of both the strengths and limitations of conventional GC/MS and the strategies presented in this chapter to overcome most of these limitations. Unfortunately, too often the technique applied to solve a problem is not always the most appropriate but the one that is available in the laboratory.

Chemical and related ionization techniques can provide substantial selectivity and sometimes significantly lower detection limits for some analytes that are efficiently ionized with these techniques. These techniques can also provide molecular weight information for some analytes that do not give a $M^{+\cdot}$ ion with electron ionization. High resolution and tandem MS can provide many benefits including enhanced analyte selectivity, information about ion composition, lower detection limits, and information to support structure determination. Fortunately, these strategies are more available than in previous decades because of advances in instrumentation design and powerful bench-top computer systems. The time-honored chemical derivatization techniques to increase analyte selectivity, lower detection limits, aid in molecular weight determination, and support structure determination should not be overlooked. Sometimes a simple derivatization approach can provide the needed information and preclude the need for an instrument that is not available in the laboratory.

REFERENCES

1. Watson, J. T. *Introduction to Mass Spectrometry*, 3rd ed.; Lippincott-Raven: Hagerstown, MD, 1997.
2. Budzikiewicz, H.; Djerassi, C.; Williams, D. H. *Mass Spectrometry of Organic Compounds*; Holden-Day: San Francisco, CA, 1967, p 29.
3. McLafferty, F. W.; Turecek, F. *Interpretation of Mass Spectra*, 4th ed.; University Science Books: Sausalito, CA, 1993.
4. Field, F. H.; Franklin, J. L.; Lampe, F. W. *J. Am. Chem. Soc.* **1957**, *79*, 2419.
5. Munson, B. *Anal. Chem.* **1971**, *43*, 28A–43A.
6. Munson, B. *Anal. Chem.* **1977**, *49*, 772A–778A.
7. Munson, M. S. B.; Field, F. H. *J. Am. Chem. Soc.* **1966**, *88*, 2621–2630.
8. Mitchum, R. K.; Korfmacher, W. A. *Anal. Chem.* **1983**, *55*, 1485A–1499A.
9. Ghaderi, S.; Kulkarni, P. S.; Ledford, E. B.; Wilkins, C. L.; Gross, M. L. *Anal. Chem.* **1981**, *53*, 428–437.
10. Brodbelt, J. S.; Louris, J. N.; Cooks, R. G. *Anal. Chem.* **1987**, *59*, 1278–1285.
11. Harrison, A. G. *Chemical Ionization Mass Spectrometry*, 2nd ed.; CRC Press: Boca Raton, FL, 1992.
12. *NIST Chemistry Web Book*; NIST Standard Reference Database Number 69—February 2000 release, URL http://webbook.nist.gov/chemistry. Also Hunter, E. P. L.; Lias, S. G. *J. Phys. Chem. Ref. Data* **1998**, *27*, 413–656.
13. Buchanan, M. V. *Anal. Chem.* **1982**, *54*, 570–574.
14. Field, F. H. *J. Am. Chem. Soc.* **1969**, *91*, 6334–6341.
15. Rudewicz, P.; Munson, B. *Anal. Chem.* **1986**, *58*, 2903–2906.
16. Hunt, D. F.; Gale, P. J. *Anal. Chem.* **1984**, *56*, 1111–1114.
17. Hunt, D. F.; Harvey, T. M; Brumley, W. C.; Ryan, J. F.; Russell, J. W. *Anal. Chem.* **1982**, *54*, 492–496.
18. Simonsick, W. J., Jr; Hites, R. A. *Anal. Chem.* **1984**, *56*, 2749–2754.
19. Simonich, S. L.; Hites, R. A. *Environ. Sci. Technol.* **1994**, *28*, 939–943.
20. Horning, E. C.; Horning, M. G.; Carroll, D. I.; Dzidic, I.; Stillwell, R. N. *Anal. Chem.* **1973**, *45*, 936–943.
21. Dzidic, I.; Carroll, D. I.; Stillwell, R. N.; Horning, E. C. *Anal. Chem.* **1976**, *48*, 1763–1768.
22. Hunt, D. F.; Stafford, G. C.; Crow, F. W.; Russell, J. W. *Anal. Chem.* **1976**, *48*, 2098–2105.
23. Ong, V. S.; Hites, R. A. *Mass Spectrom. Rev.* **1994**, *13*, 259–283.
24. Milley, J. E.; Boyd, R. K.; Curtis, J. M.; Musial, C.; Uthe, J. F. *Environ. Sci. Technol.* **1997**, *31*, 535–541.
25. Dougherty, R. C.; Weisenberger, C. R. *J. Am. Chem. Soc.* **1968**, *90*, 6570–6571.
26. Tannenbaum, H. P.; Roberts, J. D.; Dougherty, R. C. *Anal. Chem.* **1975**, *47*, 49–54.

27. McLuckey, S. A.; Glish, G. L.; Kelley, P. E. *Anal. Chem.* **1987**, *59*, 1670–1674.
28. Laramee, J. A.; Kocher, C. A.; Deinzer, M. L. *Anal. Chem.* **1992**, *64*, 2316–2322.
29. Laramee, J. A.; Deinzer, M L. *Anal. Chem.* **1994**, *66*, 719–724.
30. Laramee, J. A.; Eichinger, P. C. H.; Mazurkiewicz, P.; Deinzer, M. L. *Anal. Chem.* **1995**, *67*, 3476–3481.
31. Oehme, M.; Stöckl, D.; Knöppel, H. *Anal. Chem.* **1986**, *58*, 554–558.
32. Laramee, J. A.; Arbogast, B. C.; Deinzer, M. L. *Anal. Chem.* **1986**, *58*, 2907–2912.
33. Stemmler, E. A.; Hites, R. A. *Biomed. Environ. Mass Spectrom.* **1988**, *15*, 659–667.
34. Stemmler, E. A.; Hites, R. A.; Arbogast, B.; Budde, W. L.; Deinzer, M. L.; Dougherty, R. C.; Eichelberger, J. W.; Foltz, R. L.; Grimm,C.; Grimsrud, E. P.; Sakashita, C.; Sears, L. J. *Anal. Chem.* **1988**, *60*, 781–787.
35. Arbogast, B.; Budde, W. L.; Deinzer, M.; Dougherty, R. C.; Eichelberger, J.; Foltz, R. D.; Grimm, C. C.; Hites, R. A.; Sakashita, C.; Stemmler, E. *Org. Mass Spectrom.* **1990**, *25*, 191–196.
36. Ong, V. S.; Hites, R. A. *J. Am. Soc. Mass Spectrom.* **1993**, *4*, 270–277.
37. Marquis, P. J.; Hanson, R. L.; Larsen, M. L.; DeVita, W. M.; Butterworth, B. C.; Kuehl, D. W. *Chemosphere* **1994**, *29*, 509–521.
38. Rostad, C. E. *Environ. Sci. Technol.* **1997**, *31*, 1308–1312.
39. Stemmler, E. A.; Hites, R. A. *Electron Capture Negative Ion Mass Spectra of Environmental Contaminants and Related Compounds*; VCH Publishers: New York, 1988.
40. Dougherty, R. C. *Anal. Chem.* **1981**, *53*, 625A–636A.
41. Dougherty, R. C.; Roberts, J. D.; Biros, F. J. *Anal. Chem.* **1975**, *47*, 54–59.
42. Dzidic, I.; Carroll, D. I.; Stillwell, R. N.; Horning, E. C. *Anal. Chem.* **1975**, *47*, 1308–1312.
43. Hunt, D. F.; McEwen, C. N.; Harvey, T. M. *Anal. Chem.* **1975**, *47*, 1730–1734.
44. Smit, A. L. C.; Field, F. H. *J. Am. Chem. Soc.* **1977**, *99*, 6471–6483.
45. Crow, F. W.; Bjorseth, A.; Knapp, K. T.; Bennett, R. *Anal. Chem.* **1981**, *53*, 619–625.
46. Oehme, M. *Anal. Chem.* **1983**, *55*, 2290–2295.
47. Hunt, D. F.; Crow, F. W. *Anal. Chem.* **1978**, *50*, 1781–1784.
48. Mitchum, R. K.; Korfmacher, W. A.; Moler, G. F.; Stalling, D. L. *Anal. Chem.* **1982**, *54*, 719–722.
49. Beynon, J. H. *Mass Spectrometry and its Applications to Organic Chemistry*; Elsevier: New York, 1960.
50. Russell, D. H.; Edmondson, R. D. *J. Mass Spectrom.* **1997**, *32*, 263–276.
51. Marshall, A. G.; Grosshans, P. B. *Anal. Chem.* **1991**, *63*, 215A–229A.
52. Amster, I. J. *J. Mass Spectrom.* **1996**, *31*, 1325–1337.
53. Guilhaus, M. *J. Mass Spectrom.* **1995**, *30*, 1519–1532.
54. Schwartz, J. C.; Syka, J. E. P.; Jardine, I. *J. Am. Soc. Mass Spectrom.* **1991**, *2*, 198–204.
55. Price, P. *J. Am. Soc. Mass Spectrom.* **1991**, *2*, 336–348.
56. Bartmess, J.; Little, J.; Price, P.; Voyksner, R.; Wang-Iverson, D. *ASMS "Terms and Definitions"—An Opportunity for Participation*; The 45th ASMS Conference on Mass Spectrometry and Allied Topics, Palm Springs, CA, 1–5 June 1997.
57. Lide, D. R., Ed.-in-Chief. *CRC Handbook of Chemistry and Physics,* 78th ed.; CRC Press: Boca Raton, FL, 1997/1998, pp 1–10.
58. Edmondson, R. D.; Russell, D. H. *J. Am. Soc. Mass Spectrom.* **1996**, *7*, 995–1001.
59. March, R. E. *J. Mass Spectrom.* **1997**, *32*, 351–369.
60. Guan, S.; Marshall, A. G.; Scheppele, S. E. *Anal. Chem.* **1996**, *68*, 46–71.

61. Tong, H. Y.; Giblin, D. E.; Lapp, R. L.; Monson, S. J.; Gross, M. L. *Anal. Chem.* **1991**, *63*, 1772–1780.
62. Grange, A. H.; Donnelly, J. R.; Brumley, W. C.; Billets, S.; Sovocool, G. W. *Anal. Chem.* **1994**, *66*, 4416–4421.
63. Yost, R. A.; Enke, C. G. *Anal. Chem.* **1979**, *51*, 1251A–1264A.
64. Cooks, R. G.; Glish, G. L *Chem. Eng. News* 30 November 1981, 40–52.
65. McLafferty, F. W., Ed. *Tandem Mass Spectrometry*; John Wiley: New York, 1983.
66. Hoffmann, E. de *J. Mass Spectrom.* **1996**, *31*, 129–137.
67. Cody, R. B.; Burnier, R. C.; Freiser, B. S. *Anal. Chem.* **1982**, *54*, 96–101.
68. Louris, J. N.; Cooks, R. G.; Syka, J. E. P.; Kelley, P. E.; Stafford, G. C.; Todd, J. F. J. *Anal. Chem.* **1987**, *59*, 1677–1685.
69. Johnson, J. V.; Yost, R. A.; Kelley, P. E.; Bradford, D. C. *Anal. Chem.* **1990**, *62*, 2162–2172.
70. Hunt, D. F.; Shabanowitz, J.; Harvey, T. M.; Coates, M. *Anal. Chem.* **1985**, *57*, 525–537.
71. *NIST/EPA/NIH Mass Spectral Database*; National Institute for Standards and Technology: Gaithersburg, MD, 1992.
72. Chess, E. K.; Gross, M. L. *Anal. Chem.* **1980**, *52*, 2057–2061.
73. Tondeur, Y.; Niederhut, W. N.; Campana, J. E.; Missler, S. R. *Biomed. Environ. Mass Spectrom.* **1987**, *14*, 449–456.
74. Reiner, E. J.; Schellenberg, D. H.; Taguchi, V. Y. *Environ. Sci. Technol.* **1991**, *25*, 110–117.
75. Knapp, D. R., Ed. *Handbook of Analytical Derivatization Reactions*; John Wiley: New York, 1979.
76. Drozd, J.; Novak, J. P. *Chemical Derivatization in Gas Chromatography*; Elsevier Scientific: New York, 1981.
77. Frei, R. W.; Lawrence, J. F., Eds. *Chemical Derivatization in Analytical Chemistry*; Plenum Press: New York, 1981.
78. Felice, L. J. *Anal. Chem.* **1982**, *54*, 869–872.
79. Later, D. W.; Lee, M. L.; Wilson, B. W. *Anal. Chem.* **1982**, *54*, 117–123.
80. Howdeshell, M. J.; Hites, R. A. *Environ. Sci. Technol.* **1994**, *28*, 1691–1697.
81. Glaze, W. H.; Koga, M.; Cancilla, D. *Environ. Sci. Technol.* **1989**, *23*, 838–847.
82. Budzikiewicz, H.; Djerassi, C.; Williams, D. H. *Mass Spectrometry of Organic Compounds*; Holden-Day: San Francisco, CA, 1967, p 399.
83. Budzikiewicz, H.; Djerassi, C.; Williams, D. H. *Mass Spectrometry of Organic Compounds*; Holden-Day: San Francisco, CA, 1967, p 383.

6

Compounds Not Amenable to Gas Chromatography

The successful separation, identification, and measurement of compounds with GC/MS requires target or other analytes that are thermally stable at the temperatures required to achieve appropriate separations in a reasonable analysis time. Similarly, analytes must have vapor pressures of about 10^{-8} Torr or greater at injection port, GC column, and ion source temperatures that are attainable and practical with standard instrument systems. While exceptions exist, most GC columns and commercial instrument systems are designed to operate at temperatures up to about 300 °C. Finally, analytes that are thermally stable and volatile must not be so polar, acidic, basic, or reactive that they adhere strongly to surfaces or other substances and, therefore, are not amenable to GC. These requirements were defined and discussed in more detail in Chapter 2 in the section entitled "Limitations of Mass Spectrometry". Chapters 3 and 4 contain hundreds of examples of organic compounds amenable to GC. Thermally stable, volatile, and nonreactive compounds are often target analytes because they are more persistent in the environment and not difficult to detect and measure using the well-developed GC/MS methods described in Chapters 3 and 4. Compounds that are thermally unstable, nonvolatile, or reactive are rarely amenable to GC.

In reality, however, most condensed-phase environmental samples and sample extracts probably contain a mixture of thermally stable, volatile, and unreactive compounds and some that are thermally unstable, nonvolatile, or reactive. When samples or extracts containing substances that are not amenable to GC are analyzed by conventional GC/MS, they leave residues of thermal decomposition products, unvaporized sample, and reaction products in GC injection port liners, GC columns, and the MS ion source. These residues and carbonaceous deposits can have deleterious effects on conventional GC/MS systems as described in Chapter 2. Analysts often look with disdain on *dirty* samples which contaminate instrumentation and increase the need for maintenance and GC column replacement. This problem has led to considerable research to develop methods for preliminary separations to remove undesirable components, that is, to *clean-up dirty* samples before GC/MS.

On the other hand, the thermally unstable, nonvolatile, and reactive *undesirable substances* can include many compounds of considerable interest. Second and later generation insecticides and herbicides frequently contain nitrogen, oxygen, phosphorus, or sulfur, and combinations of these elements and are designed to be more biodegradable than the first-generation persistent chlorinated hydrocarbon pesticides. However, these newer products are often more polar, thermally unstable, and nonvolatile and often they cannot be determined by conventional GC/MS. Some of these compounds are of long-standing environmental concern, but efficient techniques for their determination, comparable to GC/MS, have not been available. In addition, macromolecules of biological origin are of considerable interest because they are associated with pathogenic bacteria, protozoa, or viruses in environmental media.

The purpose of this chapter is to consider strategies for the determination of this diverse class of thermally unstable, nonvolatile, or reactive compounds. Historically, a major limitation of MS has been the inability to determine most compounds of this type. However, significant research efforts over the last 25 years have produced a variety of techniques that can be used to identify and measure analytes that previously were beyond the scope of any form of MS. While these techniques are not as broad in scope or as well developed as conventional GC/MS for broad spectrum environmental analyses, great progress toward developing them has been made in recent years. In the next section of this chapter the structures and some properties of representative members of several classes of thermally unstable, nonvolatile, or reactive analytes are briefly described. Analytical strategies are considered including methods for the separation of these substances in the condensed-phase. Finally, techniques are described for interfacing condensed-phase separation systems with a mass spectrometer and introducing thermally unstable, nonvolatile, or reactive analytes.

Classes of Compounds and Their Structures and Properties

Chart 6.1 shows the structures and names of a variety of compounds of environmental interest, which are representative of classes of analytes that are not amenable to GC/MS or are marginal analytes with GC/MS. Some are thermally unstable or nonvolatile to varying degrees, and others are just too polar, acidic, basic, or reactive to permit separation by GC. The structures shown in Chart 6.1, and the groups of compounds described briefly in the following sections, are not intended to be a comprehensive listing of all types of compounds in this category, but they are illustrative of the structural types of some important compounds of environmental interest. These compounds and classes of compounds are frequently used as examples in subsequent sections of this chapter.

Table 6.1 lists compounds that are not amenable to GC/MS, or are marginal analytes, and which have maximum contaminant levels (MCLs) in drinking water in the United States. The MCL is a human health risk-based value and, therefore, is an indication of the presently known potential for exposure to these compounds and their hazardous nature.

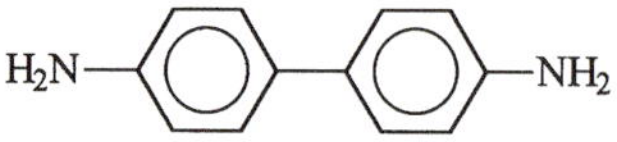

Benzidine

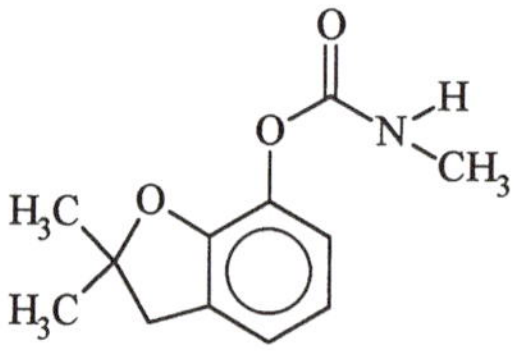

Carbofuran

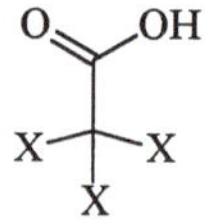

Haloacetic Acids
X = Cl, Br, or H

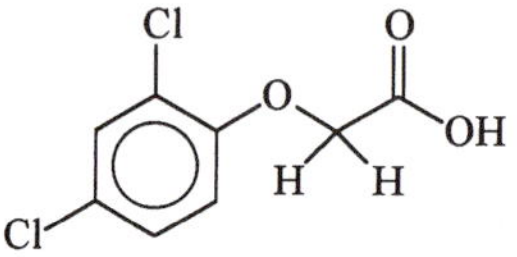

2,4-Dichlorophenoxyacetic acid (2,4-D)

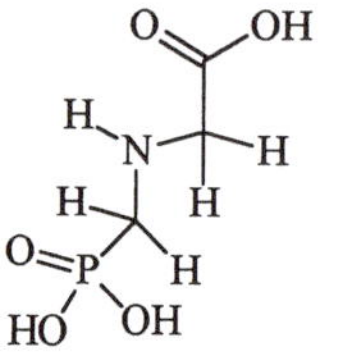

Glyphosate

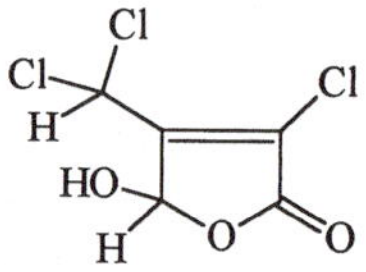

3-Chloro-4-(dichloromethyl)-5-hydroxy-2(5H)-furanone (MX)

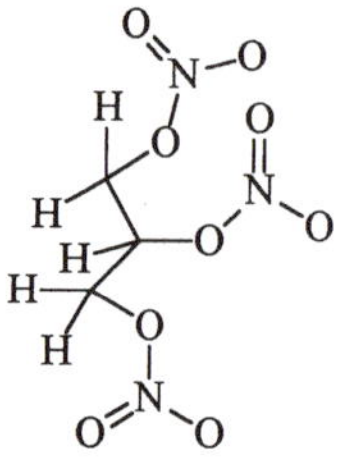

Nitroglycerine

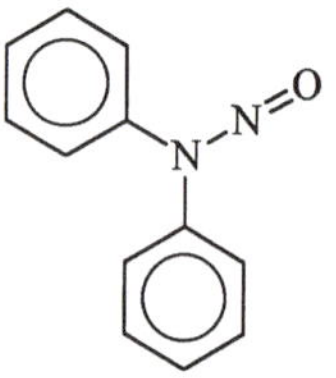

N-Nitrosodiphenylamine

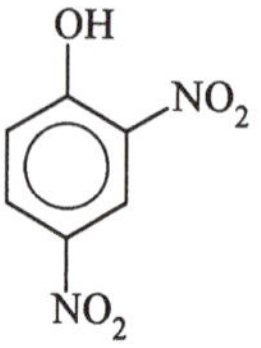

2,4-Dinitrophenol

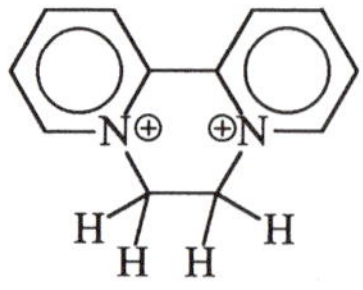

Diquat

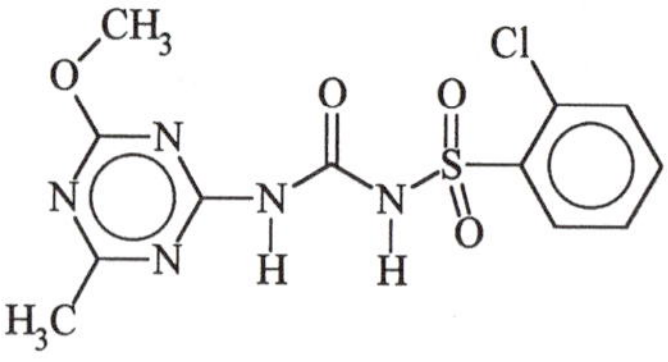

Chlorsulfuron

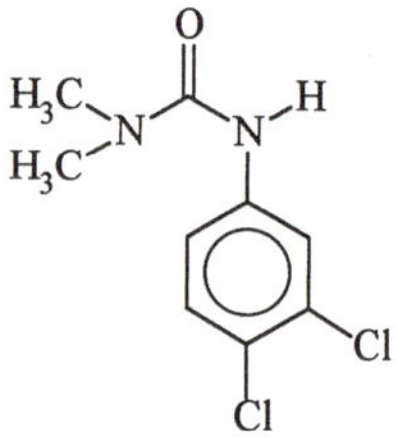

Diuron

Chart 6.1

Table 6.1 Compounds not amenable to GC/MS or marginal GC/MS analytes, their uses or source, and their maximum contaminant levels (MCLs) in drinking water in the United States

Compound	Use or source	MCL in drinking water (μg/L)
4-Amino-3,5,6-trichloropicolinic acid (picloram)	Herbicide	500
Bromoacetic acid	Disinfection by-product	[a]
Carbofuran	Insecticide	40
Chloroacetic acid	Disinfection by-product	[a]
Dibromoacetic acid	Disinfection by-product	[a]
Dichloroacetic acid	Disinfection by-product	[a]
2,4-Dichlorophenoxyacetic acid (2,4-D)	Herbicide	70
2,2-Dichloropropionic acid (dalapon)	Herbicide	200
Diquat	Herbicide	20
Glyphosate (Roundup™)	Herbicide	700
2-(1-Methylpropyl)-4,6-dinitrophenol (dinoseb)	Herbicide	7
Oxamyl	Insecticide	200
7-Oxobicyclo[2.2.1]heptane-2,3-dicarboxylic acid (endothall)	Herbicide	100
Pentachlorophenol[b]	Wood preservative	1
Trichloroacetic acid	Disinfection by-product	[a]
2-(2,4,5-Trichlorophenoxy)propionic acid (2,4,5-TP, silvex)	Herbicide	50

[a]60 μg/L for the sum of the concentrations of the five haloacetic acids cited in the table.[1]
[b]Pentachlorophenol is a marginal GC/MS analyte.
Source: Data from Ref. 2.

Aromatic Amines

Aromatic amines are defined and discussed in Chapter 4 (under "Classes of Semivolatile Organic Compounds"). These important compounds, especially the primary amines, are marginal GC/MS analytes, and alternative strategies are appropriate for many applications, particularly when long-term reliability and low detection limits are needed. Benzidine (left end of top row in Chart 6.1), various substituted benzidines (Table 4.1), and the other aromatic primary diamines listed in Table 4.1 are best determined using the techniques described in this chapter.

Carbamate Pesticides

A carbamate is an ester of carbamic acid ($H_2N–COOH$), which is the half-amide of carbonic acid (H_2CO_3). Carbamate pesticides, which are mostly used as insecticides, are a large and diverse group of substances that are esterified with a variety of organic groups. In addition, most carbamate pesticides have one or two organic groups substituted for the hydrogens of the nitrogen. For example, carbofuran (center of top row in Chart 6.1) has an *N*-methyl group. Many carbamate pesticides have the *N*-methyl structure because they are often prepared by the reactions (Chart 6.2) of a variety of substituted phenols (Reaction 6.1) or oximes (Reaction 6.2) with methyl isocyanate. The products are called *N*-methylcarbamates and *N*-methyl oxime carbamates or *N*-methylcarbamoyloximes, respectively. Some carbamates are restricted to use by certified applicators in the United States, and the uses of carbofuran are severely restricted in the United States.[3, 4] Carbofuran also has a MCL in drinking water in the United States (Table 6.1).

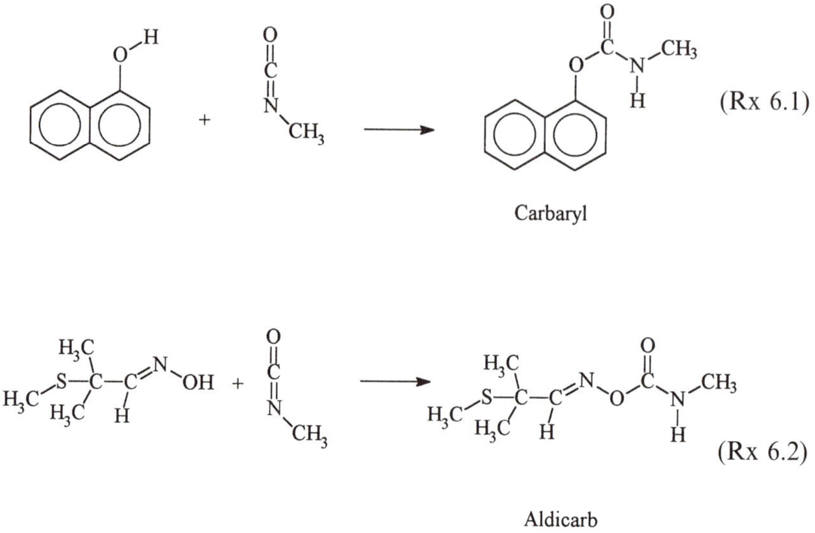

(Rx 6.1)

Carbaryl

(Rx 6.2)

Aldicarb

Chart 6.2

While a few carbamates are amenable to separation by GC, for example, chlorpropham in Table 4.1, most are not and either show signs of thermal decomposition or fail to elute from the GC column.[5] Although not a general approach, very short GC columns and low injector and column oven temperatures do allow the separation of some carbamates by GC.[6] The thermal instability of many *N*-methylcarbamates seems to derive, at least in part, from the relative ease of reversal of Reaction 6.1 when these compounds are heated in a mass spectrometer sample introduction system.[7] The electron ionization (EI) mass spectra of the carbamate pesticides in the NIST/EPA/NIH database were probably measured with sample introduction by direct insertion probe.[8] The *N*-methylcarbamate spectra generally show very low abundance ($<15\%$) or no $M^{+\cdot}$ ions and very abundant $(M-57)^{+\cdot}$ ions corresponding to the loss of methyl isocyanate, which could occur either before or after ionization. Most of the carbamates prepared from oximes, except aldicarb (Chart 6.2) and its oxidation products, show a similar tendency to lose methyl isocyanate. However, the oxime derivatives are less thermally stable and usually have no $M^{+\cdot}$ ions in their probe EI spectra. Some carbamates have organic groups other than methyl substituted for the hydrogens on the nitrogen, and these are more thermally stable and some have $M^{+\cdot}$ ions in their probe EI spectra.

Substitution of a sulfur for one or both oxygens of a carbamate gives a thio- or a dithio-carbamate. Some thiocarbamates are sufficiently thermally stable for GC and are semivolatile analytes listed in Table 4.1 and briefly discussed in Chapter 4 under "Nitrogen-based Pesticides". Substitution of both carbamate oxygens by sulfur gives dithiocarbamates and some *N*,*N*,*S*-trialkyl dithiocarbamates containing small alkyl groups are amenable to GC, but more complex substituents are likely to cause thermal decomposition.[9]

Carboxylic and Other Acids

Carboxylic acid functional groups (–COOH) are contained in an enormous variety of structures of organic compounds, but only a few of them, for example, acetic acid and propionic acid, are amenable to separation by GC. Most of these acids tend to decarboxylate on heating or they may vaporize but adhere to basic sites in injection port liners and other surfaces and give broad tailing GC peaks with variable areas and poor signal/noise. Carboxylic acids of environmental interest include disinfection by-products (DBPs), herbicides, degradation products of some pesticides, and other commercial products. Sulfonic and phosphonic acids, which contain carbon–sulfur and carbon–phosphorus bonds, respectively, are also found in some products such as azodyes, herbicides, and their degradation products. These acids are generally stronger than carboxylic acids and they are not amenable to GC.

The haloacetic acids containing chlorine or bromine or both (right end of top row in Chart 6.1) are DBPs whose formation in drinking water is described in the next section. The U.S. Environmental Protection Agency (USEPA) has classified dichloroacetic acid as a probable human carcinogen and trichloroacetic acid as a possible human carcinogen.[1] A MCL of 60 μg/L in drinking water is established in the United States for the sum of the concentrations of the five haloacetic acids listed in Table 6.1.[1] The haloacetic acids are strong acids with pK_a values generally below 3.[10] Chloroacetic acid, which is also listed as a hazardous air pollutant in Table 3.1, is volatile and amenable to GC. However, strong interactions between these acids and surfaces or basic sites in the GC system and a tendency toward thermal decomposition of the brominated acids precludes the use of GC for the separation of the haloacetic acids.

The EI spectra of five of the nine possible chloro-, bromo-, and bromo-chloro-acetic acids are contained in the NIST/EPA/NIH Mass Spectral Database.[8] For these measurements the acids were probably introduced into the mass spectrometer through a small orifice from a heated reservoir or with a direct insertion probe. The spectra of the chloro-, dichloro-, and trichloro-acetic acids either have no $M^{+\cdot}$ ions or very low abundance $M^{+\cdot}$ ions. The most abundant ions in their spectra correspond to the loss of CO_2 from the $M^{+\cdot}$ ions. The bromo- and dibromo-acetic acids have similar spectra except that the $M^{+\cdot}$ ion of bromoacetic acid has a relative abundance of 50%. While these five acids have sufficient vapor pressures to allow measurements of their mass spectra with a reservoir or probe inlet system, drinking water analyses require an on-line separation of mixtures of acids, and an alternative to GC is required. A structurally related herbicide, 2,2-dichloropropionic acid (dalapon), has an established MCL in drinking water in the United States (Table 6.1) and is not amenable to GC.

Many carboxylic acids have been or are used as herbicides on a variety of crops and for general weed control. A very common example is 2,4-dichlorophenoxyacetic acid (left end of second row in Chart 6.1). Other closely related herbicides are 2,4,5-trichlorophenoxyacetic acid (2,4,5-T) and 2-(2,4,5-trichlorophenoxy)propionic acid (2,4,5-TP or Silvex), which are banned in the United States but might be used in other countries.[4] A related group of herbicides are derivatives of benzoic and other aromatic acids, for example, 2-methoxy-3,6-dichlorobenzoic acid (dicamba). Table 6.1 gives the names of the carboxylic acid herbicides that have MCLs in drinking water in the United States. The EI spectra of many of these and related acids are available in the NIST/EPA/NIH database and some of their spectra have very abundant $M^{+\cdot}$ ions.[8] These compounds were very likely introduced into the mass spectrometer with a direct insertion probe and some of them have sufficient thermal stabilities and vapor pressures to allow measurements of their spectra with a probe inlet system. However, analyses of environmental samples require on-line separations, and GC is unsuitable for the separation of these herbicides because of strong interactions between the acids and surfaces or basic sites in injection port liners and standard GC columns and thermal decomposition at elevated temperatures in a GC system.

The herbicide glyphosate (middle of second row in Chart 6.1) is a tribasic acid containing both carboxylic acid and phosphonic acid groups. The spectrum of glyphosate in the NIST/EPA/NIH database shows clear signs of thermal decomposition even with the probable use of a direct insertion probe for sample introduction.[8] The spectrum has a 10% $M^{+\cdot}$ ion at m/z 169 and a base peak at m/z 102, but most of the ion abundance is distributed among a large number of low m/z fragments below m/z 83. Glyphosate and a dicarboxylic acid herbicide, endothall, both have MCLs in drinking water in the United States (Table 6.1). Neither these acids nor the imidazolinone herbicides, which also contain a carboxylic acid group, are amenable to separation by GC.[11]

Azodyes have the general structure Ar−N=N–Ar′ in which one or more azo groups link together a variety of aromatic (Ar) ring systems including phenyl, naphthyl, biphenyl, and others. Many compounds in this diverse group of dyes contain one or more carboxylic acid or sulfonic acid substituents, which impart water solubility and provide linking groups for the attachment of the dyes to fibers and other materials. Compounds in this group are of environmental interest because of the potential for anaerobic reduction of the azo group to produce carcinogenic aromatic primary amines. The azodye carboxylic and sulfonic acids are not amenable to GC.

Disinfection By-products

Disinfection by-products (DBPs) are incidentally produced when chlorine, ozone, and other oxidants are added to water to control bacteria, protozoa, and viruses. Disinfection is mainly associated with drinking water, but municipal and industrial wastewaters are often disinfected during treatment before discharge into a local river or lake. Chlorine reacts with humic substances present in many ground and surface waters to produce a variety of chlorinated organic compounds including the trihalomethanes (Table 3.1) and the haloacetic acids (preceding section and Table 6.1).[12] When bromide ion is present in the source water, as it is in most surface and ground waters, chlorine oxidizes bromide to generate very reactive bromine atoms, which produce brominated and mixed bromochloro DBPs.[13] When the bromide ion concentration is relatively high, as it is in some coastal areas and ground waters, brominated compounds dominate the distribution of DBPs.[12] Ozonation of water containing humic substances produces formaldehyde, acetaldehyde, and other aliphatic aldehydes.[14]

Although many DBPs are amenable to GC and were discovered with GC/MS, some, including the haloacetic acids and aldehydes, are not. Another DBP that is not amenable to GC is 3-chloro-4-(dichloromethyl)-5-hydroxy-2(5*H*)-furanone (right end of second row in Chart 6.1). This compound is a potent bacterial mutagen that was called mutagen X before its structure was known and is still frequently abbreviated to MX. It was discovered in the early 1980s in chlorinated wastewater from pulp mills and subsequently found in chlorinated drinking waters in Finland, Great Britain, and the United States.[15] The compound and similar structures are not amenable to separation by GC.

Nitric Acid Esters and N*-Nitro Compounds*

The nitric acid esters of glycerol (1,2,3-trihydroxypropane) and pentaerythritol [tetrakis(hydroxymethyl)methane] are among the most well known of this group of compounds. Glyceryl trinitrate, commonly called nitroglycerine (left end of third row in Chart 6.1), is a familiar explosive. The tetranitrate ester of pentaerythritol is another explosive often abbreviated as PETN. The two possible glyceryl nitrates, two glyceryl dinitrates, glyceryl trinitrate, and the corresponding esters of pentaerythritol are potential environmental pollutants in soil, streams, and ground water in the vicinity of munitions manufacturing and processing plants. These compounds are also of interest in forensic investigations, but they are thermally unstable and shock sensitive and are not amenable to separation by GC.

The EI mass spectrum of nitroglycerine in the NIST/EPA/NIH database was likely measured with a direct insertion probe and does not have a $M^{+\cdot}$ ion at *m/z* 227.[8] An ion at *m/z* 76 (10%) is $N_2O_3^+$ and the base peak at *m/z* 46 corresponds to NO_2^+. Therefore, decomposition is complete probably even with mild conditions in the direct insertion probe. Similarly, the EI spectrum of PETN has no $M^{+\cdot}$ ion at *m/z* 316, a base peak at *m/z* 76, and an 80% ion at *m/z* 46. *N*-Nitro compounds are also explosives and two well-known compounds are *N,N,N*-trinitrohexahydrotriazine (RDX) and *N,N,N,N*-tetranitro-octahydrotetrazocine (HMX). The EI spectra of these compounds in the NIST/EPA/NIH database were surely measured with a direct insertion probe. The spectra of RDX and HMX do not have $M^{+\cdot}$ ions at 222 and 296, respectively. The highest *m/z* ion present in the spectrum of RDX is *m/z* 128 (25%) and the base peak is *m/z* 46 (NO_2^+). The spectrum of HMX has an ion at *m/z* 148 (10%) and a base peak at *m/z* 46.

N-*Nitroso Compounds*

N-Nitroso compounds are defined and described in Chapter 4 (under "Classes of Semivolatile Organic Compounds"). One of the compounds mentioned in that section is *N*-nitrosodiphenylamine (middle of third row in Chart 6.1), which is formed by the reaction of NO with diphenylamine. Diphenylamine has been used for many years as a stabilizer in propellants where it captures nitrogen oxides that form by slow decomposition of these materials. The N–NO bond in this class of compounds is weak and many *N*-nitroso compounds decompose on heating and are not amenable to separation by GC.

Organometallic Compounds

Organometallic compounds contain a metal to carbon bond and a few of them have been or are important industrial products and/or environmental contaminants. A few elements which are not strictly metals, for example, As and Se, are present in many organic compounds and are included in this category of analytes. Examples from this class of compounds are listed below and the structures of some environmentally important organometallic compounds are shown in Chart 7.1.

- Tetra-alkyllead gasoline additives used in some countries.
- Methylcyclopentadienyl manganese tricarbonyl used as a gasoline additive in some countries.
- Organomercury halides produced naturally in the environment from elemental mercury and mercury compounds used as pesticides.
- Organoarsenic compounds produced naturally in the environment from inorganic arsenic and arsenic compounds used as pesticides.
- Organoselenium compounds produced naturally in the environment.
- Tri-*n*-butyltin chloride formed in sea water by the leaching of tri-*n*-butyltin compounds from antifouling paints on the hulls of boats and ships.

Some organometallic compounds are amenable to separation by GC, especially those in which the metal is bonded to only small nonpolar organic groups. Examples of compounds in this category are diethylmercury, trimethylarsine, and tetraethyllead. However, this type of compound is not common in the environment because they are often reactive and are oxidized or hydrolyzed to compounds having one or two fewer carbon–metal bonds. The more probable environmental contaminants are organometallic compounds in which the metal is also bonded to one or more carboxylate anions, halides, or some other inorganic anion. Examples of compounds in this category, which are usually not amenable to GC, are phenylmercuric chloride, dialkyllead acetates, and tri-*n*-butyltin chloride.

Some compounds of arsenic, cadmium, mercury, lead, and thallium are either banned or severely restricted from use as pesticides or gasoline additives in the United States and some other countries.[4] The presence of organometallic compounds of these elements in the environment is the result of past utilization and disposal, continuing use of some products, or methylation in the environment, which is a natural detoxification mechanism. Elements such as arsenic, mercury, selenium, and others continue to be widely dispersed by emissions from fossil fuel-burning power plants and metal smelting and processing operations.

Tri-*n*-butyltin compounds are biocides that have been widely used in many products including coatings applied to the hulls of ships and boats.[16] A variety of tri-*n*-butyltin compounds, for example, the fluoride or methacrylate, are used in antifouling paints. In sea water, tri-*n*-butyltin chloride, which is very toxic to aquatic organisms, is slowly leached from the coating and kills the organisms that attach to hulls and cause metal corrosion and drag. Because of these toxic effects some tri-*n*-butyltin compounds are restricted to use by certified applicators in the United States.[3] The concentrations of tri-*n*-butyltin chloride in sea water and aquatic organisms are of considerable interest and there have been attempts to separate tributyltin chloride by GC. However, the polarity of the Sn–Cl bond causes considerable difficulties due to adsorption on injection port liners and columns and reactions with active sites in GC columns. This results in broad tailing GC peaks, variable peak areas, and poor signal/noise. The EI mass spectra of the tri-*n*-butyltin halides and the oxide are contained in the NIST/EPA/NIH database and they were probably measured with a direct insertion probe.[8] These spectra do not have $M^{+\cdot}$ ions and the highest m/z fragments are caused by loss of one of the butyl groups.

Besides the techniques discussed in this chapter, the mass spectrometric strategies for elemental analyses discussed in Chapter 7 are also applicable to organometallic compounds.

Phenols

Phenols are defined and discussed in Chapter 4 (under "Classes of Semivolatile Organic Compounds"). While many phenols are amenable to separation by GC, phenols with multiple electron-withdrawing nitro, fluoro, and chloro groups are acidic and often give broad tailing GC peaks with variable peak areas and poor signal/noise. Pentachlorophenol (Table 6.1), dinoseb (Table 6.1), and 2,4-dinitrophenol (right end of third row in Chart 6.1) are difficult to separate and measure with accuracy, precision, and low detection limits by GC/MS. This is probably a result of strong interactions between these acids and surfaces or basic sites in GC injection port liners and standard columns. The herbicide dinoseb, which is banned for use in the United States,[4] is a derivative of 2,4-dinitrophenol with a *sec*-butyl group attached to the ring adjacent to the phenolic hydroxyl group. Both pentachlorophenol, a widely used wood preservative, and dinoseb have drinking water MCLs in the United States (Table 6.1).

Polycyclic Aromatic Hydrocarbons

Polycyclic aromatic hydrocarbons (PAHs) are defined and discussed in Chapter 4 (under "Classes of Semivolatile Organic Compounds"). As mentioned there, PAH with molecular weights $>\sim 300$ and/or with more than six fused rings, including several classified as reasonably anticipated to be human carcinogens,[17] are generally not amenable to separation by GC. Even somewhat smaller PAHs, for example, benzo[*a*]pyrene and benzo[*g,h,i*]perylene (Chart 2.1) or indeno[1,2,3-*c,d*]pyrene (Table 4.1), can have retention times of 25–30 minutes or longer with typical GC columns and conditions. This can result in broad GC peaks with variable areas and poor signal/noise.

Quaternary Ammonium Compounds

Quaternary ammonium compounds have a nitrogen with a positive charge because all the valence electrons of the nitrogen are engaged in bonding to other atoms. Frequently, the nitrogen is bonded to four organic groups as in, for example, the tetramethylammonium cation. Generally, these cations are not volatile without decomposition and they are not amenable to separation by GC. Diquat (left end of last row in Chart 6.1) and the closely related compound paraquat are herbicides that have two quaternary nitrogens and are doubly charged cations with, usually, chloride or bromide counter ions. Diquat has a MCL in drinking water in the United States (Table 6.1) and paraquat is restricted to use by certified applicators in the United States.[3] The EI spectra of these compounds are in the NIST/EPA/NIH database and were very likely measured with a direct insertion probe for sample introduction.[8] The spectra show very weak, that is, $<2\%$, ions at m/z 92 and 93, which could be the doubly charged molecular cations or thermal decomposition products. The principal ions are at m/z 156 (100%) for diquat and 156 (100%), 171 (10%), and 186 (10%) for paraquat. These principal ions are singly charged and are probably produced by the reductive decomposition of the dications.

Sulfonylurea Herbicides

Sulfonylureas are urea (the diamide of carbonic acid) derivatives with an RSO_2 group substituted for one of the hydrogens on a nitrogen. This family of herbicides, which has a sulfur–nitrogen bond, is represented by chlorsulfuron (middle of last row in Chart 6.1). These compounds are polar and generally thermally unstable at typical GC injection port and column temperatures.[18] Compounds in this family invariably contain one heterocyclic ring with two or three nitrogen atoms and a substituted phenyl or another heterocyclic group. The multiplicity of polar structures within these compounds and the weakly acidic hydrogens on the urea nitrogens provide abundant opportunities for intermolecular hydrogen bonding, which probably contributes to their lack of volatility. The 1992 release of the NIST/EPA/NIH database does not contain the EI mass spectra of any of 18 sulfonylurea herbicides whose structures have been published.[18–20]

Urea Derivatives and Related Pesticides

Substituted ureas, as illustrated by diuron (right end of last row in Chart 6.1), are widely used herbicides. In some derivatives the carbonyl oxygen is substituted by a sulfur which is then a thiourea. With typical GC operating conditions, the substituted ureas diuron, siduron, and *N*-(1-naphthyl)thiourea failed to elute from the GC column.[5] Linuron and 2-chlorophenylthiourea gave tailing or distorted GC peak shapes, multiple GC peaks, or other evidence of partial thermal decomposition.[5] However, one urea derivative, tebuthiuron, is listed as a GC/MS analyte in Table 4.1. The EI spectra of some substituted urea herbicides are in the NIST/EPA/NIH database and these spectra were probably measured using a direct insertion probe for sample introduction.[8] Generally, these spectra contain $M^{+\cdot}$ ions in the 1–40% relative abundance range, but have few fragment ions corresponding to losses of small groups or single atoms. The mass spectra tend to be dominated by ions with $m/z < 100$, which suggests that the compounds, or their $M^{+\cdot}$ ions, are unstable.

Strategies for Broad Spectrum Analyses

The broad spectrum (BS) analysis strategy was defined and described in Chapter 2. This approach to chemical analyses is appropriate when the goal is to determine what compounds are present in a sample without a predetermined list of target analytes. This goal is realistic for general exploratory studies or in the case of a specific environmental effect that could be caused by many different substances, for example, a fish kill. The BS strategy can be successfully implemented because a large number of compounds are amenable to a few of the versatile GC/MS methods described in Chapters 3 and 4. These methods for air, water, and solid samples include techniques for sampling, sample preparation, GC separation, and measurement. The scope of each method was established by analyses of various sample matrices fortified with a variety of compounds. This process established lists of target analytes for the methods, but many untested compounds can also be detected and identified within each method. However, as broad in scope as these GC/MS methods are, they include only part of the universe of organic compounds of potential interest. A truly BS analysis must also include compounds not amenable to GC and compounds that are not even soluble in a typical organic solvent used in GC/MS.[21]

Broad spectrum analytical strategies for compounds not amenable to GC, such as those in the groups described in the previous sections of this chapter, are essentially undeveloped compared to the GC/MS approaches in Chapters 3 and 4. This is probably due to a combination of factors including the following:

- The classes of compounds not amenable to GC contain a variety of reactive functional groups, and often several functional groups per compound. These compounds have a broad range of thermal stabilities, volatilities, water solubilities, organic solvent solubilities, and chemical reactivities and cannot be categorized into a few broad classes as was possible with the GC/MS analytes.

- The condensed-phase separation techniques for compounds not amenable to GC are either less well developed or less versatile than capillary column GC. These techniques tend to accommodate smaller and more well-defined groups of analytes rather than very broad classes of compounds.
- The interfacing strategies used to connect condensed-phase separation techniques and the mass spectrometer are generally more complex, but less versatile and less well developed than the fundamentally simple and versatile techniques for open tubular column GC/MS briefly described in Chapter 2.

Because of these constraints, the goal of developing a few BS analytical approaches for compounds not amenable to GC has not been realized. Analytical strategies are usually limited to small groups of analytes whose physical and chemical properties are influenced by a few similar functional groups. The target analyte approach is widely practiced with this category of compounds.

However, the potential for broader approaches exists and this has been explored to some extent.[22] In any organic solvent extract of an environmental sample, which is prepared for the purpose of a BS GC/MS analysis, there may be compounds, for example, carbamate pesticides, larger PAHs, or urea herbicides, that are not amenable to GC/MS, but that could be separated and analyzed by using one of the techniques described in this chapter. Similarly, after a sample is exhaustively extracted with an organic solvent, the original sample phase may still contain compounds, for example, carboxylic acids or quaternary ammonium salts, that can be partially or completely separated, identified, and measured by using the techniques described in this chapter. Many samples can be concentrated without excessive heating and analyzed for compounds not amenable to GC/MS. These concepts do not appear to have been applied in many environmental studies, but they have considerable potential as condensed-phase separation/MS techniques are further developed and become more widely used. Some investigators are beginning to apply both GC/MS and condensed-phase separation/MS methods to analyses of environmental samples.[23]

Desorption Ionization and Related Probe Techniques

The direct insertion probe (DIP), which was briefly described in Chapter 2 under "Limitations of Conventional GC/MS", expanded greatly the range of compounds that could be easily introduced into the mass spectrometer. The nominal 70-eV EI DIP spectra of some compounds not amenable to GC were discussed in this chapter in the section entitled "Classes of Compounds and their Structures and Properties". However, a DIP sample introduction system requires both a reasonably pure sample and sufficient sample, perhaps a few micrograms, to permit manual sample handling. Since most environmental and many other types of samples consist of very dilute mixtures or solutions of microgram to nanogram quantities of compounds which require on-line separation, the DIP is of very limited use in most analyses.

Many other probe and closely related techniques have been developed in the last 25 years and these have enormously expanded the applicability of MS to thermally unstable, nonvolatile, higher molecular weight, and chemically reactive substances. While there have been few, if any, environmental applications of these techniques to date, again because on-line separations are usually not possible, they are often very convenient and can provide rapid results and valuable information. Furthermore, the tandem MS techniques described in Chapter 5 can be used with samples that consist of several components and this obviates, to a limited extent, the on-line separation requirement if sufficient sample in a concentrated form is available. Because these probe techniques have considerable potential in environmental analyses, several are briefly described in this section with references to reviews and research papers for more complete information. In general, the probe techniques discussed are more appropriate for preliminary screening and identifications of compounds than for quantitative analyses because calibration for quantitative analysis is difficult. Potential environmental applications include rapid screening and analyses of bulk materials, solid samples, and fractions collected in preliminary separations, and the identification of biological molecules of environmental interest.

Direct Exposure Probe

The direct exposure probe (DEP) is a DIP that is extended so that a surface coated with sample is within the ion source and is directly exposed to ionizing electrons from a filament or the reagent ions of a chemical ionization reagent gas. The DEP is a simple modification of a DIP that is more effective at producing ions, including $M^{+\cdot}$ ions, at lower temperatures from smaller samples of thermally unstable, nonvolatile, or reactive substances.[24] When used with chemical ionization, this technique is called desorption chemical ionization (DCI). The DEP is a moderately priced accessory that is usually available for mass spectrometers that are or can be equipped with a standard DIP. If an unknown organic solid is collected from a waste drum and presented for analysis, a DCI experiment is a reasonable first step to rapidly gain some preliminary information about the nature and purity of the substance. Similarly, a solid residue from the evaporation of solvent from a separated liquid chromatography fraction may be rapidly examined by DCI. The DCI spectra of many compounds have been measured and reported in the literature, including the spectra of eight sulfonylurea herbicides.[25]

Field Desorption

Field desorption (FD) is a process in which a sample coated on an emitter is placed within a few millimeters of an electrode with an 8–12 kV potential difference, and sample molecules are desorbed into the gas phase and ionized.[26] The emitter is a thin tungsten wire mounted on a probe, and the wire is usually coated with activating microcrystalline needles of pyrolytic carbon or silicon. There are several mechanisms of ionization in FD including the gentle removal of an electron by the high field to produce high-abundance $M^{+\cdot}$ ions with little or no fragmentation. The emitter may be electrically heated to enhance volatilization, ionization, or fragmentation, and a wide range of polar and nonvolatile molecules, including lower molecular weight polymers, have been examined using this technique. Field desorption is not commonly available on many mass spectrometers, but it has been available on most models of magnetic deflection instruments. If available, FD is a technique that can be applied to the types of environmental samples described under the DEP.

Fast Atom Bombardment

Fast atom bombardment (FAB) is the bombardment of a sample on a probe with accelerated atoms, usually Ar or Xe, which energize the sample and dislodge into the gas phase various atoms, molecules, positive ions, and negative ions created by multiple collisions within the sample.[27] The sample is contained in a viscous supporting matrix such as glycerol, and the beam of high-energy atoms is created by ionizing the gas to generate Ar^+ or Xe^+ ions, accelerating these ions through a potential difference of 3–10 kV, then neutralizing the ions to form a beam of atoms which retain their kinetic energy. The device which produces the beam of keV energy atoms is called a fast atom gun. The FAB technique provided a major advance in MS technology and allowed the measurement, for the first time, of gas phase ions from a great variety of higher molecular weight, polar, thermally unstable, and nonvolatile molecules containing multiple acidic, basic, and other functional groups. These measurements included peptides, small proteins, organometallic compounds, pharmaceuticals, natural pigments, carbohydrates, and synthetic polymers. The FAB ionization process produces positive and negative ions and polar organic molecules that are usually observed as their protonated $(M + H)^+$ or deprotonated $(M - H)^-$ ions. In the presence of Na^+ or K^+, high abundances of molecules associated with alkali metal cations are observed, and some fragmentation of these molecular species is usually present in FAB spectra.

Unlike most other probe techniques, considerable efforts were made to couple FAB with flow injection and microbore (1 mm ID or less) liquid chromatography (LC). The combined technique, which is called continuous flow FAB, allows the introduction of compounds in a flowing stream and on-line LC separations of mixtures of compounds including peptides and other biomolecules.[28,29] Continuous flow FAB, coupled with tandem MS, has been used for the flow injection quantitative analysis of alkylbenzenesulfonate surfactants in water and sludge samples from a sewage treatment plant.[30] Because of the constraints of continuous flow FAB interfaces, flow rates are restricted to about 1–5 µL/min, and LC columns are either very narrow bore open tubular (10 µm ID) or packed micro capillaries (50 µm ID).[28] The FAB interface is a moderately priced accessory that has been commercially available on a number of types of mass spectrometer. However, because of other developments in desorption techniques and other LC/MS interfaces, which are discussed later in this chapter, FAB systems are not as readily available on new mass spectrometers as they have been in the past.[31] The basic FAB probe technique, if available, can be used for rapid screening of environmental samples as described under the DEP, or continuous flow FAB can be used with flow injection or microbore LC for more detailed analyses. This technique also can be used for the examination of biological molecules of environmental interest, for example, small peptides, deoxyribonucleic acid (DNA) fragments, carbohydrates, and natural pigments associated with pathogenic or indicator organisms.

Liquid Secondary Ion Mass Spectrometry

Liquid secondary ion MS (LSIMS) is essentially the same as FAB except the desorption and ionizing agent is a beam of keV ions, typically Cs^+ ions.[31–33] The device that produces the beam of ions is called an ion gun. The two terms, FAB and LSIMS, have been used in the literature to distinguish between the type of bombarding particle, but the resulting mass spectra are often very similar or the same. The ion gun technique has advantages over the fast atom gun including higher sensitivity and the ability to focus the ion beam on a small target or a portion of a sample. Other issues that can contribute to practical differences between the techniques include background contamination and ion source pressure. The ion gun technique has been used with a similar broad range of higher molecular weight, polar, thermally unstable, and nonvolatile molecules containing multiple acidic, basic, and other functional groups. The availability of this accessory and the potential applications are the same as described under FAB.

Plasma Desorption

Plasma desorption (PD) was among the earliest desorption techniques described and applied to nonvolatile organic compounds.[34] In this technique the sample is bombarded with very high energy (MeV) particles from a particle accelerator or from the spontaneous nuclear fission of ^{252}Cf. Plasma desorption mass spectrometers were, and still may be, commercially available and several hundred or more were constructed primarily for research on biomolecules, especially small proteins and natural products.[35] Assuming the availability of such systems, the PD technique may have environmental applications similar to those described for the other desorption ionization techniques.

Matrix Assisted Laser Desorption

Perhaps the most popular and widely used desorption technique of the 1990s is matrix assisted laser desorption ionization (MALDI).[36] In this technique, analyte molecules are placed on a probe in a liquid or solid matrix consisting of small molecules that strongly absorb at the frequency of the laser radiation. The absorbed energy is transferred from the matrix molecules to the analyte molecules, which are ejected as ions into the gas phase. The matrix has several functions including absorbing nearly all the radiant energy, preventing thermal decomposition of the analyte molecules, and preventing undesirable chemical interactions between the analyte molecules. This technique has been enormously successful for the determination of very large protein and other biomolecules with molecular weights of 10^3–10^6 Da. Typically mono-, di-, and a few higher protonated molecules are observed with little or no fragmentation, but adduct ions from the matrix are observed. The most likely and potentially significant application of MALDI in environmental analyses is the identification of biomolecules, for example, proteins, associated with pathogenic or indicator organisms. A tutorial review that summarizes the mechanics and mechanisms of desorption ionization in FAB, LSIMS, PD, and MALDI has been published.[37]

Desorption Ionization and Microorganisms

While the environmental applications of these primarily probe techniques are very limited, some, particularly MALDI, are likely to grow in importance as more instrumental techniques are used to assess human exposure to pathogenic environmental microorganisms including bacteria, protozoa, and viruses. Classical microbiological techniques require slow culturing and depend heavily on microscopic visual inspection for identification and measurement. Newer biochemical techniques, such as the polymerase chain reaction,[38] can rapidly amplify the quantity of a characteristic DNA fragment and provide sufficient material for a rapid and decisive mass spectrometric analysis. The recognition of characteristic peptides or proteins from microorganisms should be attainable as sensitivity and mass spectrometric resolving power increases. The difficulty of detecting the protozoa *Cryptosporidium*, which caused a 1993 outbreak of waterborne disease in Milwaukee, is an indication of the need for new biochemical/instrumental approaches to microbiological analyses.[39] Desorption ionization is emphasized in a major review of the capabilities of MS for the identification of microorganisms.[40]

Chemical Derivatization

Chemical derivatization is the conversion of a compound, or a group of structurally related compounds, into another compound, or group of compounds, by reaction with a chemical reagent or reagents. This technique was discussed in Chapter 5 as a strategy to enhance selectivity and lower detection limits for certain types of analytes. Derivatization has been used for many years to facilitate GC separations of compounds that either are not amenable to GC or are marginal GC analytes. This approach is most successful with compounds that have reactive functional groups that are readily converted into derivatives that are amenable to separation by GC. For example, esterification of carboxylic acids changes the functional group from one that often decomposes on heating, or is adsorbed on to basic sites on surfaces, to a less reactive and thermally stable structure. A complete description of chemical derivatization is beyond the scope of this book, and other reference works should be consulted for more complete information.[41–44] In this section, several types of derivatives that are used in environmental and other analyses are briefly described. Some limitations and disadvantages of derivatization, which have led investigators to seek alternative strategies, are also presented.

Carboxylic Acids and Phenols

The haloacetic acid disinfection by-products and acid herbicides shown in Chart 6.1 and Table 6.1 are frequently determined as their methyl ester derivatives, which are amenable to GC separation. Diazomethane is one of the most effective and convenient methylating agents (Reaction 6.3) although other reagents are used. Reactive methylating reagents, such as diazomethane, also convert acidic phenols, for example, 2,4-dinitrophenol (Chart 6.1) and pentachlorophenol (Table 6.1), into volatile, thermally stable, and neutral methyl ethers. The more reactive reagents, including diazomethane, are very sensitive to water and, therefore, analytes must be transferred into an aprotic solvent, which is then thoroughly dried before the reagent is added. A major concern with diazomethane is that it is a very toxic gas that is subject to violent explosions under certain conditions. Another powerful methylating reagent, dimethyl sulfate, is also extremely hazardous and is on the reasonably anticipated to be a carcinogen list.[17]

$$R\text{–}COOH + H_2C{=}N{=}N \longrightarrow R\text{–}COOCH_3 + N_2 \qquad \text{(Rx 6.3)}$$

Some methylating reagents are less hazardous and function in the presence of moisture or even in the aqueous phase. However, these methylating reagents, which include sulfuric acid in methanol or BF_3 in methanol, are not always as effective as diazomethane and may require longer reaction times and heating and may not produce high percentage conversions of all analytes. Individual analyte reactivities with various reagents can very widely as a function of analyte structure and reaction conditions. The mutagenic disinfection by-product MX (Charts 6.1 and 6.3), which is not amenable to separation by GC, is determined by GC/MS after methylation of the hydroxy group with sulfuric acid in methanol for 1 h at 70 °C.[15] High-resolution MS with SIM is often used to provide enhanced selectivity for the derivative (Chapter 5, "High-Resolution Mass Spectrometry"). With EI the methyl derivative of MX does not give a $M^{+\cdot}$ ion, and the two fragment ions used for SIM and quantitative analysis have m/z values that differ by <0.02 Da (Chart 6.3).

Carboxylic acids and phenols are ubiquitous and often important compounds, and many other techniques have been investigated for the preparation of their volatile and thermally stable esters and ethers. Pentafluorobenzyl esters and ethers are described in Chapter 5 as useful derivatives to increase selectivity and lower detection limits for phenols, thiols, carboxylic acids, and other weak acids (Chart 5.5). Resin acids, which are toxic to fish, have been identified by conventional GC/MS in a pulp mill effluent as their pentafluorobenzyl esters.[45] The esters were prepared by treating the dry residue of the sample extract with pentafluorobenzyl bromide, which is a much less hazardous reagent than diazomethane or dimethyl sulfate. An advantage of this approach is that methyl and other esters indigenous to the environmental sample can be distinguished from acids that are converted into pentafluorobenzyl esters. Some herbicides are commercially formulated as esters and these may be distinguished from the free acids if the sample is not treated with strong base first to hydrolyze the esters.

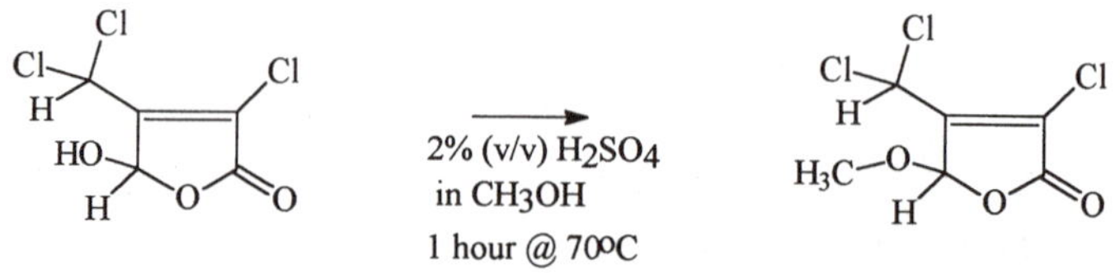

2% (v/v) H_2SO_4 in CH_3OH
1 hour @ 70°C

3-Chloro-4-(dichloromethyl)-5-hydroxy-2(5H)-furanone (MX)

MX methyl ether (MW 230)

$(M-OCH_3)^+$ ions at 198.9120, 200.9091, 202.9061

$(M-HCO)^+$ ions at 200.9277, 202.9247, 204.9218

Chart 6.3

Phase transfer derivatization is a process in which the reagents are placed in the sample and/or an immiscible solvent, and the process of derivatization facilitates the transfer of the derivative from the aqueous phase to the solvent. The technique has been used with several reagents and solvents including supercritical carbon dioxide.[46] Several chlorinated phenoxyacetic acid herbicides are methylated in high yield by methyl iodide and a tetra-alkylammonium salt catalyst, and the methyl esters are transferred to supercritical carbon dioxide in the process.[47] Methyl esters of some acids may be too volatile and can be lost during solvent evaporation or other processing steps. Lower molecular weight acids in wet precipitation and aerosol samples were identified as their *n*-butyl esters by treatment of the residue from the evaporation of the sample with BF_3 and *n*-butanol.[48]

Microorganisms, especially pathogenic organisms, are of enormous environmental significance, but are usually not considered analytes in analytical chemistry or MS. Nevertheless, fatty acids that originate in the cell membrane phospholipids of bacteria are frequently used to identify bacteria by GC separation of their methyl esters.[40,46] The cell membrane fatty acids are in the C_{14}–C_{18} or higher range and the distribution of acids is characteristic of various bacteria. Free fatty acids in this molecular weight range are not amenable to GC separation.

Other Analytes Not Amenable to GC

Acetylation and perfluoroacetylation of aromatic amines, for example, benzidine (Chart 6.1), was described in Chapter 5 as a strategy to increase selectivity and lower detection limits for this class of compounds. These derivatives are also more amenable than the more basic aromatic amines to GC separation. The quaternary ammonium salt herbicide paraquat, which has a structure similar to that of diquat (Chart 6.1), has been derivatized for GC separation by conversion into a mixture of partially saturated cyclic amines by reduction with sodium borohydride in sodium hydroxide solution. Reduction gave mono and diene products, which were extracted into an organic solvent before GC analysis. The production of two, and possibly other products, complicates the quantitative analysis and reduces sensitivity because the signal is divided among several derivatives.[49]

Organometallic compounds such as tri-*n*-butyltin chloride can be determined by GC after phase transfer reduction of the Sn–Cl group to Sn–H by sodium borohydride in the aqueous sample containing methylene chloride.[50] This is very effective because the hydride is immediately partitioned, as it is formed, into the methylene chloride layer where it is protected from hydrolysis by water. The tri-*n*-butyltin hydride is sufficiently volatile and thermally stable for separation by GC. Alternatively, the Sn–Cl group can be converted into a Sn–R group, where R is a small alkyl group such as methyl, ethyl, or propyl. This technique requires the extraction of the tri-*n*-butyltin chloride into an organic solvent, which must be thoroughly dried before reaction with a Grignard reagent, R–Mg–X, or an organolithium reagent.[51]

Thermally stable, nonreactive, and volatile trimethylsilyl derivatives of mono- and poly-functional alcohols have been widely used to permit GC separation of a variety of compounds that are not amenable to GC.[52,53]

Limitations and Disadvantages

In the foregoing discussion of derivatives, some of the types of structures shown in Chart 6.1 were not mentioned. Readily prepared, thermally stable, and volatile derivatives are not available for carbamate pesticides, nitric acid esters, *N*-nitroso compounds, sulfonylurea herbicides, urea pesticides, and some other classes of compounds. Derivatization requires the analyte to have a chemically reactive functional group and some classes of compounds either are not chemically reactive or do not have an easily derivatized functional group. On the other hand, some derivatives are readily prepared, but their mass spectra are sometimes dominated by ions characteristic of the derivative and not of the parent molecule. This occurs with some trimethylsilyl derivatives whose spectra have high abundances of the trimethylsilyl cation (Me_3Si^+) at m/z 73, which is not helpful for the identification of the parent compound.

Searching for new methods of producing volatile derivatives is not attractive because of the variability and complicating by-products often encountered in these reactions, and the additional time, skill, and cost required for derivatization during sample preparation. These considerations have reduced the use of derivatization in some studies even when acceptable techniques were available. The limitations and disadvantages of chemical derivatization have tended, in recent years, to increase research to find condensed-phase separation techniques that can be connected directly to a mass spectrometer to create the equivalent of GC/MS for thermally unstable, reactive, and nonvolatile compounds.

Pyrolysis Mass Spectrometry

Pyrolysis is the deliberate thermal decomposition of chemical substances. The technique of pyrolysis MS (Py/MS) was probably developed following unsuccessful attempts to separate by GC or obtain the mass spectra of thermally unstable compounds. The volatile pyrolysis products from the thermal decomposition were analyzed and found to contain useful analytical information. A variety of thermal degradation techniques have been developed, but there have been only a few applications to environmental analyses.[54] Samples may be very rapidly heated to a specific temperature, then held at that temperature as the volatile pyrolysis products are analyzed by MS or collected for GC/MS. Alternatively, the sample may be heated slowly and the volatile products analyzed continuously or trapped for subsequent GC/MS analysis. A number of variations of these techniques have been used including the entrainment of volatile products in a flowing inert gas stream. Several types of Py/MS sample introduction systems have been commercially available and some have the ability to maintain precise temperature control.

Pyrolysis MS techniques are important for the determination of the nature of volatile compounds produced from organic polymers at elevated temperatures, for example, in an incinerator. Brominated biphenyl flame retardants, which are reasonably anticipated to be carcinogens,[17] have been detected in plastics with Py/MS.[55] The Py/MS technique was used to classify beeswax, honey, pollen, and bee feces that were implicated in yellow rain occurrences in southeast Asia during the Vietnam war.[56] Fourteen sulfonylurea herbicides (see chlorsulfuron in the middle of the last row in Chart 6.1) were pyrolyzed at 800 °C for 5 s and the volatile decomposition products were analyzed by conventional GC/MS. The neutral fragments formed gave mass spectra that could be used to identify sulfonylurea herbicides.[18] Microorganisms have been studied by Py/MS and this technique may be useful for the identification of some organisms.[40, 57]

With the development of the desorption ionization techniques described in previous sections, and condensed-phase separation/MS techniques described later in this chapter, the Py/MS technique probably will find only occasional use in environmental analyses. Nevertheless, given the availability of the equipment in the laboratory, Py/MS may be a viable choice for the rapid examination of apparently nonvolatile solids collected from a waste site. Similarly, a solid residue from the evaporation of solvent from a separated liquid chromatography fraction may be rapidly examined by Py/MS. The identification of high molecular weight polymers, and their additives, in waste streams could be an important environmental application of this technique.

Condensed-Phase Separation Techniques

The development of high-performance condensed-phase separation techniques over about the last 25 years is one of the most significant advances in the history of separation science. These techniques, which are now routinely available on commercial instruments, allow the separation of thermally unstable, nonvolatile, or reactive compounds with an analytical performance that sometimes is the equivalent of, or even superior to, GC. A few condensed-phase separation techniques are used in environmental analyses and applications of these techniques with on-line mass spectrometers are likely to increase in the future. A thorough review of all the condensed-phase separation techniques that conceivably could be used with MS is beyond the scope of this book. In this section, several separation techniques are briefly described with emphasis on those aspects that are relevant to on-line separations with a mass spectrometer. Other reference books and reviews are cited which provide detailed information about these separation techniques. All of the techniques discussed are considered *high performance* and this descriptor, which is frequently used in the literature, is not used in subsequent discussions.

Liquid Chromatography

Modern analytical liquid chromatography (LC) is conducted in a closed system consisting of a metal tube (column) packed with very fine particles, a sample injection device, a high pressure pump to force the mobile phase solvent or solvent mixture through the packed tube, and a detector. Wall-coated open tubes with very small inside diameters (IDs), that is, 50 μm or less, are also used for some applications. Figure 6.1 is a diagram showing the major components of a LC/MS instrument. These major components are briefly reviewed in

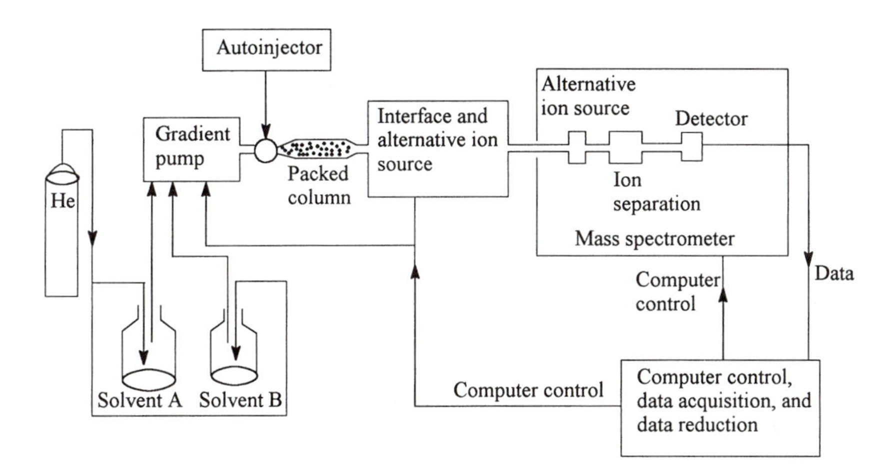

Figure 6.1 The major components of a LC/MS instrument.

this section, and various MS interfaces are discussed in greater detail in subsequent sections of this chapter.

Packed LC analytical columns vary in diameter from 4.6 mm ID, or larger, down to a fraction of 1 mm ID, and from a few centimeters in length to 30 cm or more. The LC literature frequently refers to the use of 4.6 mm ID analytical columns, but there is a trend toward the use of smaller bore columns (<0.5–2 mm ID), which provide higher column efficiencies and sensitivity, use substantially less mobile phase, and allow interfacing to a mass spectrometer with relative ease and little waste of sample.

Contemporary column packings consist mainly of silica particles, usually 3–10 μm in diameter, coated with a covalently bonded organic stationary phase. This organic phase can be nonpolar, for example, the ubiquitous saturated straight-chain C_{18} alkyl group, or more polar such as cyanopropyl groups. The organic group (R) is attached to the silica through formation of Si–O–Si–R bonds at silanol groups (Si–OH) on the silica surface. Organic polymers, such as crosslinked styrene–divinylbenzene copolymers, are also used as stationary phases in analytical separations. Analytes are retained on the column by an equilibrium process in which the dissolved molecule or ion is alternately associated with the stationary phase, through weak noncovalent bonding interactions, and the mobile phase where it is transported toward the detector. Unfortunately, commercial packed columns are expensive and are subject to rapid deterioration with raw samples, contaminated sample extracts, and repeated use. Short, inexpensive, and expendable guard columns are frequently used in series before the analytical column to protect it from sample components that shorten column lifetime.

Mobile phases vary widely from very polar water to nonpolar hydrocarbon solvents, including mixtures of miscible solvents. The type of mobile phase used depends on the type of column used and the analytes under investigation. Isocratic elution is the use of a mobile phase which has a constant composition throughout the elution of analytes from the column. Gradient elution is the gradual changing of the mobile phase composition, for example, a mixture of two solvents, as the analytes elute from the column. Gradient elution is frequently used to enhance analyte resolution and shorten the time required for a separation. Mobile phase solvents are usually purged with helium during operations to remove dissolved air and other gases, which can disrupt the smooth flow of mobile phase through the system under high pressure (Figure 6.1).

Conventional detectors for LC (not shown in Figure 6.1) are fewer in variety than those used with GC. The single-wavelength ultraviolet–visible (UV–VIS) absorption detector is probably the most widely used, but it is only applicable to analytes that absorb in the UV–VIS, and high sensitivity is limited to analytes that absorb strongly. Diode array detectors (DADs) can measure continuously the UV–VIS spectra of eluting analytes over the range of about 190–600 nm, but are similarly limited to some analytes. Preinjection or postcolumn on-line derivatizations are widely used to provide a UV–VIS absorbing or fluorescing derivative. The UV–VIS detector responds to natural background materials in some environmental samples and sample extracts and this increases background noise and reduces the reliability of the detector. In general, conventional LC detectors have significant limitations similar to those of conventional GC detectors as discussed in Chapter 2. As with GC detectors, retention times and retention indices are generally inadequate for the identification of analytes in environmental samples and sample extracts. Further abundant general and detailed information on LC history,[58] theory, terminology, equipment, micro columns,[59] mobile phases, detectors,[60] and retention mechanisms[61] has been published.[62–65]

Normal-Phase Liquid Chromatography. Normal-phase LC is the use of a more polar stationary phase with a much less polar or nonpolar mobile phase such as hexane, toluene, or methylene chloride. This form of LC takes its name from the traditional open-column class separations of, for example, aliphatic and aromatic hydrocarbons. The hydrocarbon class separations employ silica or alumina stationary phases and hydrocarbon or chlorinated hydrocarbon mobile phases. Contemporary normal-phase LC employs polar stationary phases covalently bonded to a silica support. This form of LC is definitely the second choice, after reverse-phase LC (next section), except for hydrocarbon or lipid samples that are not sufficiently resolved by, or soluble in, an aqueous reverse-phase mobile phase. Many of the classes of analytes described earlier in this chapter as thermally unstable, nonvolatile, or reactive (Chart 6.1) are not very soluble in nonpolar mobile phases. In addition, normal-phase LC does not offer the flexibility of pH adjustments, ionic strength adjustments, and other devices to fine tune retention, resolution, and selectivity.

Reverse-Phase Liquid Chromatography. The workhorse of modern LC is reverse-phase LC, which is the use of a nonpolar or slightly polar stationary phase with a polar mobile phase such as water, methanol, acetonitrile, or mixtures of these solvents. A vast array of organic compounds containing a variety of functional groups (Chart 6.1) are soluble in mixtures of water and methanol or acetonitrile. Compounds in this diverse group are often amenable to separation with stationary phases consisting of C_8, C_{18}, phenyl, trimethylsilyl, or other nonpolar organic groups chemically bonded to the silica surface. The pH of typical mobile phases can be adjusted over a reasonable range to improve resolution and selectivity. Various buffer substances, salts, and ion pair reagents can be used to control pH and ionic strength and facilitate the separation of ionic analytes. Reverse-phase LC is more complex than GC because of the number of operational parameters including column dimensions, column packings, mobile phases, gradient elution, and various mobile phase additives. Most analyses that employ a condensed-phase separation use reverse-phase LC and gradient elution with mixtures of water and methanol or acetonitrile. Acetonitrile has a viscosity lower than that of methanol, which is an advantage with narrow bore columns.

Ion Exchange Chromatography. Ion exchange is a traditional form of LC which uses a stationary phase consisting of a crosslinked synthetic organic polymer, often called a resin, with $-SO_3H$ or $-NH_2$ groups attached to phenyl or other aromatic rings on the polymer backbone. Mobile phases are usually water or water and a miscible organic solvent. In basic solution the SO_3H groups of a cation exchange resin are ionized and consist of SO_3^- groups and associated counter ions, for example, Na^+. Cationic analytes, for example, diquat in Chart 6.1, are retained on the cation exchange column by displacing the resin counter ions in an equilibrium process. The cationic analytes are subsequently eluted with a mobile phase containing a high concentration of counter ions or counter ions of a higher charge, for example, Ca^{2+}. Isocratic and gradient elutions are used with gradients in both solvent composition and counter ion concentration. Mobile phase pH is a very important operational parameter, which can have a significant impact on the retention or elution of various ionizable analytes from a resin.

In an acid solution the amino groups of an anion exchange resin are protonated and exist as NH_3^+ groups and associated counter ions, for example, Cl^-. Analyte anions, for example, haloacetate or phenoxyacetate anions (Chart 6.1), are retained on the column by displacing the resin counter ions in an equilibrium process. The anion analytes are subsequently eluted with a higher concentration of counter ions or some type of gradient elution. Anion exchange chromatography is widely used for the separation of inorganic anions and chelated metal anions in aqueous samples. With some natural or industrial water samples that contain high concentrations of ions such as Na^+, Ca^{2+}, Mg^{2+}, Cl^-, and SO_4^{2-}, cation and anion analytes may not be retained on the column and this can result in poor analyte recoveries.

Supercritical Fluid Chromatography

Supercritical fluids are not condensed-phases, but supercritical fluid chromatography (SFC) is considered along with the condensed-phase separation techniques because it is used for the separation of some of the same classes of analytes. Supercritical fluids are substances heated above their critical temperatures where they cannot be condensed regardless of the pressure. The supercritical region exists over a range of pressures above the critical pressure, and supercritical fluids have physical properties similar to both liquids and gases.[66, 67]

The properties of supercritical fluids that make them attractive as chromatography mobile phases include favorable mass transfer characteristics and readily variable solvent strengths. Supercritical fluids are less viscous than liquid solvents, and solutes diffuse more rapidly than in liquid solvents. This property can potentially provide more rapid and higher resolution separations with SFC than with LC. Supercritical fluids have solvent strengths comparable to liquid solvents. The solvent strength of a supercritical fluid can be controlled by temperature, just as with liquid solvents, and pressure, which is not an option with noncompressible liquid solvents. Lower pressures, and therefore lower densities, favor less polar analytes and higher pressures, and higher densities favor more polar analytes. Pressure gradient elution is an important capability of SFC.

The nature of the supercritical fluid is a major factor contributing to its value as a mobile phase. Many substances have been evaluated and used in SFC including low molecular weight hydrocarbons, fluorinated methanes, alcohols, ammonia, water, and nitrogen oxides. However, for all practical purposes, CO_2 is the only substance that is widely used. All of the other substances either have too high a critical temperature (methanol, water), are flammable (ethane, ethene, and so on), expensive, or otherwise undesirable (ammonia, fluorinated methanes, SF_6, N_2O, Xe). Carbon dioxide has a low critical temperature (31 °C) and critical pressure (72.9 atmos). It is chemically unreactive, nonflammable, nontoxic, inexpensive, and available in high purity.

Unfortunately, supercritical CO_2 is also rather nonpolar with a solvent strength, depending on the pressure, in the range between fluorinated aliphatic hydrocarbons and hexane.[66] This tends to limit SFC with pure CO_2 to less polar compounds such as some polycyclic aromatic hydrocarbons, polychlorinated biphenyls, chlorinated hydrocarbon pesticides, alcohols, and fatty acid esters of glycerol. As noted in Chapter 4, some of the higher molecular weight compounds in these groups are difficult to separate by GC because of their low vapor pressures at typical GC column temperatures. Several of the classes of analytes shown in Chart 6.1, for example, some carbamates, *N*-nitrosodiphenylamines, and urea herbicides, are amenable to separation by SFC with a pure CO_2 mobile phase.[68–70] Small portions of more polar solvents (modifiers), such as methanol, are often added to the CO_2 to enhance the solubility of more polar compounds.

For SFC, the solvent supply and helium purge in Figure 6.1 are replaced by a source of pure CO_2, the pump and injector are modified for supercritical fluid operation, and the column is placed in an oven to maintain the temperature above the critical temperature. Both packed and open tubular columns have been used for SFC, but current trends in commercial instruments favor the packed-column approach.[71] Open tubular columns used with SFC/MS usually have IDs of ~50 μm, limited solute capacities of ~10 ng or less per component, and require small injections (~ 10–100 nL) of relatively concentrated solutions (10–100 μg/mL). These and other factors can result in long analysis times for complex mixtures.[70] Packed columns (1–4.6 mm ID) with standard LC stationary phases overcome many of these difficulties.

Standard GC and LC detectors are used with SFC, but suffer from the usual deficiencies with real-world environmental or other samples (Chapter 2, "Alternatives to Conventional GC/MS"). Mass spectrometer interfaces used with SFC are very similar to those used in LC/MS and they are discussed along with the LC interfaces later in this chapter. There have been few environmental analytical applications of SFC, but potential applications, in addition to those analytes mentioned previously, include polysiloxanes, ethoxylated alcohol surfactants, and low molecular weight polymers.[72] In general, SFC with a pure CO_2 mobile phase is most appropriate for higher molecular weight analytes that have vapor pressures too low to allow separation by GC, but are relatively nonpolar and soluble in the mobile phase. In practice the SFC technique is more like normal-phase LC than reverse-phase LC.

Capillary Electrophoresis

Capillary electrophoresis (CE) is the separation of charged analytes in solution in narrow bore (~50–100 μm ID) fused silica tubes by application of a large potential difference to the ends of the tube.[73,74] A high positive voltage, for example, 25 kV, is applied to an electrode in contact with an electrolyte solution, the anode, in a source reservoir, and one end of the tube is immersed in the electrolyte solution. The other end of the tube is immersed in an electrolyte solution in a destination reservoir, the cathode, which is usually grounded but could be at some smaller positive or a negative potential. The mobile phase is usually water, or mixtures of water and a miscible organic solvent, and contains the supporting electrolyte which is usually a pH buffer. A sample is injected into the capillary in the source reservoir, and the separation of ions is accomplished by the different rates of migration of different ions in the electric field. Figure 6.2 is an exaggerated illustration of a simple CE system, which shows the migration of ions under the influence of the high potential difference. The fused silica tube is usually uncoated, the positive ions in solution migrate away from the anode and toward the cathode, and negative ions migrate toward the anode. This process is called electrophoretic migration and is illustrated by the movement of groups of positive and negative ions near the bends in the greatly exaggerated tube in Figure 6.2.

A second and very important ion migration mechanism is caused by the interaction of supporting electrolyte ions in solution with silanol (Si–OH) groups on the surface of the fused silica tube. Silanol groups are acidic and are ionized in a rapid equilibrium process when in contact with an aqueous solution, as shown in Reaction 6.4:

$$\text{Surface–Si–OH} + H_2O \rightleftarrows \text{Surface–Si–O}^- + H_3O^+ \qquad \text{(Rx 6.4)}$$

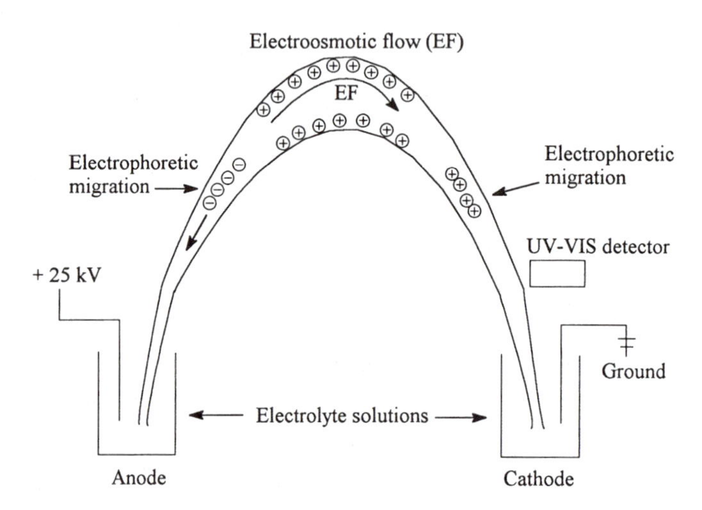

Figure 6.2 Exaggerated diagram of a simple capillary electrophoresis apparatus.

The degree of ionization and the resulting charge on the wall surface is a function of the solution pH, the percentage of nonaqueous solvent, and other factors. Even in moderately low pH solutions the fused silica walls are charged negatively and attract a layer of positive ions from the supporting electrolyte, as shown in the central part of Figure 6.2. These electrostatically bound, but still quite mobile, positive ions tend to move away from the high positive voltage anode and toward the cathode and, in the process, drag along the bulk liquid in the narrow bore tube. This phenomenon is called electro-osmotic (EO) flow and it is a major factor in the migration of ions over a wide pH range. The EO flow is sufficiently strong in more basic solutions to overcome the electrophoretic migration of negative ions toward the anode and drag them with the bulk liquid toward the cathode. The electrophoretic migration of positive ions toward the cathode is supported by the EO flow and this accounts for the rapid separations observed with CE.

Conventional single-wavelength UV–VIS detectors are widely used with CE and are placed at transparent windows in the fused silica tube, as shown in Figure 6.2. As ions which absorb in the UV–VIS pass the window they are detected as very narrow bands with very high resolution. There are many variations on the basic CE experiment. The polarity of the high voltage can be reversed, which reverses the electrophoretic migration of ions and the EO flow. This technique is not usually used to separate and detect negative ions because they would have to overcome the strong EO flow toward the negative electrode, which would result in long migration times for the negative ions. Negative ions are usually separated by using the conventional arrangement (Figure 6.2) and allowing the strong EO flow toward the cathode to overcome their electrophoretic migration toward the anode. The charge on the fused silica tube walls can be reversed by derivatizing the silanol groups with reagents that contain free amino groups that attract protons from the solution and become positively charged.[75] This creates a mobile layer of negative electrolyte ions along the tube walls, and the EO flow can be reversed without reversing the polarity of the electrode. Surfactants added to the mobile phase create charged micelles which partition neutral analytes into their nonpolar ends and also have electrophoretic mobilities. This technique, called micellar electrokinetic capillary chromatography, can be used to separate neutral compounds and ionic compounds in the same solution.[76] The fused silica tube can be packed with C_{18}–silica particles, and neutral compounds can be separated by reverse-phase LC by using the EO flow, rather than high pressure, to force the mobile phase through the tube.[77] This technique is called capillary electrochromatography.

The CE technique has emerged as a very important separation technique because of the very high resolving power that can be achieved and the high speed of separations compared to LC. The technique is limited in the sense that injection volumes (tens of nanoliters) and flow rates (nanoliters per minute) are exceedingly small and the detector must be very sensitive to analyte mass or the sample solution must be more concentrated (i.e., 10–100 μg/mL). These limitations can be overcome by sample stacking techniques that effectively combine multiple injections into a single analysis, or by some type of traditional concentration of an environmental extract. A major advantage of CE, compared to other condensed-phase packed column separation techniques, is the very low cost of CE columns, which are expendable and easily replaced if contaminated or damaged.

The vast majority of reported applications of CE are concerned with the separation of biological molecules such as proteins, peptides, DNA fragments, and drugs. There have been few environmental applications, but the potential for separation of charged analytes of environmental interest is high, especially when CE is on-line to a mass spectrometer. Negatively charged polysulfonated azodyes have been separated and detected with picomole to femtomole sensitivity[78] and the herbicides paraquat and diquat (Chart 6.1) were separated in about 10 min by CE.[79] The exit of a CE column is connected to a mass spectrometer using one of several interfaces that are modifications of LC/MS interfaces. The CE/MS technique is discussed along with the SFC/MS and LC/MS techniques in the next section.

Mass Spectrometer Interfaces

The condensed-phase separation techniques discussed in the previous section were developed to provide high resolution separations of compounds not amenable to GC. However, conventional detectors, such as ultraviolet absorption, cannot provide sufficient information to support high confidence identifications of target and other analytes in complex mixtures (Chapter 2, "Alternatives to Conventional GC/MS"). For this reason a significant amount of research over the last 25 years has been directed to the development of LC/MS, SFC/MS, and CE/MS. The connection of a condensed-phase separation technique to a mass spectrometer requires an interface (Figure 6.1). The interface is defined as the device that receives the condensed-phase flow from the separation system and converts this flow into gas phase analyte molecules or ions suitable for injection into the high vacuum of the mass spectrometer. The design of this interface is considerably more challenging than the design of an interface between a mass spectrometer and a gas chromatograph equipped with a narrow-bore open tubular column, which has a very small flow of helium gas. Interfaces are divided in this section into two general types:

- Most of the earlier interface designs, roughly between 1973 and 1984, used a strategy that was a natural extension of the approach used with GC/MS. The function of the interface was to separate as much as possible of the neutral mobile phase from the dissolved neutral analytes, which were then introduced into the vacuum of the mass spectrometer where they were ionized by standard techniques. These designs were directed toward normal and reverse phase LC and SFC.
- Several early and the most recent interface designs convert the analytes into ions in the interface, or separate ions directly as in CE. These designs use the differences in the properties of ions and neutral molecules to separate the bulk of the neutral mobile phase from the

analyte ions and introduce mainly the analyte ions into the high vacuum of the mass spectrometer.

Several of the interface designs based on the earlier strategy are currently in use and have been used for a number of years. These designs are discussed first with brief descriptions of those techniques that are mainly of historical interest. The second design strategy is covered in the last part of this chapter. Because of this division, the individual interface descriptions are not in the chronological order of their development. Citations to the original literature can be used to trace the chronological, and often concurrent, developments and the exchange of ideas among various instrument developers. Examples of environmental analyses are given with most designs, and citations to publications are provided for additional details and information.

Direct Liquid Introduction

Direct liquid introduction (DLI) was the earliest technique used to introduce the flow from a LC column into a mass spectrometer.[80] The DLI strategy is to split the flow from the chromatograph, typically about 1 mL/min from a standard 4.6 mm ID column, and direct roughly 1%, or 10 μL/min, through a narrow bore capillary into the chemical ionization (CI) source of a mass spectrometer. With microbore LC columns (1 mm ID or less), up to 100% of the flow of 10–50 μL/min can be directed into the mass spectrometer.[81] The ion source end of the capillary is drawn to a very small opening, as small as ~5 μm diameter, to control the flow of liquid into the source. The mobile phase is usually methanol, pentane, acetonitrile, or some other fairly volatile solvent or a mixed solvent which can include water. The capillary may be heated to vaporize the mobile phase and analytes, or heat is provided by the hot (~250 °C) ion source. The flow into the ion source is adjusted at the splitter or by the size of the orifice to maintain an ion source pressure appropriate for CI, that is, about 0.2–2 Torr. The vaporized mobile phase molecules are present in enormous excess over the analyte molecules. Ionization of the vaporized mobile phase by filament electrons produces reagent ions which react with the analyte molecules to provide typical CI mass spectra of separated analytes (Chapter 5, "Chemical and Related Ionization Techniques").

The DLI technique extended the range of compounds that can be separated on-line and injected into a mass spectrometer. This extended range included many polar or reactive compounds that are not amenable to separation by GC. The DLI interface also evolved over time, and several variations of the basic idea were developed. One design accommodated very narrow bore (10–15 μm ID) open tubular LC columns with stationary phases chemically bonded to the tube walls.[82] The mobile phase flow rates from these columns were sufficiently low, a few nanoliters per minute or less, to permit the acquisition of standard EI spectra of the analytes. However, the best DLI devices had several deficiencies and these are worth noting because they are some of the same deficiencies that have plagued other types of interfaces.

- Many thermally unstable and nonvolatile analytes cannot be detected because vaporization of the analyte, usually with some heating, is required.
- Analyte detection limits can be rather high, for example, hundreds of nanograms, especially if most of the LC flow is diverted from the MS. Very low injection volumes (nanoliters) used with open tubular LC columns require very high mass spectrometer sensitivity or sample solutions that are more concentrated, (i.e., 10–100 μg/mL).
- Ions below about *m/z* 120 often cannot be measured because of the great abundance of reagent ions from the ionization of the mobile phase.
- Many inorganic acids, bases, buffers, and salts commonly used to optimize reverse-phase LC separations can only be used sparingly because they cause nonvolatile deposits in the interface and ion source and some of them cause corrosion of the metal parts of the mass spectrometer.
- Plugging of capillary tubes and small pinhole orifices with various deposits is a major problem.
- Chemical ionization is generally the only ionization technique available and CI requires a significant level of maintenance for the ion source, vacuum system, and other parts of the spectrometer, which become contaminated with mobile phase, carbon, and other deposits.

Several DLI interfaces were commercially available during the mid-1980s and these were designed as inexpensive accessories for mass spectrometers that are or can be equipped with a standard direct insertion probe. These devices may still be available in some laboratories and may be used for some types of analyses. Publication of DLI research and applications declined significantly after the late 1980s as new developments in interfaces captured the attention of investigators. There are few reported environmental applications of DLI probably because most DLI investigations were conducted during a period when environmental research was primarily focused on analytes that are amenable to GC/MS. Standard solutions of a few explosives,[83] carbamate pesticides,[84] and several metabolites of the herbicide trifluralin[85] have been studied by LC/DLI/MS, but little is known about the quantitative aspects of this technique. Given the availability of a DLI probe, this LC/MS technique could be of some value for the preliminary investigations of environmental samples.

Direct Fluid Introduction

The DLI interface has survived, however, as the direct fluid introduction (DFI) interface for SFC. The DFI interface is used with open tubular SFC and, after splitting the higher flows, with packed column SFC.[72] In this application, the ion source end of the capillary is drawn to a very narrow opening, ~5 μm, which restricts the flow of supercritical CO_2 and keeps sufficient pressure in the interface to maintain supercritical conditions. The exit into the ion source must be heated to an elevated temperature, 150–450 °C, to overcome Joule–Thompson cooling during the rapid free jet expansion of the CO_2 and to volatilize the analytes. As with DLI, a CI source is most appropriate for SFC/MS, and the CO_2 gas is used to produce charge exchange spectra (Chapter 5) which are similar to standard EI spectra. Alternatively, an independent source of methane or other reagent gas can be used to produce CI mass spectra of separated analytes.

Although some compounds of environmental interest have been studied by SFC/DFI/MS and their CO_2 charge exchange, methane, isobutane, or ammonia CI mass spectra measured,[68, 69] few environmental analytical applications of this technique have been reported. This is, at least partly, for the reasons given in the discussion of SFC under "Condensed-Phase Separation Techniques". Open tubular SFC columns have a limited capacity per analyte ($\sim < 10$ ng), injection volumes are low (10–100 nL), sample or sample extract concentrations must be somewhat high (10–100 μg/mL), and analysis times can be relatively long. In addition, pressure programming can affect mass spectrometer performance by increasing mobile phase gas loads in the mass spectrometer during the separation.[86] Spectrometer resolving power and signal/noise can be negatively affected although this may vary depending on the type of mass spectrometer, the analyte, reagent gas, and the ionization mechanism, that is, charge exchange, positive ion CI, electron capture, or negative ion CI.

For these and other reasons the trend of recent years is to use higher capacity packed columns. However, the higher flows with packed columns require either splitting the mobile phase before DFI, which negatively impacts analyte detection limits, or the use of some other type of mass spectrometer interface. The types of interfaces used with packed-column SFC/MS are very similar to LC/MS interfaces and are described in subsequent sections of this chapter. Among compounds of environmental interest, several amide and carbamate pesticides, urea herbicides, and organophosphorus insecticides have been studied by packed-column SFC/MS.[87, 88] These studies were conducted using 1 mm ID packed microbore columns, a CO_2 or modified CO_2 mobile phase, and a high flow rate (100–150 μL/min—liquid) interface to a mass spectrometer. The higher flow rate was accommodated by additional pumping capacity between the exit of the flow restrictor and the CI ion source.

Moving Belt Interface

Figure 6.3 shows a diagram of a moving belt (MB) interface, which was developed in the mid-1970s primarily through the efforts of Bill McFadden at the Finnigan Corporation.[89] The MB interface evolved from an earlier design based on a moving wire, which is mainly of historical interest.[90] With the MB interface, the LC effluent flows onto a moving stainless steel belt that carries the liquid film past several heaters and through vacuum locks that promote the vaporization and removal of the mobile phase. The analyte residues, separated in space on the belt, are vaporized by more intense heating, and the analyte vapors enter the high vacuum of the mass spectrometer through a small orifice. The vaporized

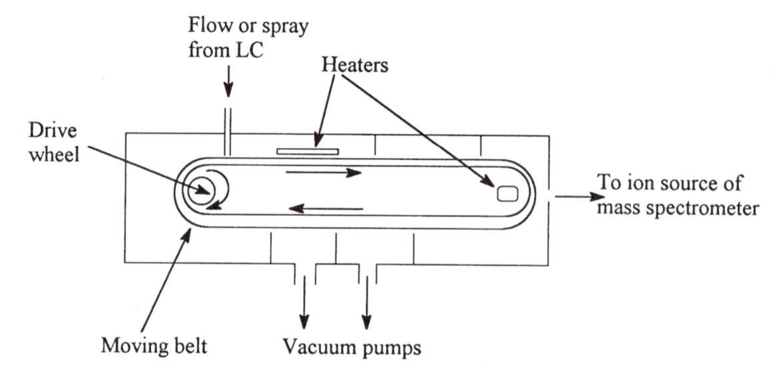

Figure 6.3 Diagram of a moving belt LC/MS interface.

compounds are ionized by conventional 70-eV electron ionization (EI) or CI in the mass spectrometer ion source. A major attraction and advantage of the MB interface is that it can be used to produce conventional EI mass spectra which contain fragment ions and their intrinsic structural information. For the types of compounds shown in Chart 6.1, MB EI spectra are usually very similar to the direct insertion probe EI spectra discussed in this chapter under "Classes of Compounds and their Structures and Properties". Several hundred or more MB interfaces were manufactured and sold commercially during the late 1970s and early 1980s. Many of these were used to evaluate the technique, develop modifications to improve the interface, and conduct research in various areas.

Analyte transfer from the LC column to the mass spectrometer was originally reported in the range 25–40%.[89] Transfer efficiency depends on several processes including the efficiency of deposition of analytes on the belt and the vapor pressures of the analytes during solvent removal and at the point of thermal desorption. Aerosol spray deposition was developed to address the problems of mobile phase splatter and uneven surface coverage on the belt.[91] Alternative techniques for vaporization of analyte residues were investigated including laser desorption[92] and fast atom bombardment.[93] By the mid-1980s the MB interface was developed to the point of providing reasonable performance for certain classes of analytes, for example, the alkylated phenols and dihydroxybenzenes found in coal gasifier condensates.[94] However, major deficiencies remained that were difficult to control and these included losses of chromatographic resolution on the MB, carryover of analytes and background noise from eluted components, losses of some analytes during solvent vaporization, the difficulty of optimizing the vaporization of each analyte in real time, applicability only to compounds with sufficient vapor pressure and thermal stability, moderate sensitivity, and the inevitable mechanical failures. Few quantitative studies were conducted and by the mid- to late 1980s other LC interfaces attracted the attention of investigators. The MB interface is probably no longer commercially available, but it may be available in some laboratories. Given the availability of the interface, this technique can be used for preliminary evaluation of samples separated by LC or SFC, and useful qualitative information, especially EI spectra, can be obtained from some analytes.

Particle Beam Interface

The particle beam (PB) interface is based on technology used in molecular beam studies and was first described by Ross Willoughby and Richard Browner in 1984.[95] The technique was initially described as a "monodisperse aerosol generation interface for combining (MAGIC) liquid chromatography with mass spectroscopy". The MAGIC acronym emphasized the important use of an aerosol spray in the interface.[96] Later designs and commercialization in the late 1980s emphasized the separation function of the interface and the name *particle beam* was widely accepted.

Figure 6.4 is a diagram of a PB interface.[97] In this particular design, the mobile phase flow from the LC column is mixed with a gas, usually helium, in the nebulizer, which converts the liquid into an aerosol that is dispersed into a heated desolvation chamber by the gas pressure. Figure 6.4 shows a concentric pneumatic nebulizer, but other types of nebulizer have been used including the monodisperse aerosol generator,[95,96] an ultrasonic pneumatic nebulizer,[98] the heated capillary (thermospray) nebulizer,[99] and a micro flow rate nebulizer suitable for use with packed microbore (1 mm ID or less) LC columns.[100,101] In the desolvation chamber, which is somewhat below atmospheric pressure, solvent evaporates from the droplets, reducing their size. The aerosol particles, dispersion gas, and solvent vapors are accelerated to a high speed as they pass through the beam collimator or nozzle into the momentum separator. Some particles, particularly the smaller ones, may strike the walls of the desolvation chamber where analyte is lost in the condensation drain.

In the momentum separator the low atomic weight nebulizer gas and low molecular weight solvent vapors diffuse rapidly and are pumped out of the system. Lower molecular weight, nonpolar, and volatile analytes and other substances in the chromatographic effluent also vaporize and diffuse out of

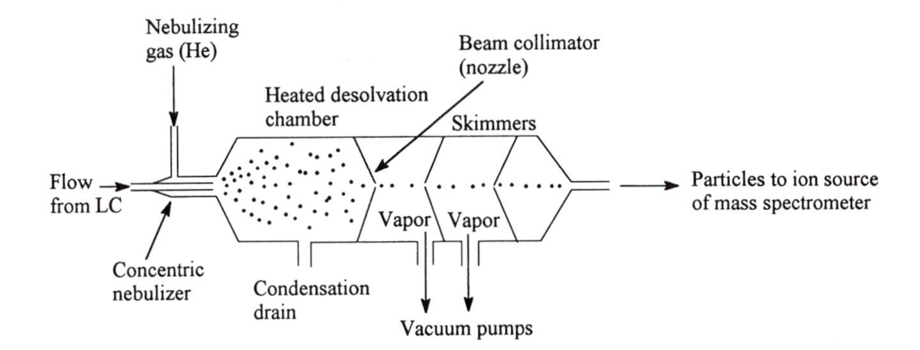

Figure 6.4 Diagram of a particle beam LC/MS interface.

the particle beam and are removed from the system. The heavy particles containing the less volatile analytes and possibly some residual solvent tend to move in a straight line through the skimmers and into the ion source of the mass spectrometer. A two-stage momentum separator (Figure 6.4) is very efficient and the pressure is reduced from somewhat below 1 atmos in the desolvation chamber to below 1 Torr at the second-stage rough pump, and to about 10^{-6} Torr in the spectrometer ion source. In one innovative design the walls of the desolvation chamber consist of a gas permeable membrane, and the exterior of the membrane is continuously swept with a counter current stream of dry helium.[99] Mobile phase and other vapors pass through the membrane and are removed from the system, which reduces the burden on the momentum separator. An important advantage of the PB interface is that, unlike the MB, there are no moving parts that can fail and the dimensions of the various opening are sufficiently large that plugging is not a major problem.

The analytes in the particles must be vaporized before ionization and this is accomplished by allowing the PB to strike a heated surface which, hopefully, will flash vaporize the analytes. The most common surface used is the stainless steel inside wall of the ion source heated to 250–300 °C, and this is a significant limitation on the performance of the PB interface. Particles enter the ion source through a small hole, and those compounds that are thermally unstable or that lack sufficient vapor pressure either decompose or are adsorbed on the heated surface. Experiments have been conducted with other surfaces including inert fluorocarbon polymer surfaces, and these give better performance with some analytes.[102] If the analytes have sufficient thermal stability and vapor pressure, they are flash vaporized and ionized with either conventional 70-eV EI or CI in the mass spectrometer ion source. A major attraction and advantage of the PB interface is that it can be used to produce conventional EI mass spectra, which contain fragment ions and their intrinsic structural information. Several hundred or more PB interfaces were manufactured and sold commercially during the late 1980s and early 1990s. Many of these were used to evaluate the technique, develop modifications to improve the interface, and conduct research in various areas. Because the PB interface has been used to study the mass spectra of many compounds of environmental interest, and used in some quantitative studies, these spectra and quantitative results are considered in more detail in subsequent sections. A review of PB instrumentation and applications was published in 1993.[103]

Compounds Amenable to Particle Beam Sample Introduction. For the types of compounds shown in Chart 6.1, PB EI spectra are sometimes very similar to the direct insertion probe EI spectra discussed in this chapter under "Classes of Compounds and their Structures and Properties". However, the PB interface is suitable for use with only some of these classes of compounds. Several examples are given of positive and negative ion CI spectra obtained with a PB interface.

The haloacetic acids (Chart 6.1) in their acid form are either too volatile, or too thermally unstable when striking a hot surface, to be measured with this interface. Chloroacetic acid with a molecular weight of 94 vaporizes easily and is pumped from the system in the momentum separator. Bromoacetic and dichloroacetic acids behave similarly. Trichloroacetic and dibromoacetic acids are detected with poor signal/noise and have very low $M^{+\cdot}$ ion abundances, suggesting thermal decomposition. The heavier brominated and mixed chlorobromoacetic acids are more likely to be transported as particles into the ion source, but they are less thermally stable and may decompose on the hot ion source walls. A closely related compound, the herbicide dalapon or 2,2-dichloropropionic acid, was an analyte in a multilaboratory comparison of several LC/MS interfaces, but it was not detected by any of the four laboratories that used PB interfaces, even at the very high concentration of 500 mg/L.[104]

The chlorinated phenoxyacetic acids and related carboxylic acid herbicides, for example, 2,4-D (Chart 6.1), are among the most well-studied groups of compounds with the PB interface. These acids attracted attention because they are a large and very important class of environmental contaminants. Most measurements of these compounds have been made by the extraction of the acids from the sample with an organic solvent, derivatization to the methyl esters, and GC with an electron capture detector. A LC separation followed by a PB interface to a mass spectrometer offered the potential for a less complicated and more reliable analytical method. An early measurement of 2,4-D with a PB interface[105] gave an EI spectrum with a strong similarity to the spectrum in the NIST/EPA/NIH database, which was probably measured with a direct insertion probe (DIP).[8] However, the relative abundance of the $M^{+\cdot}$ ion in the PB spectrum was about 30%, which is only about half the abundance in the database spectrum. Subsequent PB measurements of 2,4-D and seven related herbicide acids confirmed that PB spectra are very similar to database DIP spectra, but the $M^{+\cdot}$ ion abundances in the PB spectra are always lower than in the database spectra.[106] It was also concluded that PB measurements on this group of acids were too variable for reliable quantitative measurements. The issue of quantitative measurements of these and other analytes by PB/MS is discussed in more detail in a subsequent section entitled "Quantitative Analysis". The herbicide glyphosate (Chart 6.1) gives no response with the PB interface as would be expected from an examination of the NIST/EPA/NIH database spectrum.[8]

Nitroglycerine (Chart 6.1) is one of the best tests of the ability of an interface to handle thermally unstable compounds. With a microbore LC column, a microflow nebulizer, a custom designed Teflon™ PB target in the ion source, and negative-ion CI, nitroglycerine did not produce an $M^{-\cdot}$ ion by electron capture, but the base peak was an $(M + NO_3)^-$ adduct ion.[107] The abundances of the adduct and NO_3^- ions were dependent on the amount of sample injected, and thermal decomposition of nitroglycerine was significantly greater with a commercial PB interface. The compound *N*-nitrosodiphenylamine (Chart 6.1) could not be detected by PB/MS.[108] It has also been observed that nitrophenols, including the herbicide dinoseb (Table 6.1), are not detected with the PB interface.[106] Compounds in this group are likely to be too volatile and are lost in the momentum separator. The herbicides paraquat and diquat (Chart 6.1) give PB spectra quite similar to those in the NIST/EPA/NIH database, which were described previously in the section entitled "Classes of Compounds and their Structures and Properties". These spectra are indicative of singly charged ions, which are probably produced by the reductive thermal decomposition of the dications.

The PB EI spectra of the sulfonylurea herbicides apparently have not been reported, but the PB EI spectra of 33 carbamates and 14 oxidation and other reaction products have.[109] Nearly all of the 22 *N*-methylcarbamates derived from substituted phenols show very low abundance (< 10%) or no $M^{+\cdot}$ ions and very abundant and characteristic $(M - 57)^{+\cdot}$ ions, which correspond to the loss of methyl isocyanate. This decomposition, which could occur either before (thermally) or after ionization, is the reverse of the synthetic process shown in Chart 6.2 (Reaction 6.1). The *N*-methylcarbamate spectra usually have three or four lower *m*/*z* fragment ions, which are useful for structural confirmation or diagnosis. In general, the PB spectra are very similar to the spectra in the NIST/EPA/NIH mass spectra database, which were probably measured with a DIP inlet system.[8] The PB EI spectra of the *N*-methylcarbamoyloximes, which are derived from various oximes, are usually devoid of $M^{+\cdot}$ ions and often do not have the characteristic $(M - 57)^{+\cdot}$ ions.[109] For example, aldicarb (Chart 6.2) has no $M^{+\cdot}$, and the major fragment ions are the result of hydrogen rearrangements and losses of substructures attached to the C=N group. Other compounds in this group appear to behave similarly. The few compounds with more complex substituents than methyl attached to the carbamate nitrogen are more likely to have $M^{+\cdot}$ ions in their spectra. An exception is the fungicide benomyl which is thermally unstable and decomposes to give *n*-butylisocyanate and carbendazim (Chart 6.4). Therefore, the PB spectra of benomyl and carbendazim are practically identical.

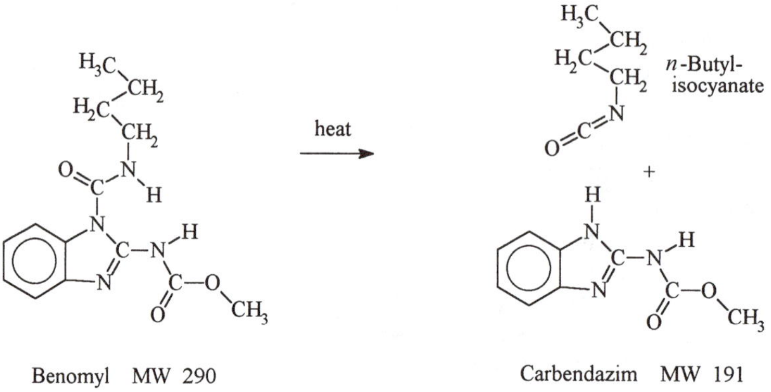

heat

n-Butyl-isocyanate

Benomyl MW 290

Carbendazim MW 191

Chart 6.4

The positive and negative ion CI spectra of this group of carbamate pesticides have also been measured with the PB interface.[109] Many carbamates gave the expected $(M + H)^+$ and $(M + NH_4)^+$ ions with methane and ammonia reagent gases. As expected there was more fragmentation with methane than with ammonia. Negative ion CI with the same reagent gases gave $(M-H)^-$ ions with only a few carbamates, and extensive fragmentation was observed with most compounds.

A group of urea derivatives (e.g., diuron in Chart 6.1) are herbicides, and the PB EI spectra of several of these compounds have been measured. These spectra are similar to the DIP spectra described in the section entitled "Classes of Compounds and their Structures and Properties". In general, the spectra are dominated by low mass base peaks, usually below *m/z* 100. The $M^{+\cdot}$ ion abundances are low, usually below 10%, and there are few, if any, other ions. Chart 6.5 summarizes the PB spectra of four representative compounds from this group and shows the fragmentations that produce the low mass ions.[5,110,111] When electron-withdrawing chlorines are substituents on the *N*-phenyl groups (monuron, diuron, and linuron), the positive charge is retained on the nitrogen on the other side of the carbonyl group. The base peaks at *m/z* 72 and 61 make these spectra less useful because the low *m/z* ions are more likely to be obscured by low *m/z* background ions. For example, ammonium acetate, which is used to improve LC performance, dissociates to give acetic acid, which has a $^{13}C\ M^{+\cdot}$ ion at *m/z* 61 that interferes with the base peak in linuron. In addition, the low *m/z* ions are not particularly informative about the structure of the compounds. Higher mass $M^{+\cdot}$ ions and fragments are either of low abundance or not present. The PB spectra of a few thiourea derivatives have also been measured and their PB spectra have much more abundant $M^{+\cdot}$ ions and some higher mass fragment ions.[5,112].

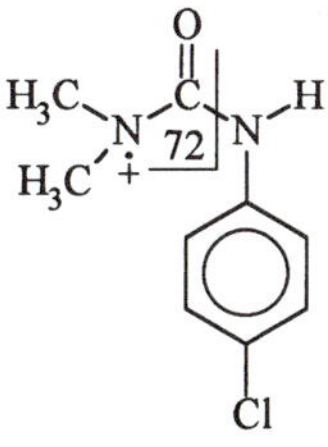

Monuron MW 198

m/z	RA %
198	10
72	100

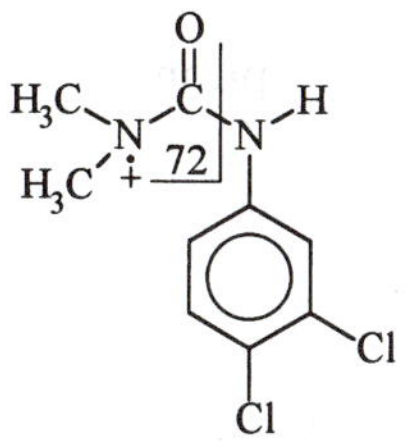

Diuron MW 232

m/z	RA %
232	12
72	100

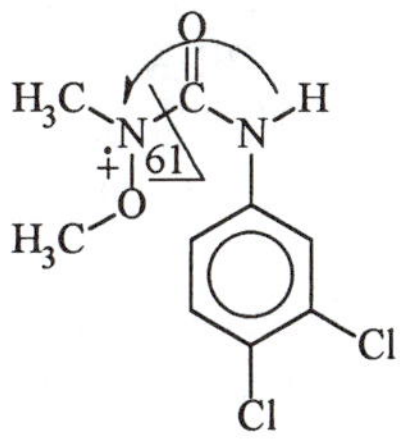

Linuron MW 248

m/z	RA %
248	7
61	100

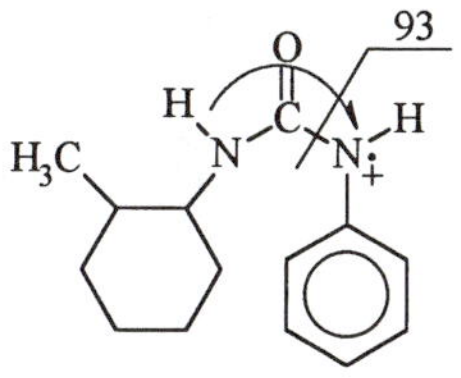

Siduron MW 232

m/z	RA %
232	10
93	100

Chart 6.5

Aromatic amines, such as benzidine (Chart 6.1), are marginal GC/MS analytes but are amenable to measurement with the PB interface.[5] Benzidine, 3,3′-dimethylbenzidine, 3,3′-dichlorobenzidine, and 3,3′-dimethoxybenzidine are compounds that have been widely used in the preparation of azodyes. These amines give either $M^{+\cdot}$ ion base peaks or very abundant $M^{+\cdot}$ ions and moderately abundant M^{2+} ions. The PB spectra of the 3,3′-substituted benzidines also contain an array of fragment ions and are similar to the spectra in the NIST/EPA/NIH database, which were probably measured with DIP sample introduction. The PB interface has also been used in the LC/MS separation and identification of numerous substituted anilines and naphthalamines formed by reduction of azodyes.[113]

Organometallic compounds such as in tri-*n*-butyltin chloride, which are usually converted into derivatives for GC separation, would appear amenable to measurement with the PB interface. However, no $M^{+\cdot}$ ions were observed in studies of the PB spectra of tetra-*n*-butyltin, tri-*n*-butyltin chloride, di-*n*-butyltin dichloride, and *n*-butyltin trichloride.[114] The most abundant ion in the spectra of the three chlorides is *m/z* 155, which is $SnCl^+$. Some lower abundance ions are observed at higher masses, which correspond to the loss of one or two *n*-butyl groups or Cl atoms from the $M^{+\cdot}$.

The PB spectra of several polycyclic aromatic hydrocarbons (PAHs), which have five or six or more fused rings, for example, benzo[*a*]pyrene, benzo[*g,h,i*]-perylene, and indeno[1,2,3-*c,d*]pyrene, have been measured. The larger PAH gave total ion signals roughly four times greater than signals from the lower molecular weight PAH that are typical GC analytes, for example, acenaphthene.[115] The lower molecular weight PAHs have vapor pressures that are too high for this interface and they are depleted in the momentum separator. The PB spectra of PAHs are very similar to the EI spectra in the NIST/EPA/NIH database.[8]

The PB spectra of several other compounds or groups of compounds have been measured. These include the natural pesticide rotenone,[116] several substituted aromatic sulfonic acids,[117,118] some degradation products of polydimethylsiloxanes in soil,[119] and the plant growth regulator Alar (the mono 2,2-dimethylhydrazide of succinic acid).[120] In one study, non-ionic surfactants, plasticizers, plastic additives, and various synthetic organic compounds were identified or tentatively identified in wastewater by LC/PB/MS.[121] Both EI and negative-ion CI were used to identify the aromatic sulfonic acids, and positive-ion CI was used to measure Alar in apple juice.

Liquid and Supercritical Fluid Chromatography. Reverse-phase LC is the most common separation technique used with the PB interface. Most investigators use 10 to 25 cm × 2 mm ID columns packed with C_{18}–silica. Mobile phase flow rates are in the 0.25–0.5 mL/min range, which are compatible with several commercial PB interfaces. Larger diameter columns and higher flow rates can be used with some interfaces.[115,121] Gradient elution is most often used and begins with a high fraction of water and a low fraction of methanol or acetonitrile in the mobile phase and subsequent programming to 100% or nearly 100% organic solvent. Some reports indicate that methanol gives more abundant ions with some analytes, especially with high fractions of methanol, but acetonitrile is often used and seems to provide more consistent analyte signals across the mobile phase gradient.[5,115] Ammonium acetate, or another volatile buffer, may be required to separate basic compounds, for example, benzidine (Chart 6.1), that are strongly retained at acidic silanol sites on some columns.[122] Ammonia blocks these sites, and basic compounds elute faster as narrow symmetrical peaks with enhanced signal/noise (however, see the next section on quantitative analyses for other effects caused by ammonium acetate and other mobile phase components). Acetic acid is often added to the mobile phase to keep acids in the molecular form and allow separations on a reverse-phase column.[104] Anion exchange columns are used with similar flow and gradient elution conditions to separate chlorinated aromatic sulfonic acids.[117,118]

The effluent from packed column supercritical fluid chromatography (SFC) has been introduced into a mass spectrometer with a PB interface.[110,111,123] The interface is either modified to introduce a make-up organic solvent after the SFC restrictor, or the supercritical carbon dioxide mobile phase is modified with an organic solvent. The organic solvent is necessary to facilitate particle formation in the interface. The packed-column SFC technique has some significant advantages, and this approach deserves more attention, especially with compounds well suited to separation by SFC. Microbore LC columns, with an ID of 1 mm or less, and a micro flow rate nebulizer, which was developed recently, may provide the highest level of performance obtained to date with the PB interface.[100–102,108,124,125]

Quantitative Analyses. Because of the generally positive potential of the PB interface and the importance of quantitative analyses in environmental and other studies, several groups of investigators focused on the issues of calibration, measurement accuracy, measurement precision, and instrument detection limits. Early studies gave mixed indications of calibration linearity as some investigators observed reasonably linear calibration plots while others observed nonlinear behavior particularly over the larger concentration ranges.[5,112,120,126] Figure 6.5 shows plots of the integrated quantitation ion abundances of benzidine and three substituted benzidines as a function of the amount injected (between 5 and 460 ng).[126] These plots, which used fairly precise data from a Hewlett-Packard PB interface, were judged nonlinear, and similar results were obtained with these compounds in the same concentration range with several other commercial PB interfaces. The plots in Figure 6.5 are similar to that reported for ethylenethiourea (between 1 and >100 ng) with a different commercial PB interface.[112] The upward curving plots are interpreted as more efficient mass transport through the PB interface and into the mass spectrometer at higher analyte concentrations.

Coeluting substances were also found to enhance the transport of analytes through the PB interface. These substances included ammonium acetate and other volatile buffers, structural analogs of the analytes, isotopically labeled analytes, and possibly column bleed.[120,122,126–128] Several explanations for these effects have been offered:

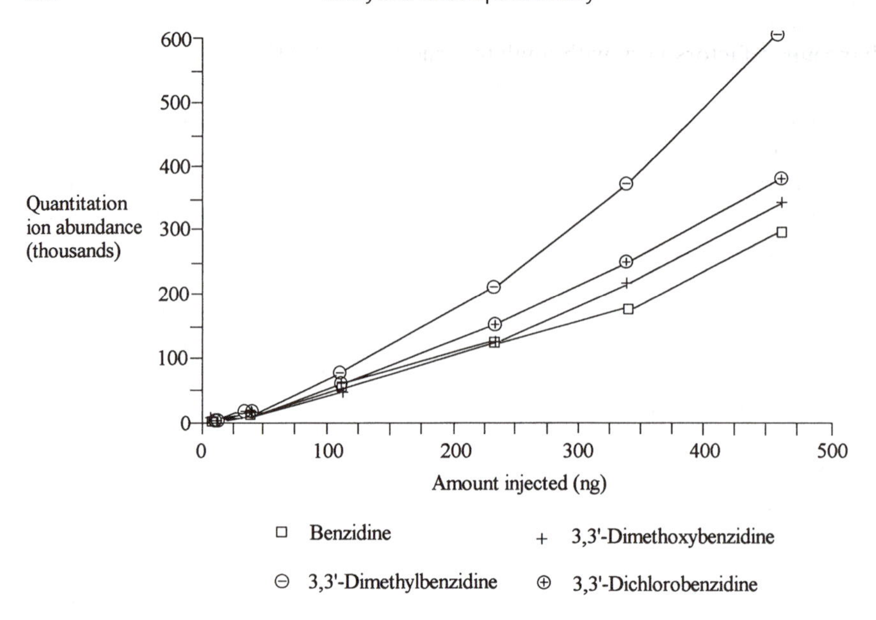

Figure 6.5 Calibration plots for benzidine, 3,3′-dimethoxybenzidine, 3,3′-dimethylbenzidine, and 3,3′-dichlorobenzidine obtained with a particle beam LC/MS instrument. (Redrawn with permission from Ref. 126. Elsevier Science Inc.)

- Higher concentrations of analytes or the presence of some coeluting substances shift the particle size distribution toward larger particles, and these particles are more efficiently transported through the interface.[126]
- While the size of the particles may remain roughly the same over a range of concentrations, they may increase in mass at higher concentrations or when coeluting substances are present. The higher density particles are more efficiently transported through the interface.[126]
- Substances that interact with analytes through hydrogen bonding in the liquid-phase droplets reduce the rate of evaporation of analytes from the particles and possibly increase the solubility of analytes in the droplets.[122]
- The rate of nucleation of solutes to form submicrometer particles may be enhanced by coeluting ammonium acetate.[115]

The first explanation is consistent with the concept that particle size is a function of concentration.[96] However, evidence exists that particle size distribution is not the major factor in transport efficiency.[129] Lower density particles may be more dispersed by turbulence or beam spreading before the skimmers in the momentum separator and have a lower probability of being transported into the ion source.[126] The possibility of chemical interactions causing more efficient transport is supported by several studies and the observation that signal enhancement factors vary with analyte structure.[122, 128] Chart 6.6 shows hydrogen bonding between caffeine, a GC/MS analyte, and ammonium acetate in solution. Caffeine is sometimes used as a LC/MS test compound and, because it is readily amenable to separation by GC, losses in the PB momentum separator are expected. The ion abundances of caffeine in PB/MS are enhanced in the presence of ammonium acetate, and this may be rationalized if interactions such as those in Chart 6.6 reduce the caffeine evaporation rate or increase its solubility.[122, 126]

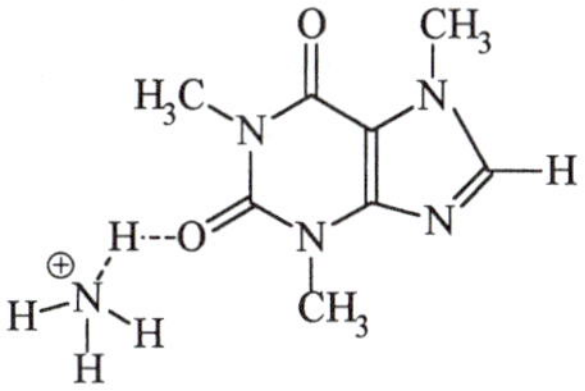

Chart 6.6

Regardless of the reasons, the efficiency of transport of an analyte through a PB interface depends at least on the design of the interface, the temperature and other operational parameters of the interface, the nature of the mobile phase, the vapor pressure of the analyte, the concentration of the analyte, and the presence of coeluting substances. It has been shown that calibration with external standards and a second-order regression curve can give very good recoveries and reasonable precision when samples are free from unexpected coeluting substances.[116] However, the frequent occurrence of unexpected coeluting substances in environmental and other samples will cause calibration errors and analytical bias. The potential for calibration errors has likely inhibited the application of LC/PB/MS to quantitative analysis. The most reliable PB calibration method is the use of coeluting isotope-labeled internal standards. These calibrations are linear because factors that enhance the transport of native analyte to the mass spectrometer also enhance, to the same extent, the transport of the coeluting labeled internal standard.[126]

Measurement precision in PB/MS has been evaluated by repetitive determinations of the integrated quantitation ion abundances of 16 potential analytes.[5] These measurements encompassed 10–14 replicate injections of calibration solutions into two commercial LC/PB/MS systems over 2 work days. The relative standard deviations (RSDs) for these measurements were in the range 7–26%. In a later study with another PB interface, RSDs for 12 compounds were in the 2.2–12.2% range.[126] Four laboratories used two different LC/PB/MS systems to measure nine carboxylic acid herbicides and the phenolic herbicide dinoseb in simulated sample extracts at three concentration levels.[104] At concentrations of 50 and 500 mg/L, recoveries ranged from 30 to 122% and RSDs ranged from 3 to 45%. At the lowest concentration of 5 mg/L only one laboratory was able to detect the analytes, and the measured concentrations were a factor of 2–3 above the true values. Assuming a 5-μL injection, these acids could not be detected at the 25 ng level. These measurements confirmed the conclusion that variability was too great for quantitative measurements of this class of compounds.[106] A later study revealed evidence for thermal decomposition of carboxylic acid herbicides in PB systems.[130] Twelve laboratories participated in a multilaboratory comparison of the measurements of the four benzidines referenced in Figure 6.5.[126] In this study, recoveries at 10 and 100 mg/L were in the 95–104% range, and the multilaboratory RSDs were in the 8.6–20% range. These results demonstrated that reasonable quantitative analyses can be obtained with carefully controlled conditions. In the latter study, enhancements of signals due to coeluting substances was also confirmed. In a single laboratory study, RSDs in the 6–20% range were determined for injected quantities of 0.5–2.5 ng of four large polycyclic aromatic hydrocarbons (PAHs) measured with selected ion monitoring (SIM).[115]

Estimated instrument detection limits (IDLs) for 16 potential analytes varied between 10 and 440 ng in one study with two different PB interfaces.[5] The IDL was defined as the injected quantity from a standard solution that gave a 3:1 signal/noise for each compound's integrated quantitation ion abundance in the complete mass spectrum. Compounds investigated included benzidine and substituted benzidines, carbamate pesticides, urea and thiourea herbicides, and the natural product rotenone. In a later study with some of the same compounds and a different PB interface, IDLs were in the range 1–130 ng.[126] An IDL of 1.3 ng was obtained for Alar by using SIM and signal enhancement from a coeluting structural analog.[120] Similarly, the IDL for ethylenethiourea was estimated at 1.25 ng by using complete mass spectral data acquisition.[112] Six aromatic sulfonic acids were determined with full mass range data acquisition and ~3:1 or better signal/noise at the 100 ng level.[117] Estimated IDLs for the larger PAHs chrysene, dibenzo[*a*,*h*]anthracene, and benzo[*g*,*h*,*i*]perylene were in the 2–4 ng range with full mass range data acquisition and 0.2–2 ng with SIM.[115] Using a recently developed micro flow rate nebulizer with the PB interface and SIM, estimated IDLs in the range 1–40 ng were reported for 45 carbamate, carboxylic acid, phenolic, triazine, urea, and other pesticides and herbicides [124] Many of these IDLs were below 5 ng and all were estimated at a signal/noise of 5. Even lower IDLs in the range 0.6–5 ng were obtained for the carboxylic acids and phenols with a microbore LC column, the micro flow rate nebulizer, an inert fluorocarbon polymer PB target, and SIM.[125]

Particle Beam Summary. Despite clear limitations on the types of compounds that can be successfully measured, the attainable detection limits, and the accuracy of quantitative analyses, the PB interface is one of the most attractive techniques for the introduction of LC or SFC separated analytes into a mass spectrometer. The flexibility of using either conventional 70-eV EI, positive ion CI, negative ion CI, or electron capture ionization is a major advantage of the PB technique. Although the number of commercial suppliers of PB interfaces declined during the late 1990s, a large number of PB interfaces are available in industrial, government, and university laboratories. These interfaces can provide important information when used with appropriate samples and classes of target compounds. Improved performance from new PB designs, especially those incorporating inert PB targets and nebulizers matched to microbore columns, may renew interest by investigators and commercial suppliers in the future.

Thermospray

The thermospray (TSP) LC/MS interface was invented and developed by Marvin Vestal in the early 1980s.[131–134] Unfortunately, the name given to this technique has several different meanings and this can be somewhat confusing in the scientific and commercial literature. Thermospray is a method of creating an aerosol spray; it is also referred to as an ionization technique, and is the name given to a particular LC/MS interface. Sometimes TSP is used just to create an aerosol spray, and a filament or corona discharge is used to produce ionizing electrons in a TSP interface. In another design, a TSP nebulizer is used as the source of an aerosol spray for a PB interface.[99] Thermospray is a combination nebulization and limited ion-separation process. In the presence of certain reagent ions, some neutral molecules can be ionized during the thermospray process.

Figure 6.6 shows a diagram of a TSP LC/MS interface. The flow from the LC column is forced through a narrow-bore stainless steel capillary that is electrically heated. A portion of the mobile phase in the heated capillary is vaporized by contact with the hot metal walls and this vapor acts as a nebulizing gas. Under proper operating conditions a fine aerosol emerges from the capillary at a very high speed. Typical capillary diameters are in the range 0.1–0.15 mm ID and they are useful for mobile phase flows of about 0.5–1.5 mL/min. Smaller diameter capillaries are required for lower mobile phase flows. For mobile phases such as aqueous acetonitrile or methanol, the capillary temperature is maintained in the range 100–200 °C at a mobile phase flow rate of about 1 mL/min.[134] Some experimentation is required to determine the proper operating temperature in an isocratic or gradient elution. Too much heat produces a dry vapor, which is ineffective, and too little heat produces undesirable droplets of mobile phase. Commercial TSP interfaces use thermocouples to monitor the temperatures of the capillary exit and vapor and other operating parameters. Feedback circuits are employed to maintain precise temperature control of the

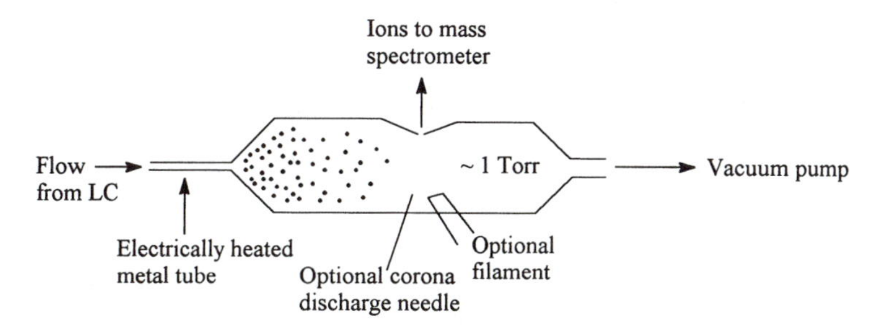

Figure 6.6 Diagram of a thermospray LC/MS interface.

capillary especially as the flow rate or mobile phase composition changes during gradient elution.[135] A mechanical vacuum pump maintains a pressure of about 1 Torr in the TSP interface. This pump removes most of the solvent vapors and other neutral substances from the interface.

The important discovery of Vestal was that when a solution containing certain analytes and ionic substances is thermosprayed, ions are produced in the gas phase without the use of a filament or a corona discharge or any other source of ionizing electrons. These ions can be drawn into a mass spectrometer (Figure 6.6) and a mass spectrum can be measured. The explanation of this phenomenon is that when a solution containing ions is thermosprayed, there is a finite statistical probability that some droplets in the TSP aerosol will contain an excess of either positive or negative ions.[133] This process is not difficult to imagine and occurs because TSP nebulization is very fast and traps some ions in droplets without a corresponding counter ion. The counter ion is trapped in another droplet, and a statistically equal number of droplets with excess positive and negative ions are formed. Most droplets probably contain an equal number of positive and negative ions and are neutral. The small number of charged droplets probably contain only one or a few excess positive or negative ions. These charged droplets release positive and negative ions into the gas phase where they can be detected in the mass spectrometer. In this sense TSP is a limited process for the separation of some positive ions from some negative ions.

Volatile salts, primarily ammonium acetate, are commonly used to generate the ions observed in the TSP process. When an aqueous mobile phase containing ammonium acetate is thermosprayed and the mass spectrum of the vapors is measured, positive ions are observed at m/z 18, 35, 36, and 54. These ions are the ammonium ion and ammonium ion solvated by ammonia, one water molecule, and two water molecules, respectively.[136] The negative-ion spectrum of ammonium acetate has an abundant ion at m/z 59 due to the acetate anion. The positive and negative ions that are observed, and others that can be formed in solution, are the excess positive and negative ions in the charged droplets from the thermosprayed solutions.

When neutral analyte molecules (M) with basic sites are present in an aqueous ammonium acetate solution, they are protonated to form $(M + H)^+$ ions. Weaker bases may associate with an ammonium ion through hydrogen bonding to form $(M + NH_4)^+$ ions (Reaction 6.5). The formation of these ions is favored by the invariably great excess of ammonium acetate over analyte molecules. When a solution containing protonated molecules or ammonium ion complexes is thermosprayed, and the mass spectrum of the vapors is measured, abundant $(M + H)^+$ and/or $(M + NH_4)^+$ ions are usually observed.

$$\underset{(\sim 0.1\ \text{M})}{M + NH_4^{+}\ {}^{-}OAc} \rightleftarrows (M + H)^+ + (M + NH_4)^+ \qquad \text{(Rx 6.5)}$$

The relative abundances of these ions probably depend, in part, on their equilibrium concentrations in solution. Adducts of $(M+H)^+$ or $(M+NH_4)^+$ with various mobile phase solvent molecules are also usually observed. Organic compounds that contain quaternary ammonium groups and exist as ions in solution do not require ammonium acetate, and TSP of these solutions produces the M^+ ions directly. Acidic compounds that form anions in basic solutions, for example, chlorinated phenols and carboxylic acids, produce $(M-H)^-$ and $(M+OAc)^-$ ions in their TSP mass spectra.

Several mechanisms for the transfer of ions from the droplets with excess charge to the gas phase have been proposed. These mechanisms were first discussed in connection with the electrospray techniques that are described later in this chapter. In one model, solvent molecules evaporate from the surface of a charged droplet in the heated TSP aerosol and reduce the size of the droplet. Analyte ions reach the surface where they also evaporate, often with one or more weakly bound solvent molecules. Alternatively, the analyte ions may be the residues in the gas phase after evaporation of all the associated solvent molecules, uncharged analytes, and ammonium acetate. Although ammonium ions and various solvated ammonium ions are observed in TSP mass spectra, most of the ammonium acetate probably evaporates as dissociated ammonia and acetic acid, which are pumped out of the interface with the mobile phase solvent molecules.[137]

Ions in the gas phase are drawn into the mass spectrometer through a small orifice (Figure 6.6) by the natural flow of gases from the higher pressure region to the vacuum of the mass spectrometer (the pressure gradient). Ions may also be drawn through the orifice by electric fields on lenses used to focus ions in the mass spectrometer. These processes are probably quite inefficient with only a small fraction of the ions formed actually entering the mass spectrometer. A retarding electrode may be used to increase the efficiency of transfer of ions into the spectrometer.[138] The bulk of the mobile phase vapors move in a straight line and are removed from the interface by the mechanical vacuum pump (Figure 6.6). The TSP interface is the first example discussed where differences in the properties of ions and neutral molecules are used to separate the bulk of the neutral mobile phase from the analyte ions and introduce mainly the analyte ions into the high vacuum of the mass spectrometer.

In addition to the solution phase equilibria, rapid re-equilibration of ions and molecules and some ionization of neutrals also occurs in the gas phase.[136, 137] Protonated molecules and solvated ions that are released from the droplets rapidly equilibrate with abundant mobile phase components (water, acetonitrile, methanol, ammonia, acetic acid, and so on) in the gas phase. The relative abundances (RAs) of the $(M + H)^+$ and $(M + NH_4)^+$ ions and solvent adducts of these ions can be modified in the interface. The spectrum of ions and RAs that are observed depend on the specific conditions in the interface. These conditions include the concentrations and proton affinities of the substances present, the temperature of the TSP vapor, the propensity of ions to fragment, the design of the interface, the solution pH, and other factors. Some vaporized neutral analyte molecules are also ionized through chemical ionization (CI) processes in which protons are transferred from gas-phase ammonium or hydronium ions to analyte molecules. Fragmentation of some analyte ions also occurs, but these ionization reactions are generally soft, low energy processes, which produce mainly protonated molecules and some adduct ions.

Impurities in reagents and mobile phase solvents may also complicate TSP spectra. In addition to the ions previously mentioned, the ammonium acetate positive-ion spectrum may also contains ions at *m*/*z* 59, 60, 77, and 119 and these have been attributed to several sources.[22,139] The *m*/*z* 59 ion may be the acetonitrile adduct of NH_4^+ in appropriate mobile phases, or a $Na(H_2O)_2^+$ ion in water. Ions at *m*/*z* 60 and 77 were attributed to protonated acetamide (MW 59) and the ammonium ion adduct of acetamide, respectively.[22] It was assumed that acetamide, a hydration product of acetonitrile, was an impurity in the mobile phase. However, *m*/*z* 77 and 119 ions were observed in solutions that did not contain acetonitrile and were attributed to an acetamide impurity in some samples of ammonium acetate.[139] The *m*/*z* 119 ion may be a cluster consisting of a proton and two acetamide molecules. Analogous background ions may be present in other mobile phases. Depending on the conditions and mobile phase options selected, a complex variety of ions, especially at masses below about *m*/*z* 120, can be observed. For this reason, TSP data acquisition often begins at some *m*/*z* in the range 120–160 to avoid the often abundant background ions.

If molecules do not contain sufficiently basic sites and have low proton affinities, ionization does not occur in either the liquid or gas phases, and these neutral substances are removed from the interface by the vacuum system. In order to extend the range of applicability of commercial interfaces, manufacturers usually include a filament and a corona discharge needle to generate ionizing electrons (Figure 6.6). When these devices are used, electrons ionize solvent molecules and form reagent ions that are capable of ionizing substances that are not ionized by ammonium ion (Chapter 5, "Chemical and Related Ionization Techniques"). Unfortunately, papers in the scientific literature are not always clear in specifying whether a volatile salt, a filament, or a discharge or some combination was used in a study. Combinations of volatile salt and ionizing electrons can modify and complicate the spectra of some analytes.

The filament is most useful in high organic mobile phases where reagent ions are readily produced and ionize sufficiently volatile analytes through ion–molecule reactions. However, a filament has a short lifetime in high aqueous mobile phases, which presents a dilemma when employing the widely used LC gradient elution starting near 100% water. The best application of filament ionization is with the little-used normal-phase LC. By contrast, discharge ionization is most effective in highly aqueous mobile phases, but the needle rapidly deteriorates in high organic mobile phases, which presents another dilemma when the LC gradient elution ends with a high percentage of organic solvent.

If a volatile salt is used for ionization of an analyte, it is not required to be in the mobile phase during LC, but can be added to the mobile phase after the LC column (postcolumn addition).[140] If an ionized analyte is already present in solution, for example, an organic quaternary ammonium ion or an acid anion, a volatile salt is not required although solution pH control may be required. Ionization initiated by electrons from a filament or discharge, which produces reagent ions from solvent molecules, is CI even though it occurs in a TSP interface. Whenever ionization using one of these auxiliary devices is discussed, the technique should be clearly identified and it will be in this book. Except for one section entitled "Auxiliary Ionization Devices", all references to TSP spectra, data, measurements, and so on, in this book imply ionization in solution or in the gas phase without the use of ionizing electrons from a filament or discharge electrode and usually with the assistance of a volatile salt such as ammonium acetate.

Compounds Amenable to TSP. Because TSP was one of the earliest LC/MS interfaces developed, and hundreds of interfaces were sold by several manufacturers, many compounds have been studied under a wide variety of conditions. A general discussion has been published of the types of compounds amenable to this technique and the operating parameters used with various mobile phases and ionization techniques.[135] Another monograph describes a large number of applications in more detail including applications in biological chemistry and biochemistry.[141] In this section, TSP spectra obtained with the use of a volatile salt, primarily ammonium acetate, are briefly reviewed. The analytes considered are the general types of compounds shown in Chart 6.1, and a few others.

As a general guideline, TSP with ammonium acetate should produce ions from analytes with basic sites, which are often analytes that have larger proton affinities (Table 5.1). Compounds in this diverse group of substances often contain nitrogen or other atoms with an available unshared electron pair. Amenable compounds include alkylated benzenes, amides, amines, amine oxides, dienes, esters, ethers, heterocyclic nitrogen compounds, ketones, phosphorus compounds, and many others (Table 5.1). Many members of these classes of compounds are also less volatile, thermally unstable, or chemically reactive compounds of environmental importance (Chart 6.1). As indicated previously (Reaction 6.5), TSP of amenable neutral compounds typically produces $(M + H)^+$ and/or $(M + NH_4)^+$ ions and other solvated $(M + H)^+$ ions. Typical TSP spectra have been reported for aromatic amines,[22,107,142] azo and anthraquinone dyes,[143] carbamate pesticides,[22,140,144,145] chloroacetanilide herbicides,[22,140,145] ethoxylate surfactants,[146] phosphate and thiophosphate insecticides,[22,147,148] phthalate plasticizers,[22] triazine herbicides,[22,145,147] thiourea and urea herbicides,[22,140,145] and many other substances.

Frequently, the $(M + H)^+$ ion is the base peak in these spectra and the $(M + NH_4)^+$ ion is sometimes the second most abundant ion, but these relative abundances are often reversed. Proton affinities may be a factor in determining whether the $(M + H)^+$ ion or the $(M + NH_4)^+$ ion is most abundant. The $(M + NH_4)^+$ ion can be thought of as a struggle between the molecule M and an ammonia molecule for a proton (see the structure in Chart 6.6). The molecule M may capture the proton and release ammonia giving the $(M + H)^+$ ion, or there may be a stand-off resulting in an abundant $(M + NH_4)^+$ ion. Some compounds, particularly aromatic amines, do not produce $(M + NH_4)^+$ ions, but have a strong tenancy to form acetonitrile, methanol, and water adducts of the $(M + H)^+$ ions.[22,142]

Thermospray with ammonium acetate is generally ineffective with compounds that are very weak bases and have low proton affinities. Compounds in this group include alcohols, aldehydes, aromatic and aliphatic hydrocarbons, chlorinated and other halogenated hydrocarbons, and nitro compounds. Electron-withdrawing Cl and F atoms and nitro groups often make these compounds too weakly basic in solution and their gas-phase proton affinities are low. The compounds tetra-*n*-butyltin, tri-*n*-butyltin chloride, di-*n*-butyltin dichloride, and *n*-butyltin trichloride do not give ions with ammonium acetate.[22] The indolium dyes Basic Red 14 and Basic Orange 21 contain positively charged quaternary nitrogens and produce M^+ ions on TSP rather than $(M + H)^+$ or $(M + NH_4)^+$ ions.[149] Negative ions are observed by TSP of solutions of chloro- or nitro-phenols, chlorophenoxyacetic acid and related carboxylic acid herbicides, and other acidic compounds at an appropriate pH. These spectra are dominated by $(M - H)^-$ and $(M + OAc)^-$ ions.[150]

Thermal Decomposition. In the TSP interface, thermal decomposition occurs with analytes that are sensitive to the heat required for nebulization of the mobile phase. For example, the fungicide benomyl gives the same TSP spectrum as its decomposition product carbendazim (Chart 6.4), primarily the $(M + H)^+$ ion.[149] If the herbicides paraquat and diquat (Chart 6.1) vaporized unchanged, they would give M^{2+} ions at m/z 93 and 92, respectively. Thermospray gives no sign of these ions, but base peaks at m/z 187 and 184, which are likely the singly charged $(M + H)^+$ and $M^{+\cdot}$ ions formed by electron transfer from a reducing agent, possibly the acetate ion.[114] Several insecticides that are esters of thiophosphoric acid, including methyl parathion, suffer some thermal decomposition in a TSP interface.[148]

The compound *N*-nitrosodiphenylamine (Chart 6.1) was previously described as a product formed in the reaction of the propellant stabilizer diphenylamine (DPA) with nitrogen oxides. Figure 6.7 shows the TSP mass spectra of DPA (lower panel) and *N*-nitrosodiphenylamine (upper panel).[108] It is apparent that the DPA spectrum is contained in the *N*-nitrosodiphenylamine spectrum. Both spectra have a base peak at m/z 170 due to the $(M + H)^+$ ion of DPA and both have an ion at m/z 211, which is likely the acetonitrile adduct of the DPA $(M + H)^+$ ion. A very low abundance m/z 199 ion appears in the spectrum of *N*-nitrosodiphenylamine and this is probably the $(M + H)^+$ ion, but it is too weak for reliable diagnostic purposes. The major ion at m/z 337 distinguishes these compounds and is attributed to one of the products of the sequence of reactions shown in Chart 6.7.

The N–NO bond in *N*-nitrosamines is weak and probably homolytically cleaves on heating to give the diphenylnitrogen radical and nitric oxide. The diphenylnitrogen radical abstracts a hydrogen atom from a solvent molecule to form DPA, which accounts for its spectrum in the spectrum of *N*-nitrosodiphenylamine. Alternatively, two of the radicals combine to form tetraphenylhydrazine, which is protonated to produce the m/z 337 ion. The sequence of reactions must occur rapidly in the liquid phase of a droplet because the radicals probably have a short lifetime and would not likely combine after dispersion in the gas

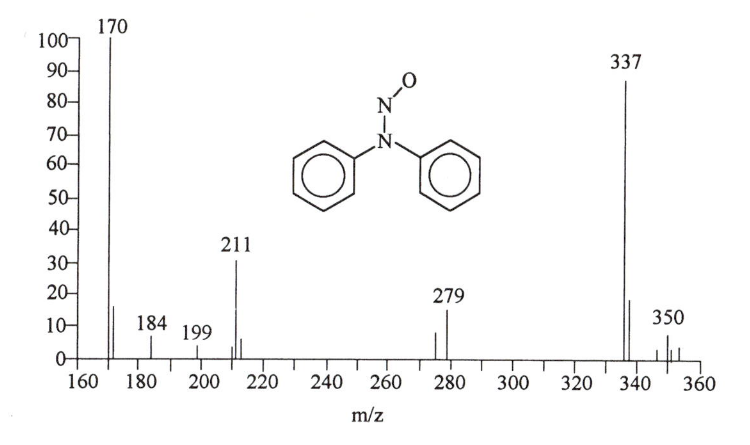

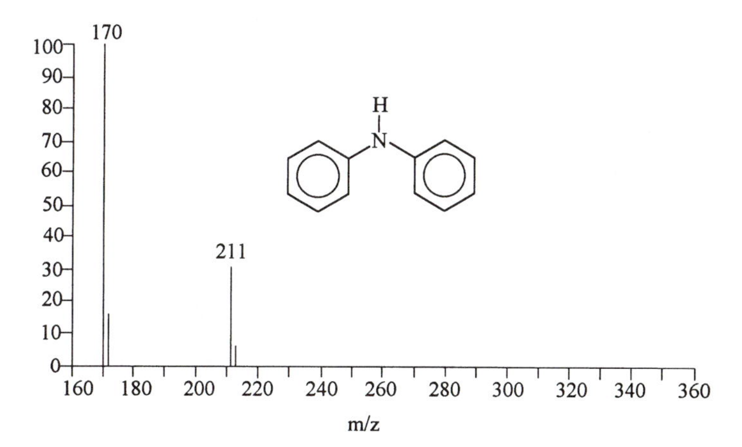

Figure 6.7 Thermospray mass spectra of diphenylamine (lower panel) and *N*-nitrosodiphenylamine (upper panel). (Redrawn from Ref. 108, published in 1990 by the American Chemical Society.)

phase. The DPA and tetraphenylhydrazine could be protonated in the liquid or gas phase as both are probably sufficiently volatile to vaporize as molecules. Diphenylamine and *N*-nitrosodiphenylamine are readily distinguished with TSP LC/MS whereas both compounds give the identical spectrum of DPA when injected into a conventional GC/MS system.

Collision Induced Dissociation. Because few fragment ions are observed in TSP spectra, there was an early recognition that collision induced dissociation (CID) and tandem MS (Chapter 5) would be valuable supplemental techniques. In a study of two dyes containing quaternary nitrogens, CID and MS/MS were used to determine the composition and structure of the compounds under investigation.[149] Thermospray of various esters of phosphoric acid, some of which are marginal GC/MS analytes, gave mainly $(M + H)^+$ and

$(M + NH_4)^+$ ions and few fragment ions.[148] With CID in a tandem mass spectrometer, a large number of fragment ions were generated that were indicative of composition and structure.

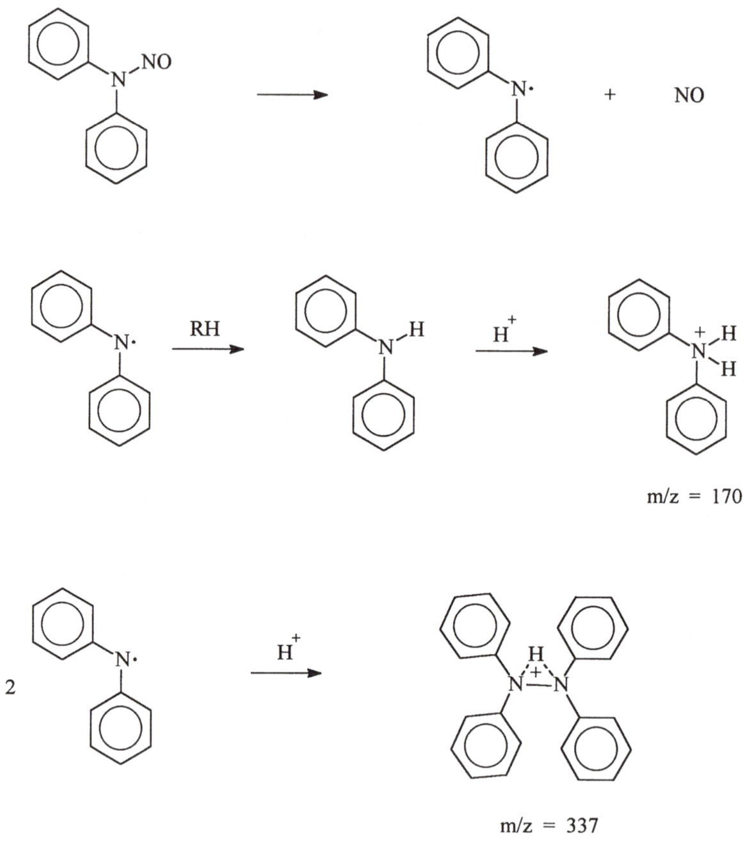

m/z = 170

m/z = 337

Chart 6.7

An ion retarding electrode is sometimes inserted close to the ion exit orifice (Figure 6.6) to increase the number of ions entering the mass spectrometer.[135] This electrode also can increase fragmentation by accelerating analyte ions before CID. Collision induced dissociation and MS/MS provide the additional information necessary to make TSP useful in structure determination. The selected reaction monitoring technique is particularly valuable in enhancing the selectivity of TSP for quantitative analysis.

Auxiliary Ionization Devices. As previously discussed, a variety of compounds of general interest are not sufficiently ionized in solution, have low gas-phase proton affinities, and are not detected by TSP with ammonium acetate. Therefore, most TSP interface designs include a filament and a corona discharge needle to generate ionizing electrons (Figure 6.6). These devices allow the TSP interface to be used as a CI reaction chamber and this has led to some interesting observations.

Thermospray of a variety of hydrocarbon-soluble azo and anthraquinone dyes produced only $(M + H)^+$ ions, and the structures of these compounds could not be determined with this limited information.[143] Because tandem MS was not available, a filament emitting 200-eV electrons was used to generate CI reagent ions in the aqueous methanol vapors. These reagent ions ionized the analytes with sufficient energy deposition to produce some modest fragmentations that allowed the identification of some structural elements. A by-product of the filament operation is the formation of thermal electrons and detection of abundant $M^{-\cdot}$ ions of some compounds. As suggested earlier, however, filaments can have a rather short lifetime of a few days under some TSP conditions and this is not a recommended general analytical technique.

The use of chloroacetonitrile, Cl–CH_2–CN, as a mobile phase additive (2% v/v) in the determination of chlorophenols and chlorinated phenoxyacetic acids has been studied.[151] In the negative ion mode with filament or discharge assisted ionization, the chloride attachment ion $(M + Cl)^-$ was often the base peak or an abundant ion. The chloride reagent ions are produced by EI of the chloroacetonitrile (Chapter 5, Reactions 5.13 and 5.14). This observation illustrates that relatively large amounts of some substances can be conducive to ion–molecule reactions that may provide some useful information, but also can complicate analyses and possibly lead to misinterpretations of spectra. In normal-phase LC with hydrocarbon solvents and filament electrons to generate reagent ions, CI processes are predominant.[152] This can be a sensitive approach, but is limited to analytes that are soluble in normal-phase solvents, for example, large PAHs, chlorinated phenols, and similar analytes. As with many CI systems, the available mass range may be limited by the high abundances of background reagent ions below about m/z 120.

The positive and negative ion spectra of 19 carbamate pesticides and 12 environmental degradation products were studied to determine the effects of mobile phase composition and volatile salts on filament and discharge CI spectra.[153] The volatile salts often cause significant changes in the abundances of fragment, adduct, and other ions in these spectra. These changes are indicative of the complexity and variability of spectra when both volatile salts and supplemental ionization devices are used. The presence of various salts may be unavoidable in some samples and extracts and cause similar effects. In another study the TSP spectra of 16 carbamate pesticides were compared to the pure CI spectra (no ammonium acetate) at three different interface temperatures with three different interface designs.[154] Typical TSP spectra were observed for the more thermally stable carbamates. As expected, somewhat more fragmentation occurred with mobile phase CI reagent ions whose conjugate bases have proton affinities lower than that of ammonia. Similar spectra were observed with the three different commercial TSP interfaces.

Liquid and Supercritical Fluid Chromatography. During the 1980s when most TSP interfaces were designed, the 4.6 mm ID LC column was a widely used industry-standard size. As a result, most commercial TSP interfaces are designed for mobile phase flow rates of 0.5–1.5 mL/min, which are used with columns having IDs of 3–5 mm. With the trend of the 1990s toward smaller diameter LC columns, that is, 2 mm ID and microbore columns of 1 mm ID or less, much lower flow rates are required and these are incompatible with most existing TSP interfaces. When using smaller diameter LC columns, the usual approach is to provide an additional pump for postcolumn addition of mobile phase, which can include the ammonium acetate, to provide a total flow rate of 0.5–1.5 mL/min into the TSP interface. Reverse-phase LC is the most common technique used with the TSP interface. Most investigators use 10–25 cm columns packed with C_{18}–silica or similar reverse-phase packing materials. Gradient elution is most often used and begins with a high fraction of water and a low fraction of methanol or acetonitrile in the mobile phase and subsequent programming to a high fraction of organic solvent.

The effluent from packed column (25 cm × 2 mm ID) supercritical fluid chromatography (SFC) also can be introduced into a mass spectrometer with a TSP interface.[70] The interface is modified so that a fused silica restrictor connected to the end of the SFC column can be placed inside the heated TSP capillary (Figure 6.6). Provision also is made to allow the introduction of methane or another CI reagent gas near the restrictor exit. The TSP capillary is used to supply the heat required to vaporize the separated analytes during the Joule–Thomson expansion of the supercritical CO_2 to a gas as it emerges from the restrictor. The interface vacuum system removes most of the CO_2 and maintains a pressure of about 1 Torr in the interface. A filament or discharge needle is used to supply ionizing electrons, which produce typical methane or other reagent gas ions for positive-ion CI. In addition to the normal reagent ions, $(HCO_2)^+$ and $[H(CO_2)_2]^+$ reagent ions are observed in high abundance. Carbon dioxide has a low proton affinity and these reagent ions readily protonate many analyte molecules. The ionizing electrons are also thermalized through numerous collisions with reagent gas and CO_2 molecules and they can be used for electron capture ionization.

The packed-column SFC/MS technique has some significant advantages and this approach deserves more attention, especially with compounds well suited to separation by SFC. The propellant stabilizer DPA and several compounds found in propellants, including 2-nitrodiphenylamine, 4-nitrodiphenylamine, and 2,6-dinitrotoluene, were separated by SFC and introduced into a mass spectrometer with the modified TSP interface.[70] These compounds gave $(M + H)^+$ ion base peaks and other typical CI ions with methane reagent gas. The nitro compounds also gave base peak $M^{-\cdot}$ anions from electron capture ionization with especially high sensitivity (subpicogram) when pure CO_2 was used as the moderating gas to produce thermal electrons. Some useful fragment ions were also observed in the negative-ion mode. The thermal decomposition of *N*-nitrosodiphenylamine (Chart 6.7) also occurred in SFC/TSP/MS because the temperature at the SFC restrictor tip was similar to the temperature used in conventional TSP. However, at a lower restrictor temperature, the protonated *N*-nitrosodiphenylamine molecule at *m*/*z* 199 was the base peak, which illustrates the potential of this technique for the detection of very thermally unstable compounds.

Quantitative Analyses. Only a few LC/MS options were available during the mid-1980s and therefore considerable attention was devoted to the evaluation of the TSP interface for quantitative analyses. These efforts were focused on calibration, measurement accuracy, measurement precision, and instrument detection limits. Table 6.2 lists TSP external standard response factors (RFs) for four urea herbicides whose structures are shown in Chart 6.5.[22] The base peak and quantitation ion for each compound is the $(M + H)^+$ ion, and two independent sets of mean RFs are listed in Table 6.2. The compounds monuron and siduron have relatively large RFs over the calibration range 40–1000 ng. These compounds have either one or no electron-withdrawing chlorine atoms in their structures. The RFs for diuron are less than 30% of those of monuron. Diuron has two chlorine atoms on the *N*-phenyl group, which reduces the electron density available for a proton. Linuron has the lowest RFs of the four—less than 3% of that of monuron. Linuron has two chlorine atoms on the *N*-phenyl group and an electronegative oxygen attached to the other urea nitrogen. The solution basicity and gas-phase proton affinity of linuron is probably the lowest of the four compounds. These RFs illustrate the wide range of sensitivities observed with TSP. Similar RF trends are observed with other groups of structurally related compounds.

An indication of the variability of TSP measurements is given by the differences in the two sets of mean RFs in Table 6.2. The differences are in the range 36–47%, and this degree of variability is not uncommon in day-to-day measurements with the same calibration solutions, LC columns, instruments, and operators. Ion abundance variations are attributed, in part, to temperature instabilities in the interface, especially during gradient elution. These variations are reduced when the interface is equipped with feedback control circuits to provide precise temperature control of the capillary as the flow rate or mobile phase composition changes.[135] Variations in ion abundances are also caused by contamination of the interface and especially the ion entrance orifice by nonvolatile residues and thermal decomposition products of various analytes. The

Table 6.2 Base peak quantitation ions and mean external standard response factors for four urea herbicides

Compound	Base peak quantitation ion	Mean response factors (peak area/ng)		Difference (%)
Monuron	199	569	396	36
Diuron	233	165	102	47
Linuron	249	15	10	40
Siduron	233	470	273	46

Source: Data from Ref. 22.

TSP interface and the vacuum pump directly attached to the interface require frequent cleaning and maintenance.

Three laboratories used TSP with three different commercial interfaces to measure 10 carboxylic acid herbicides and the phenolic herbicide dinoseb in simulated sample extracts at three concentration levels.[104] The analytes were measured as their $(M-H)^-$ anions, using external calibration standards. At concentrations of 5, 50 and 500 mg/L, recoveries ranged from 62 to 121% and relative standard deviations (RSDs) ranged from 9 to 99% when an anomalously high 150% (4% RSD) recovery of 2,2-dichloropropionic acid was neglected. This high recovery may be due to background $Na(OAc)_2^-$ ions at *m/z* 141, which is the same *m/z* as the 2,2-dichloropropionate anion. At the lowest concentration of 5 mg/L all the laboratories measured the analytes with recoveries in the 87–108% range and RSDs of 12–31% when the 2,2-dichloropropionic acid results were neglected. Assuming a 5-μL injection, these acids were measured at the 25 ng level. Good recoveries were obtained at the 5 mg/L concentration with TSP while in the same study four laboratories using particle beam (PB) interfaces either were not able to detect the analytes or the values reported were grossly inaccurate. The TSP interface provides detection limits lower than those of the PB interface for some compounds. However, the variability of measurements with the TSP interface is generally greater than with the PB interface.

Thermospray instrument detection limits for various esters of phosphoric acid were reported to be in the 0.2–30 ng range, and repetitive measurements of calibration standards gave RSDs for individual calibration points in the range 0.84–17%.[148] Thermospray measurements of a series of primary alcohol ethoxylate surfactants with quantitation ions in the *m/z* 250–1000 range gave linear calibrations over about a factor of 10 in amounts injected in the 5–500 ng range.[146] An internal standard was used to control variability, but RSDs for measurements of calibration points were not reported. Measurement of RSDs for specific substances in fortified samples, which included extraction variabilities, covered the broad range 0.9–80%. The use of selected ion monitoring with TSP is reported to give subnanogram instrument detection limits for a group of 51 carbamate pesticides and urea, chloroacetanilide, and triazine herbicides.[145]

Thermospray Summary. Thermospray is a nebulization technique and a limited ion separation process. Ionization can occur in a TSP interface either in the liquid phase or in the gas phase through various chemical ionization reactions. Supplemental ionizing electrons can also be provided by a filament or a discharge needle in a TSP interface, but these ionization techniques are independent of the TSP process.

Thermospray is effective with a variety of compounds that are sufficiently thermally stable and are either acidic or basic in solution or have larger gas-phase proton affinities. Although predominately $(M+H)^+$ and $(M+NH_4)^+$ ions are formed by many compounds, collision induced dissociation in a tandem mass spectrometer or ion-storage type mass spectrometer can be used to produce fragment ions. Both positive and negative ions are readily measured and signal/

noise for some compounds is very good, especially in high water content mobile phases. Auxiliary ionization devices can be used with TSP and they provide some additional capabilities, but may have short lifetimes and complicate the observed spectra.

The range of chromatographic conditions usable with TSP is limited, and nonvolatile inorganic buffers and salts rapidly contaminate the ion source with a significant loss of signal/noise. Important deficiencies of TSP are the abundant background ions below about m/z 120, the variability of TSP spectra with relatively small changes in experimental conditions, and the need for more frequent cleaning and maintenance of the interface and vacuum system. Analytes often give widely different absolute ion abundances, and each analyte separated in a chromatogram may require a different interface temperature for optimum signal/noise.

Thermospray has been an important technique for compounds not amenable to GC separation, but it was investigated mostly during the 1980s when the 4.6 mm ID LC column and 1 mL/min or greater mobile phase flow rates were widely used. The high mobile phase flow rates caused some of the maintenance problems associated with TSP. Lower flow rates used with microbore LC columns should be more compatible with the mass spectrometer and cause fewer problems. One advantage of TSP is that it is a simple device that is readily fabricated in laboratories and this will likely extend its uses. Thermospray interfaces may no longer be commercially available.

Atmospheric Pressure Chemical Ionization

Atmospheric pressure chemical ionization (APCI), using both positive and negative ions, was described briefly in Chapter 5 as the logical extension of standard chemical ionization in an ion source designed to maintain a pressure of 0.2–2 Torr. It was noted in Chapter 5 that the combination of GC and APCI has not been widely used for several reasons. However, during the 1970s APCI pioneers E. C. Horning and associates at the Baylor College of Medicine concluded that APCI had potential for application in combined LC/MS systems.[155,156] This conclusion was probably based on the recognition that it would be more advantageous to generate ions at atmospheric pressure and admit mainly ions into the vacuum of the mass spectrometer than to employ, what was then, the new direct liquid introduction (DLI) technique. With DLI part or all of the LC flow is vaporized into a chemical ionization (CI) ion source designed to operate at 0.2–2 Torr. The vapors are partly ionized by electron ionization to form CI reagent ions, and the bulk of the neutral mobile phase vapors must be pumped from the spectrometer. The DLI approach limits the flow of mobile phase into the mass spectrometer, may require a high capacity vacuum system, and is subject to the other limitations described earlier in this chapter. Atmospheric pressure chemical ionization is much less restrictive on mobile phase flow rates, but, depending on the design, might also require a high capacity vacuum system.

An APCI LC/MS instrument was not commercialized until the early 1980s for reasons that are described by another APCI pioneer, Bruce Thomson of the

Sciex Corporation.[157] A major problem was that chemical ionization produced few fragment ions, which are needed for reliable analyte identification and structure determination. Collision induced dissociation and tandem mass spectrometers were not commercially available during the 1970s, but this impediment was removed with the commercial introduction of the triple quadrupole mass spectrometer at the end of the decade. Another problem was the absence of a sample introduction technique for heat sensitive compounds. This was partially addressed by the development of a heated pneumatic nebulizer to convert the entire mobile phase flow into an aerosol spray in the APCI ion source of a triple quadrupole spectrometer.[157–159] Figure 6.8 is a diagram of a heated pneumatic nebulizer and an APCI ion source with a corona discharge electrode. Actual commercial devices, which are now available from most mass spectrometer manufacturers, vary in the details of design from that in Figure 6.8, but they often employ many of the same features. Figure 6.8, does not show various ion focusing lenses and other devices commonly used in commercial interfaces.

The flow from the LC column, which may be about 1 mL/min or less, is usually nebulized with pure nitrogen or air. Additional gas is heated to the range 120–150 °C and mixes with the aerosol spray droplets to promote vaporization of the dissolved analytes and the mobile phase solvents. A corona discharge needle is inserted into the region where essentially all of the spray has been converted into vapor-phase analytes and gaseous mobile phase. The discharge needle is maintained at a voltage of 2–6 kV with the same polarity as the ions to be measured, that is, positive to create positive ions and negative to create negative ions. In the positive mode, electrons are drawn from the vaporized molecules to the needle and positively charged reagent ions are created in the vapor. Essentially, all the reagent ions are formed from mobile phase solvent molecules, which are present in enormous excess over analyte molecules. In the negative mode, electrons are emitted from the discharge needle and are captured by the vaporized species to form negative reagent ions or $M^{-\cdot}$ ions.

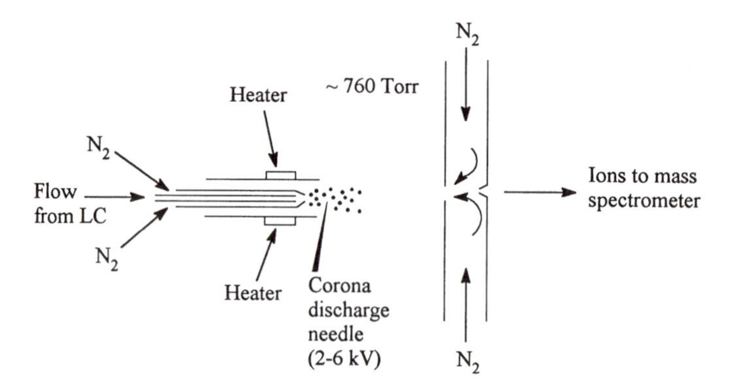

Figure 6.8 Diagram of an atmospheric pressure chemical ionization LC/MS interface with a heated nebulizer and a corona discharge electrode.

Because the LC mobile phase is usually a mixture of water and methanol or acetonitrile, $(H_3O)^+$ ions are important positive reagent ions which ionize analyte molecules by protonation. As in other chemical ionization systems, the proton affinities of the analytes determine whether $(M + H)^+$ ions are formed. Ionized nitrogen and other species also react with analytes in charge exchange reactions to form $M^{+\cdot}$ ions (Chapter 5). At atmospheric pressure the frequency of collisions between reagent ions and analyte molecules is very high and the mixture rapidly reaches equilibrium with a high probability of protonation or some other type of ionization of most analyte molecules. In the negative mode, electrons undergo numerous collisions with the gases in the interface, which reduces their energies to thermal levels where they can be captured by analytes. Electrons also react with water or oxygen to form negative reagent ions including the hydroxide ion (Chapter 5). If ammonium ions, or similar ions, are present in the mobile phase, they can also contribute to the ionization of some analytes.

As a result of the relatively high concentrations of ions and molecules in the ion source, adducts of reagent and analyte ions with various mobile phase molecules are produced. These cluster ions, for example, $(M + H + S_2)^+$ where S is a solvent molecule, create a high background and distribute the analyte molecules among a number of ions of different m/z, thus reducing analyte signal/noise. A technique was developed to reduce the abundance of cluster ions by using a counter current flow of a pure dry gas, usually nitrogen, as shown in Figure 6.8. Cluster and other ions are propelled toward the small orifice leading to the mass spectrometer by the natural flow of gases into the spectrometer vacuum system (pressure gradient) and by potential differences (not shown in Figure 6.8) in the vicinity of the orifice. As these ions pass through the dry nitrogen counter flow, collisions occur and solvent molecules are released from the weakly bound cluster ions. Depending on the design of the system and the potentials applied in the vicinity of the orifice, some fragmentation of molecular and related ions by collision induced dissociation (CID) also occurs (see the subsequent section on electrospray for more on this process). The declustered analyte ions are drawn through the orifice into the mass spectrometer whose vacuum system rapidly removes uncharged molecules that diffuse through the orifice. Ions are focused and detected in the usual way with the optional use of CID in a tandem mass spectrometer to cause fragmentation as needed.

Compounds Amenable to APCI. Because a heated nebulizer and a warm drying gas are used to vaporize the entire mobile phase effluent, analyte requirements are thermal stability and reasonable vapor pressure at a typical operating temperature of 120–150 °C. Some of the compounds and classes of compounds shown in Chart 6.1 have appropriate properties and are amenable to this sample introduction and ionization technique. Others are insufficiently volatile or thermally unstable and unsuitable for use with this interface and ion source. The LC/APCI spectra of several classes of compounds have been reported including some carbamate pesticides,[144] mono- and di-sulfonated azodyes,[160] aromatic heterocyclic sulfur compounds,[161] phosphate and thiophosphate pesticides,[162] phenols,[150, 163] triazine and urea herbicides,[164] and miscellaneous compounds.[165] Most of these compounds are sufficiently thermally stable and volatile to allow the measurement of their conventional EI spectra by using a direct insertion probe as discussed earlier in this chapter. Also, some of the phosphate and thiophosphate pesticides, phenols, triazine herbicides, and miscellaneous compounds are GC/MS analytes (Chapter 4).

The positive-ion LC/APCI spectra of most amenable compounds are readily obtained and the $(M + H)^+$ ion is the base peak in many of these spectra. The carbamates show a propensity to fragment with some having very abundant or even base peak $(M + H - 57)^+$ ions, which correspond to the loss of methyl isocyanate from the $(M+H)^+$ ion. This fragmentation is analogous to the reverse of Reaction 6.1 (Chart 6.2) in their particle beam spectra. The seven aromatic heterocyclic sulfur compounds studied produce $(M + H)^+$ ions that are generally very abundant, but the base peaks in their spectra are $M^{+\cdot}$ ions. Charge exchange ionization to remove a single electron from the sulfur or aromatic system must be energetically favorable in these compounds. Only three of 11 phosphate and thiophosphate pesticides studied have $(M + H)^+$ ion base peaks, although this ion is usually present in their spectra. Most of the phosphorus pesticides have several structurally significant fragment ions. All except one of the 14 triazine and urea herbicides studied have a single $(M + H)^+$ ion in their spectra. Aromatic amines, such as benzidine (Chart 6.1), also give abundant $(M + H)^+$ ions in their APCI spectra.

Acidic compounds produce abundant negative ions. All but one of the six monosulfonated azodyes studied have $(M - H)^-$ ion base peaks and most have several or more fragment ions of low to medium relative abundance. As might be expected, disulfonated azodyes are more difficult to measure with this technique. Although several give $(M - H)^-$ ions, the base peaks are fragment ions and most produce several or more fragment ions that may be the result of thermal decomposition in the heated nebulizer. A very useful fragment ion is the $(SO_3)^{-\cdot}$ ion at *m/z* 80, which is common to all sulfonated azodye spectra. A precursor ion scan (Chapter 5, "Tandem Mass Spectrometry") measuring *m/z* 80 in the second stage of a tandem mass spectrometer generates a chromatogram of all compounds that have ions in their spectra that undergo collision induced dissociation to produce *m/z* 80. This technique is a very effective way of detecting the elution of sulfonated azodyes from a LC.[160]

The more acidic phenols, for example, the polychlorinated and nitrated, invariably have $(M - H)^-$ ion base peaks and some fragment ions. The relative abundances of the fragment ions depend on the voltages used to accelerate ions in the vicinity of the entrance to the mass spectrometer. Less acidic compounds such as phenol itself and alkylated phenols do not produce detectable negative ions. Negative ion spectra are also reported for 11 phosphate and thiophosphate pesticides. Three of 11 compounds studied produce $M^{-\cdot}$ ions by electron capture, but these are not base peaks. The base peaks are invariably fragment ions and several or more are observed.

Liquid and Supercritical Fluid Chromatography. Reverse-phase LC is the most common technique used with APCI, and most investigators use 10–25 cm columns packed with C_{18}–silica or similar reverse-phase packing materials. Gradient elution is most often used and begins with a high fraction of water and a low fraction of methanol or acetonitrile in the mobile phase and subsequent programming to a high fraction of organic solvent. Flow rates of about 1 mL/min or less are used with LC/APCI. There are indications that methanol solutions are effective at nebulizer temperatures lower than those used with acetonitrile solutions, but acetonitrile may produce less chemical background in the form of adduct ions than would methanol.[144] Some spectra reported in the literature are measured using injection of a sample into a mobile phase without a LC column. Under flow injection conditions, especially with polar and ionic compounds at low concentrations, background metal ions, for example, Na^+, may form adducts with analytes that contribute to chemical background and reduce analyte signal/noise. With a LC column and an appropriate mobile phase pH, these background ions are separated or often not observed and analyte signal/noise is improved. Various buffer materials such as ammonium acetate and other salts are often used to control pH in LC separations and these can have deleterious effects on signal/noise. With respect to formation of negative ions, formic acid buffers should be avoided and acetic acid buffers are favored to control mobile phase pH.[166]

The APCI interface also provides a technique to introduce the effluent from packed-column supercritical fluid chromatography (SFC) into a mass spectrometer.[159] A SFC flow restrictor is attached to the end of the analytical column and terminates in the heated nebulizer device. The heater provides the energy necessary to overcome Joule–Thompson cooling during the rapid free jet expansion of the CO_2 and to volatilize the analytes. The CO_2 gas is used to produce charge exchange spectra (Chapter 5), which are similar to standard EI spectra. Alternatively, methane or another reagent gas can be used to produce CI mass spectra of separated analytes. The CO_2 also can serve as an excellent moderating gas for production of thermal electrons for electron capture ionization.[70]

Quantitative Analysis. The heated pneumatic nebulizer/APCI interface is a widely used technique for quantitative analysis, especially in bioanalytical chemistry and pharmaceutical research, but is not nearly as widely used for environmental analyses. This lower use may be because lower cost commercial instruments offering this interface are a development of the late 1990s or because of the emphasis on traditional GC/MS analytes in environmental analyses (Chapters 3 and 4). With selected reaction monitoring (SRM) in a tandem mass spectrometer (Chapter 5), linear external standard calibrations over a factor of several thousand in the amount injected are reported for a group of aromatic heterocyclic sulfur compounds.[161] Instrument detection limits for this group of compounds are in the range 200–600 pg.

Linear external standard calibrations were also obtained for five substituted phenols over the range 0.5–100 ng injected, using selected ion monitoring (SIM) of the abundances of the $(M-H)^-$ ions.[150] The relative standard deviations

(RSD) from repetitive injections and measurements at the 100 ng level were in the range 12–17%. Instrument detection limits for 14 substituted phenols are reported to be in the range 3–180 ng for full scan APCI and 1–85 pg for SIM. This approach was extended to other phenols, and linear calibrations were used to determine 16 substituted phenols in river water and measure signal/noise-based limits of detection (S/N-LOD).[163] The SIM S/N-LOD for 100 mL water samples were in the range 0.1–25 ng/L and the continuous measurement of spectra (CMS) S/N-LOD for 100 mL samples were in the range 0.1–5 μg/L.

Calibration linearity and detection limits for the phosphate and thiophosphate pesticides have been studied using SIM and both positive and negative ions.[162] Over factors of about 40–80 in the amount injected, external standard calibrations were linear and S/N-LOD were in the range 2–25 ng/L for both positive and negative ions. In the positive-ion mode the RSDs of the measurements at the S/N-LOD were in the range 28–45%. An internal standard calibration approach was used to determine a group of triazine and urea herbicides in sediments and water.[164] These calibrations were linear over a factor of 20 in concentration and the S/N-LOD were in the 1–5 ng/L range.

Atmospheric Pressure Chemical Ionization Summary. The heated nebulizer/APCI interface between LC or SFC and MS systems may be the technique of choice for analytes that are not amenable to conventional GC/MS, but which have adequate thermal stability and vapor pressure. The heated nebulizer/APCI provides the flexibility to measure both positive and negative ions with good signal/noise. Some classes of compounds produce fragment ions which support reliable identifications and structure determinations. Although some other classes of compounds produce mainly $(M + H)^+$ or $(M - H)^-$ ions, collision induced dissociation with a tandem mass spectrometer can be used to produce fragment ions.

Calibrations for quantitative analyses are linear, using either external or internal standards, and generally reasonable measurement precision and instrument detection limits are attainable. However, general experience with compounds of environmental interest and environmental samples is very limited and no detailed analytical methods' descriptions are available. The technique has not been widely accepted as part of a standard environmental analytical method and may not be for some years. There are indications of the need for frequent maintenance caused by deposition in the interface of nonvolatile material found in environmental samples.[150]

Electrospray

Of the interfaces discussed in this chapter, only TSP has been clearly identified as having the capability of vaporizing ions that exist in solution. With some of the other interfaces discussed, ions in solution conceivably might be vaporized, but are more likely left as residues during the vaporization of neutral analytes. Electrospray (ES), which is implemented in several variations with sometimes different names, is the single most important technique available to date to vaporize ions existing in solution. Since many nonvolatile or thermally unstable compounds have acidic or basic functional groups, and can be protonated or deprotonated in solution, ES may be the most versatile of all the LC/MS interfaces for this broad class of substances. Furthermore, since capillary electrophoresis (CE) is designed to separate ions in solution, ES is probably the ideal interface for combined CE/MS.

Electrospray is the most popular LC/MS interfacing and sample introduction technique of the 1990s. This is largely due to its ability to vaporize from solution high molecular weight ($\sim 10^3$–10^6) multiprotonated and multideprotonated biological molecules such as peptides, proteins, ribo- and deoxyribonucleic acids, carbohydrates, and many others. This outstanding capability, which will likely revolutionize biochemical research, has stimulated the instrument development research community and mass spectrometer manufacturers to develop a variety of ES interface designs. These devices are being evaluated with a wide range of small molecules and macromolecules with varying degrees of polarity, thermal stability, nonvolatility, and chemical reactivity. Electrospray has a long and interesting history and many reviews have been published which describe in great detail the fundamentals, theory, interface designs, and applications of this technique to a wide variety of problems.[157–159, 167–172] This section focuses on the essentials needed to understand the technique and on applications to small molecules of environmental and related interest.

Bruce Thomson at Sciex and coworkers appear to have hit on the basic idea of ES when they moved the discharge needle of the APCI ion source (Figure 6.8) close to the outlet of the nebulizer in an effort to volatilize polar, thermally sensitive, and ionized compounds in solution that were not amenable to vaporiztion with the heated pneumatic nebulizer.[173] Analytes were dissolved in water and the pH was adjusted to ensure ionization in solution. The solutions were pumped through a narrow stainless steel tube at 1 mL/min and nebulized with air, but without the application of heat. The discharge needle was placed close to the end of the sample delivery tube and maintained at 2–3 kV with a polarity opposite to that of the desired analyte ions. This opposite polarity was in sharp contrast to the polarity of the discharge needle used with APCI, which was the same as the ions to be detected. As the liquid effluent emerged from the tube the combination of nebulizing gas and the high potential electric field produced a fine spray of charged droplets at atmospheric pressure. The nebulizer gas flow propelled the spray toward the orifice leading to the mass spectrometer as solvent evaporated from the droplets and reduced their size distribution, just as in the heated nebulizer/APCI interface (Figure 6.8).

However, in contrast to the heated nebulizer/APCI interface, ions were detected in the mass spectrometer that were likely the same as the solvated ions in the liquid phase. It was proposed that, as ions in solution reached the surface of the charged droplets, they evaporated along with bound and unbound solvent molecules. At this stage, conditions in the Thomson interface are similar to conditions in the heated nebulizer/APCI interface. Rapid equilibration of ions and molecules occurs through many collisions at atmospheric pressure, and some of the solvent molecules bound to ions dissociate in the counter current flow of dry nitrogen. Ions are drawn toward the orifice leading to the mass spectrometer by the flow toward the spectrometer vacuum (pressure gradient) and by potential differences (not shown in Figure 6.8) in the vicinity of the orifice. Collisions with gaseous molecules in this region may cause some analyte ions to fragment. Ions are drawn into the mass spectrometer where they are separated by m/z and optionally fragmented in the collision cell of a tandem mass spectrometer. This sample introduction technique was called *ion evaporation* and it was explored somewhat during the 1980s.[158, 160] Ion evaporation produced primarily $(M + H)^+$ or $(M - H)^-$ ions with generally no or just a few low abundance fragment ions. This interface design did not gain widespread use or popularity for several reasons, including relatively high instrument detection limits with some compounds,[157, 160] but probably mainly because other related designs enjoyed rapid development and great success.

John Fenn and coworkers at Yale, who were influenced by the research of Malcolm Dole and associates during the 1960s, developed the design that eventually sparked a major advance in LC/MS technology.[167] They dispensed with the nebulizing gas used by Thomson, adopted a 50% (v/v) aqueous methanol solvent instead of pure water, employed a very low flow rate in the range 5–20 μL/min, and applied the high voltage directly to the stainless steel needle that delivered the liquid flow. This arrangement produced a fine stable spray of charged droplets from which ions were produced in the gas phase. After some consideration Fenn and associates decided that it was desirable for the needle to be at ground potential for convenience and safety. In their later design they surrounded the needle with a cylindrical electrode at a potential 3–4 kV above or below ground and, like Thomson's design, opposite in polarity to the ions being detected.[167, 174]

Figure 6.9 is a diagram of the later design used by Fenn and associates. While this design had elements similar to that used by Thomson, it probably was more effective for several reasons. The cylindrical electrode may have provided a higher potential at the liquid surface than that provided by the discharge needle close to the outlet of the liquid delivery tube. Perhaps more importantly, the mobile phase flow rate was much lower; an aqueous methanol mobile phase has a surface tension lower than that of pure water and is much more readily dispersed into a fine spray. In a still later design, another electrode ("End electrode" in Figure 6.9) was inserted near the entrance to the glass capillary and was maintained at a voltage similar to that of the cylindrical electrode.[175] A counter current flow of warm dry nitrogen was again used to promote vaporization of solvent from the droplets and to dissociate solvated ions as described for the APCI interface.

A dielectric glass capillary was used to deliver ions from the atmospheric pressure spray region to the first-stage vacuum of the mass spectrometer. The glass capillary was coated at both ends with a conductive noble metal. The atmospheric pressure end of the dielectric capillary was maintained at about 3–4 kV ("Cap ent" in Figure 6.9) with a polarity opposite to that of the ions being transmitted. The exit end of the capillary was maintained at a potential in the range 30–120 V ("Cap ex" in Figure 6.9) and with the same polarity as the

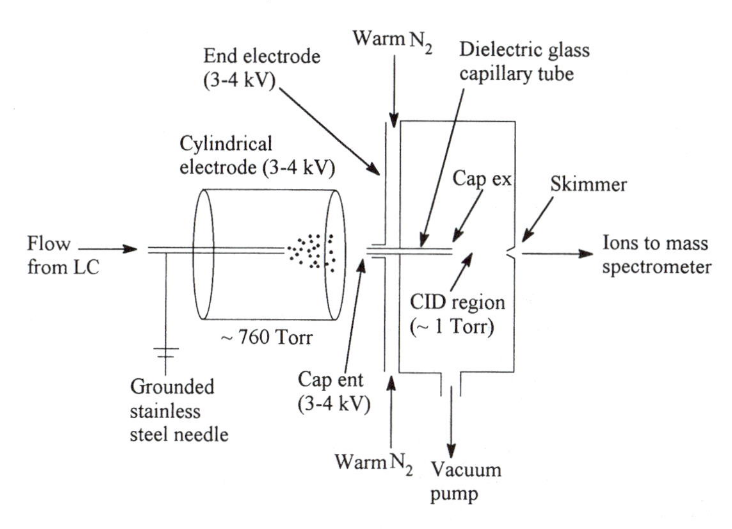

Figure 6.9 Diagram of an electrospray LC/MS interface with a grounded spray needle.

ions being transmitted. Ions were drawn into the capillary by the electric fields and the flow of gases into the lower pressure part of the interface. Ions emerged from the capillary into the gap before the skimmer (Figure 6.9) where the pressure was about 1 Torr and collisions with gaseous molecules produced some additional desolvation. However, if the potential difference between the capillary exit ("Cap ex") and the first skimmer was increased, ions were accelerated into this region and the more energetic ion–molecule collisions produced some fragment ions.[176] This type of collision induced dissociation is not as versatile as that occurring in a collision cell in a tandem mass spectrometer (Chapter 5), but the induced fragmentation can provide additional support for compound identifications. Furthermore, "Cap ex–skimmer" fragmentation can be implemented on a single-stage quadrupole mass spectrometer. Some later commercial electrospray and APCI interfaces utilize similar designs to provide similar capabilities.

Jack Henion and associates at Cornell University quickly developed an alternative design, which retained the nitrogen nebulizing gas used by Thomson, but applied the 2–3 kV potential directly to the stainless steel capillary tube.[177] Figure 6.10 is a diagram of this design. The polarity of the potential applied to the tube was the same as that of the ions being detected, that is, positive for positive ions and negative for negative ions. Suitable electrical insulation was provided to protect the equipment and workers. The sample solution was delivered in a fused silica capillary (not shown in Figure 6.10) within the stainless steel tube, and the high-voltage connection to the liquid flow was made at the exit where the liquid came into contact with the steel tube. Use of the nebulizing gas was considered a major advantage because it allowed mobile phase flow rates of up to 0.2 mL/min and higher fractions of water in the mobile phase. However, better signal/noise was obtained at about 40 μL/min, which is typical of flows used with 1 mm ID microbore LC columns. The design in Figure

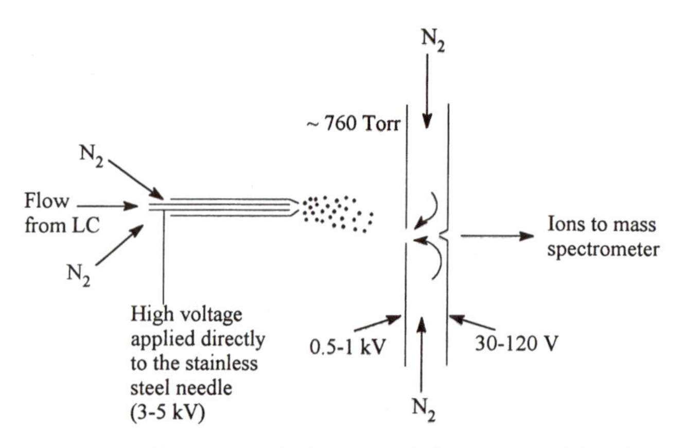

Figure 6.10 Diagram of a pneumatically assisted electrospray LC/MS interface with a high voltage on the spray needle.

6.9 gave a stable spray only at flow rates less than about 20 μL/min and with a mobile phase consisting of at least 50% (v/v) methanol in water. The design in Figure 6.10 was pneumatically assisted electrospray and was called *ion spray* by its developers.[177] In another innovation, the mobile phase delivery tube, which was at 2–3 kV potential, was offset from the orifice leading to the mass spectrometer, which had the same ion entrance system as shown in Figure 6.8. The offset position of the liquid delivery tube prevented the droplets from the higher flow rate spray from directly entering the ion sampling orifice to the mass spectrometer. Figure 6.10 also shows the approximate range of potentials applied in the vicinity of the ion entrance orifice. These and other potential differences in this vicinity can cause ion acceleration and collision induced dissociation to produce fragment ions in a manner similar to the "Cap ex-skimmer" potential difference in Figure 6.9.

Several other designs were soon developed which used some of the features shown in Figures 6.9 and 6.10, but also introduced new ideas. Dick Smith and colleagues at the United States Department of Energy Pacific Northwest National Laboratory in Richland, WA, used a design similar to that in Figure 6.10, but dispensed with the nebulizing gas and used a sheath flow of organic solvent between the stainless steel tube and the fused silica capillary delivering the sample solution.[168] Because samples were introduced by flow injection and not by chromatography, the 5–20 μL/min flow rate was acceptable. The sheath liquid, which was usually methanol or acetonitrile, allowed operation with highly aqueous sample solutions and provided an electrical connection between the high voltage on the stainless steel tube and the sample solution.

Brian Chait and the Rockefeller University group used a stainless steel sample delivery needle at a 4–6 kV potential without a nebulizing gas. Their innovation was a heated stainless steel tube (0.5 mm ID × 20 cm) to deliver the ions from the atmospheric pressure region to the mass spectrometer.[176] This metal tube, which is analogous to the glass capillary in Figure 6.9, is heated to 85–90 °C to promote desolvation of droplets and ions. A counter current flow of warm dry gas to further desolvate ions is not used in this design, but the tube is maintained at a potential of 50–380 V. Ion fragmentation in the capillary exit–skimmer region (similar to Figure 6.9) can be controlled by adjusting the potential difference (ΔV) between the metal tube and the skimmer. With a relatively low ΔV, ions emerging from the heated metal tube are further desolvated by low energy collisions with background gases. At a higher ΔV, ions are accelerated into this region and more energetic collisions produce useful fragment ions.

An ES interface was designed by Marvin Vestal, which employed a 2–3 kV potential on a stainless steel sample delivery needle.[178] This design differs from the others by using a stream of laboratory air around the outside of the needle which, however, is not used as a nebulizing gas. The gas flow entrains the spray droplets and vapors through a nozzle followed by a two-stage momentum separator similar to that used in the particle beam interface shown in Figure 6.4. A wall of the momentum separator chamber is heated to about 250 °C, which raises the temperature in the spray chamber to about 90 °C. This heat promotes evaporation of solvent from the droplets and desolvation of ions, and, as in the design of Chait and associates, a counter current flow of warm dry gas is not used for this purpose. In this design, air and neutral vaporized solvents are largely removed in the momentum separator, and ions are focused into the mass spectrometer. The nozzle and momentum separator skimmers carry various potentials to maximize ion transmission into the mass spectrometer. Some ion fragmentation is accomplished by adjusting the potential difference between the skimmers in the momentum separator, but this may not be as effective as in other designs.

In important developments, both ES and microbore LC/ion spray were demonstrated as compatible and effective with an ion-trap mass spectrometer.[179, 180] This permitted application of these techniques with a moderately priced bench-top mass spectrometer, which had capabilities for additional collision induced dissociation and tandem MS. Contemporary commercial ES and ion spray interfaces, which are available from nearly all mass spectrometer manufacturers, utilize some of the designs described in this section. However, the latest commercial systems also incorporate new devices and design details that are believed to provide improved performance. During the 1990s some of these designs have been or currently are protected by various patents, but as patents expire, manufacturers are usually quick to incorporate the best features into their products. Other innovations are expected to improve this technology substantially in future years. Designs that do not perform reliably or that are not competitive in signal/noise, flexibility, and convenience of use will soon disappear from the market. Commercial APCI interfaces also utilize some of the design features developed for ES, for example, the glass and heated metal capillary tubes to deliver ions from the atmospheric pressure region of the interface to the lower pressure regions leading to the mass spectrometer. These permit more flexibility in APCI including collision induced dissociation in the ion source region as described for ES.

Electrospray Mechanisms. In this and subsequent sections, ES and pneumatically assisted electrospray (ion spray) are not differentiated because the same principles generally apply to each approach. Electrospray is often called an ionization technique, but it is mainly an ion separation technique, that is, a process that separates positive ions from negative ions so they can be separately detected and measured. Conditions are almost always established so that ionization occurs in solution, although some ionization may occur via ion–molecule reactions in the gas phase after ES. Only under special conditions are a few neutral substances converted into ions during the ES process.[181]

Ions are formed in solution, usually in equilibrium processes, by reactions of analytes with volatile acids such as acetic acid and bases such as ammonium hydroxide. As in TSP, the typically large excess of reagent acid or base over analytes in a sample solution favors the formation of at least small equilibrium concentrations of ions from weakly basic or acidic analytes. Some strong volatile acids, for example, trifluoroacetic acid and hydrochloric acid, are not recommended for ES because they can cause signal suppression and spray instability under some conditions.[182] Ions also are formed in solution by reactions of analytes with metal ions, for example, Na^+ and Li^+, anions such as acetate, and oxidizing or charge transfer reagents.[183, 184] Some substances, for example, compounds with quaternary nitrogen or phosphorus atoms, exist as ions and no ionizing reagent is necessary.

The ES process can be envisioned as an on-the-fly electrolysis in which electrodes at a high potential difference are used to separate positive ions from negative ions. The designs of Thomson or Fenn and associates (Figure 6.9) employ an electrode at a high potential and a spray needle at ground potential. If the electrode is at a high positive potential, negative ions in the liquid at the tip of the sprayer are strongly attracted into an incipient droplet. As the incipient droplet acquires ever more excess negative ions, the strong attraction to the nearby high positive potential causes the droplet to break away from the body of the liquid. Meanwhile, positive ions migrate away from the exit of the sprayer and toward the inner surface of the needle at ground where adsorption can occur. The positive ions can be discharged by a surface reduction reaction that releases metal or other cations to balance the flow of negative ions toward the mass spectrometer. This accounts for the eventual deterioration of spray needles as the inside surface becomes pitted or coated with foreign matter.

If the nearby electrode is at a high negative potential with respect to ground, droplets with excess positive ions break away from the tip of the needle, and negative ions are repelled toward the surface of the needle where they are oxidized or the metal surface is oxidized releasing metal ions into solution. Oxidation creates a flow of electrons to ground to balance the flow of positive ions toward the mass spectrometer. The physical picture of droplets emerging from a liquid surface is often described as a Taylor cone.[169, 171] The efficiency of this process is not known, but it is probably relatively low. Therefore, charged droplets contain both positive and negative ions, but an excess of one or the other depending on the polarity of the electrode. The ES process is very fast and produces a spray with an average particle size of about 12 μm.[177] By contrast, TSP produces generally much larger particles.

Electrospray designs that apply the high voltage directly to the spray needle use a polarity opposite to that of an electrode near a grounded needle to produce the same polarity of charged droplets. A high positive potential applied to the needle in Figure 6.10 strongly attracts negative ions to the surface where they are oxidized or the metal surface is oxidized releasing metal ions into solution. Meanwhile, positive ions in solution are repelled by the high positive potential on the needle and are carried by the liquid flow toward the tip of the needle. An incipient droplet accumulates an excess of positive ions and breaks away when the forces keeping it in the liquid phase are overcome by the repulsion from the high nearby positive potential. If the potential applied to the needle is negative, positive ions are attracted and reduced on the surface, and spray particles with excess negative ions are produced. Again, charged droplets are likely to contain both positive and negative ions, but an excess of one or the other, and the process of forming the particles is very fast. The left side of Figure 6.11 shows a representation of a charged particle with analyte cations (M^+), counter anions (A^-), and solvent molecules (S) where the number of analyte cations exceeds the number of anions.

As mentioned in the TSP discussion, several mechanisms have been proposed and discussed to account for the emergence of ions in the gas phase from droplets containing excess positive or negative ions.[167–172] All ES designs incorporate some technique to promote the evaporation of solvent from the charged droplets. As solvent evaporates (left side of Figure 6.11), droplets become smaller and ions accumulate on the surface. Repulsive forces between excess ions of the same charge eventually overcome the forces holding the particle together

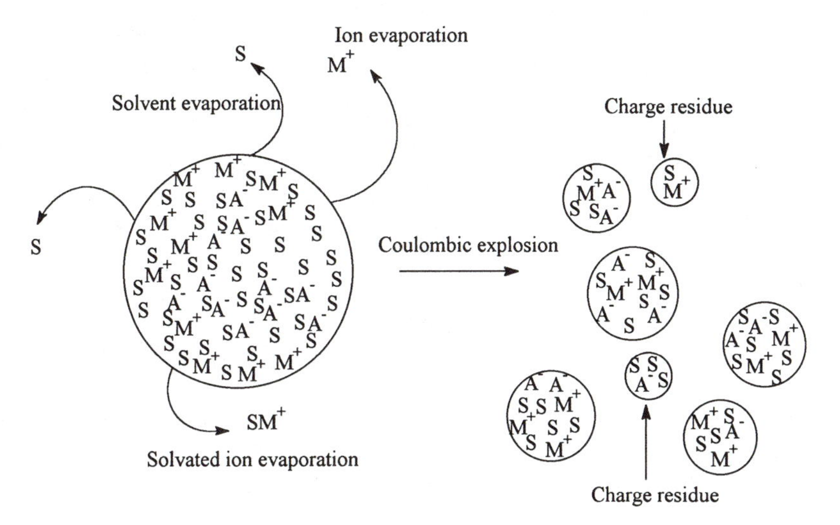

Figure 6.11 Diagram of solvent and ion evaporation from a charged electrospray droplet and a Coulombic explosion to produce smaller droplets.

(hydrogen bonding, dipole–dipole interactions, and so on) and a so-called Coulombic explosion produces several or more smaller particles which repeat the same process (Figure 6.11). Analyte ions (M^+) on the surfaces of droplets, repelled by other ions of the same charge, may also evaporate with associated solvent molecules (left side of Figure 6.11). These solvated ions in the gas phase equilibrate with gas-phase molecules and ions and are eventually drawn into the mass spectrometer where they are separated by m/z and measured. This is often called the *ion evaporation* model. An alternative mechanism is the *charge residue* model, which suggests that solvent evaporation and Coulombic explosions continue until only one ion remains. This ion, and a few associated solvent molecules, is the residual charge in the gas phase (several solvated ions on the right side of Figure 6.11). Regardless of which model is correct, and both processes may be involved, ions are produced in the gas phase and desolvated to a greater or lesser extent.

One practical problem occurs more with negative ions than with positive ions in an ES interface. There is a tendency for the spray to become unstable because of corona discharge to a nearby surface and this occurs at lower potentials with negative ions than with positive ions. Disruption of the ES substantially decreases the flow of charged particles with a great loss in analyte signal. The practical solution to this problem is to add a flow of an electron-scavenging gas, for example, oxygen, to absorb stray electrons emitted from negatively charged electrode surfaces.

Compounds Amenable to Electrospray. Many classes of compounds have appropriate properties and are amenable to ES. As a general guideline, ES should produce ions from acidic solutions of analytes with basic sites, which are often analytes that have larger proton affinities (Table 5.1). Compounds in this diverse group of substances often contain nitrogen or other atoms with an available unshared electron pair. Amenable compounds include amides, amines, amine oxides, esters, ethers, heterocyclic nitrogen compounds, ketones, phosphorus compounds, and many others. Many members of these classes of compounds are also less volatile, thermally unstable, or chemically reactive compounds. Negative ions are observed by ES of basic solutions of analytes with acidic protons including chloro- or nitro-phenols, carboxylic acids, sulfonic acids, and other acidic compounds. Other substances that are ionic, for example, inorganic cations and anions, are readily amenable to ES.

Electrospray mass spectra of most classes of compounds shown in Chart 6.1 and several others have been investigated by appropriate adjustment of mobile phase pH with volatile acids or bases. These compounds include aromatic amines,[142] benzo[*a*]pyrene metabolites,[185] carbamate pesticides,[144,186–189] chloroacetanilide herbicides and some degradation products,[145,190] *N*-nitro and nitric acid ester explosives,[191] hydrocarbon-soluble azodyes,[189] phenols,[150,163,192] phenoxyacetic acid herbicides,[193,194] phosphorus pesticides,[195] sulfonated azodyes,[160,196] sulfonylurea herbicides,[19,145] surfactants,[197,198] triazine herbicides and degradation products,[189,199] and urea herbicides.[145,186,187] Naturally ionic substances investigated by ES include inorganic anions,[200,201] metal cations,[200] the Ni(II) and Cu(II) chelates of ethylenediaminetetra-acetic acid,[202] quaternary ammonium compounds including the herbicides diquat and paraquat,[79,187,203] and several sulfonate and sulfate anions.[203] Tri-*n*-butyltin chloride produces an abundant $(n\text{-butyl})_3Sn^+$ ion with ES.[204] Chemical derivatization also has been explored as a technique to introduce ionizable functional groups into compounds that do not form ions in solution with acids or bases.[184,205]

Electrospray mass spectra of acidic or basic organic compounds are frequently dominated by $(M-H)^-$ or $(M+H)^+$ ions, respectively. Similarly, quaternary ammonium compounds and other ionized substances give primarily molecular ions in their ES spectra. As mentioned previously, Na^+, NH_4^+, and similar ions, if present, can form adducts with neutral analytes. Generally, little or no fragmentation is observed because ions are transferred from the liquid phase to the gas phase in low energy processes. Any excess energy is rapidly dissipated by frequent gas-phase collisions with other molecules at atmospheric pressure. Fragmentation to smaller ions, however, is observed with some compounds, for example, the fragile *N*-nitro and nitric acid ester explosives.[191]

The interpretation of reported ES spectra is complicated because potential differences in the vicinity of the ion entrance orifice or the capillary exit cause varying degrees of fragmentation. When assessing the degree of fragmentation of an ion, experiments should be conducted at a series of potential differences to determine the dependence of ion abundances on this operating parameter. Depending on the objectives of an analysis, it may be desirable to adjust the potential difference to maximize the abundance of the $(M-H)^-$, $(M+H)^+$, or M^+ ion for optimum signal/noise or selectivity. Alternatively, it may be important to increase the potential difference to cause decomposition of adducts or ion fragmentation to produce the optimum number of structurally significant ions. Unfortunately, optimum conditions for both alternatives are usually different for each analyte and a compromise is necessary for multicomponent analyses. No standard ES conditions or standard reference ES spectra have been proposed and none is likely for many years, if ever. Therefore, the development of standard databases of ES mass spectra, analogous to databases of electron ionization mass spectra (Chapter 2), is unlikely.

Liquid Chromatography and Capillary Electrophoresis. The LC mobile phase flow rate, electrical conductivity, and composition are significant variables in ES/MS. Without pneumatic assistance, very low flow rates of less than about 20 μL/min are required to produce a stable ES. Even with pneumatic assistance, flow rates in the 40–200 μL/min range are often used although one interface design allowed a flow rate of 2 mL/min.[187] If the conductivity of the mobile phase is too high or too low, the stability of the ES can be adversely effected. Ammonium acetate and various phosphate buffers, which are often used to improve chromatographic resolution, may cause unacceptably high conductivity and an unstable ES. Pure aqueous solutions, which are often used at the beginning of a reverse-phase gradient elution, are difficult to ES even at very low flow rates. Strong hydrogen bonding and dipole–dipole interactions in water create a high surface tension, which may not produce a stable spray even with a high voltage. By incorporating a miscible organic solvent, usually methanol or acetonitrile, either in the mobile phase or in a coaxial flow, the surface tension at the spray needle exit is lower, a stable spray is maintained at a lower voltage, and higher analyte ion abundances are often observed. Many reported ES mass spectra are measured with flow injection sample introduction where low flow rates are readily accommodated and the mobile phase composition can be controlled and maintained constant. However, when an on-line separation is required, as it is in most real-world analyses, flow rate and mobile phase characteristics are significant factors that must be considered.

Most on-line LC separations use a packed reverse-phase column and gradient elution beginning with a high percentage of water in the mobile phase and subsequent programming to 100% or nearly 100% organic solvent. Some investigators use standard 4.6 mm ID columns and split the flow after the column to divert a small percentage to the ES interface. This approach, which sacrifices substantial sample and therefore detection limits, is likely to decline in use because a variety of packed reverse-phase columns with IDs in the 0.5–3 mm range are commercially available. These smaller diameter columns not only provide higher efficiencies, but also allow, depending on the ES interface design, up to 100% of the sample flow to enter the interface.

Capillary electrophoresis (CE) has a major advantage over reverse-phase LC for separations of ionic substances. Reverse-phase LC stationary phases are nonpolar or slightly polar, and ionic substances are not well retained and elute with little or no separation. While analytes subject to ionization in solution certainly can be separated by reverse-phase LC, this usually requires manipulation of mobile phase pH or other techniques that add to the complexity of the analysis and may not be particularly effective. For example, carboxylic acids, which are ionized at moderate to high pH, must be separated at low pH to repress ionization and allow the neutral molecules to partition between the nonpolar stationary phase and the mobile phase.[193] However, unless these acids also have basic sites, positive ions are not formed in solution and they cannot be detected by positive-ion ES. The acids can be detected as negative ions if the mobile phase pH is changed by postcolumn addition of a solution of a volatile base. This technique is feasible, but requires an additional LC pump and suitable plumbing connections, which complicate the analysis and increase the cost. An alternative approach is to use an ion pairing reagent, typically a tetraalkylammonium salt, whose large cation pairs with the carboxylate group in solution to neutralize the anion and allow separation on the nonpolar LC phase.[194] Ion pairing reagents are not always effective in providing high resolution separations and again complicate the analysis. Capillary electrophoresis provides a simple alternative in which the carboxylate anions can be separated with high resolution and detected by negative-ion ES.

Because CE is designed to separate ions in solution, and ES transfers ions from the liquid phase to the gas phase, ES/MS is probably the perfect CE detector. In addition, the low flow rates used in CE, that is, a few nanoliters per minute, are compatible with high signal/noise ES. Several types of ES interfaces have been adapted to CE. In one early design, a coaxial flow of organic solvent, usually methanol, surrounds the CE capillary and provides an electrical connection to the high voltage on the ES needle at the exit.[206] The voltage on the ES needle establishes both the potential difference across the CE capillary and the high voltage necessary for ES. The coaxial or sheath flow of organic solvent also supplements the CE flow, which in some designs is too small to maintain a stable ES. This type of CE interface has also been used with an ES design that uses a grounded spray needle (Figure 6.9).[175] In another design, a liquid junction interface is used to connect the end of the CE column to the ES needle.[207] Several modifications of these and other interface designs are described in a review[208] and other publications.[209, 210]

There are some reported environmental applications of CE/ES/MS. The quaternary ammonium herbicides paraquat and diquat (Figure 6.1) were separated in about 10 min and their positive ion ES mass spectra were studied.[79] A number of metal ions and inorganic anions were separated by CE and detected with positive or negative ion ES/MS.[200] The Ni and Cu chelates of ethylenediaminetetra-acetic acid were also separated by CE and measured by negative ion ES/MS.[202] A group of 16 carboxylic acid and phenol herbicides, including picloram, 2,4-D, dinoseb, pentachlorophenol, and 2,4,5-TP (Table 6.1), were separated as anions by CE in less than 40 min and measured by negative ion ES/MS.[211] The target analytes were added to drinking water, which was adjusted to pH 10 and analyzed without further processing.

Quantitative Analysis. Both LC and, to a much lesser extent, CE have been used with ES for quantitative analysis, and linear calibrations over a factor of 100 or more in concentration have been reported.[19, 150, 194, 195, 201, 202] The precision of ion abundance measurements from repetitive injections has been studied by a few investigators and appears to be in the range of 4–35% relative standard deviation (RSD) depending on the design of the interface, the techniques used, and other factors.[19] Instrument detection limits (IDLs) have been reported by several investigators and these vary widely depending on the analytes, the type of instrumentation used, whether data acquisition was by CMS, SIM, or SRM, and the definition of detection limit used.

Atrazine and six of its degradation products gave IDLs of 0.8–6 ng with CMS data acquisition and 30–70 pg with SIM.[199] Carbamate pesticides gave IDLs of 0.4–1.8 ng with SIM data acquisition,[144] but several carbamates, hydrocarbon-soluble azodyes, and a triazine herbicide gave IDLs of 20–40 pg with an ion storage (ion trap) type of mass spectrometer.[189] For 18 phenols, IDLs are reported to be in the range 25–600 ng with CMS data acquisition and 0.05–6 ng with SIM.[150] As expected, the lowest IDLs were obtained with the most acidic polychloro- and nitro-phenols. This approach was extended to other phenols, and linear calibrations were used to determine phenols in river water. The signal/noise-based limits of detection (S/N-LOD) for 100 mL water samples and SIM were in the range 12–60 ng/L.[163] Several phosphorus pesticides were detected in the 10–200 pg range with SIM data acquisition,[195] and some phenoxyacetic acid herbicides were detected in the 0.3–1 ng range with CMS data acquisition.[194]

The 16 carboxylic acid and phenol herbicides separated by CE and measured by negative ion ES/MS gave recoveries in the 91–124% range with an internal standard and RSDs in the range 3–10%.[211]

Summary

Compounds not amenable to GC are widespread in the environment, but they are not nearly as frequently determined as the volatile and semivolatile compounds described in Chapters 3 and 4. Much more attention is given to nonvolatile, thermally unstable, acidic, basic, and reactive analytes in some related areas of chemical analysis. Considerable research has been conducted to develop on-line condensed-phase separation and mass spectrometric measurement techniques for this very large class of compounds. A variety of techniques, which were described in this chapter, are now available and they allow the separation and determination of virtually any compound or ion regardless of its structure or molecular weight. Currently, however, just a few particle beam (PB) and thermospray TSP methods have been published by the USEPA and these are not required by any environmental regulation.[212,213]

For the near future when an on-line condensed-phase separation is required, the atmospheric pressure chemical ionization (APCI) and electrospray (ES) techniques probably will receive the most attention and be applied to most problems. At present, ES is probably the premier LC/MS and CE/MS technique because, compared with APCI, it is applicable to a wider variety of compounds and ions in solution. Electrospray or APCI are applicable to essentially all the analytes previously determined with TSP, and the development of these techniques is probably a major factor in the decline of interest in TSP. Electrospray has an advantage over TSP in using minimal heating and allowing the determination of many less thermally stable and ionic analytes. The TSP and PB techniques, however, still have some positive attributes, and with modifications and new designs, may evolve and continue to be used.

The applications of tandem MS (Chapter 5) and collision induced dissociation (CID) with APCI and ES, and perhaps TSP, should expand rapidly. These technologies are now available to a wide variety of laboratories because of the mid-to-late 1990s development of lower-cost bench-top mass spectrometers with these capabilities. The tandem and CID techniques can add considerable versatility and provide the fragment ions that are often not present with APCI and ES. Additional developments are likely, and entirely different techniques and strategies for compounds not amenable to GC probably will be discovered and developed in the twenty-first century.

REFERENCES

1. *National Primary Drinking Water Regulations: Disinfection Byproducts*; Final Rule, *Federal Register* 16 December 1998, Vol. 63, 69390-69476; Title 40 Code of Federal Regulations Parts 9, 141, and 142.
2. Code of Federal Regulations, Title 40, Part 141.61.

3. *Restricted Use Products Report* URL http://www.epa.gov/opppmsd1/RestProd/.
4. *List of Pesticides Banned and Severely Restricted in the U.S.* URL http://www.epa.-gov/opppsps1/piclist.htm.
5. Behymer, T. D.; Bellar, T. A.; Budde, W. L. *Anal. Chem.* **1990**, 62, 1686–1690.
6. Trehy, M. L.; Yost, R. A.; McCreary, J. J. *Anal. Chem.* **1984**, 56, 1281–1285.
7. Thomson, J. B.; Brown, P.; Djerassi, C. *J. Am. Chem. Soc.* **1966**, *88*, 4049–4055.
8. *NIST/EPA/NIH Mass Spectral Database*; National Institute for Standards and Technology: Gaithersburg, MD, 1992.
9. Onuska, F. I.; Boos, W. R. *Anal. Chem.* **1973**, *45*, 967–970.
10. Lide, D. R., Ed.-in-Chief *CRC Handbook of Chemistry and Physics*; 78th ed.; CRC Press: Boca Raton, FL, 1997/1998.
11. Feng, R.; Czajkowska, T. *Proceedings of the 45th American Society for Mass Spectrometry Conference on Mass Spectrometry and Allied Topics*; 1–5 June 1997, Palm Springs, CA, p 45.
12. Krasner, S. W.; McGuire, M. J.; Jacangelo, J. G.; Patania, N. L.; Reagan, K. M.; Aieta, E. M. *J. Am. Water Works Assoc.* **1989**, *81*, 41–50.
13. Cowman, G. A.; Singer, P. C. *Environ. Sci. Technol.* **1996**, *30*, 16–24.
14. Glaze, W. H.; Koga, M.; Cancilla, D. *Environ. Sci. Technol.* **1989**, *23*, 838–847.
15. Kronberg, L; Holmbom, B.; Reunanen, M.; Tikkanen, L. *Environ. Sci. Technol.* **1988**, *22*, 1097–1103.
16. Clark, E. A.; Sterritt, R. M.; Lester, J. N. *Environ. Sci. Technol.* **1988**, *22*, 600–604.
17. *Annual Report on Carcinogens*; 1998, URL http://ntp-server.niehs.nih.gov.
18. Galletti, G. C.; Dinelli, G.; Chiavari, G. *J. Mass Spectrom.* **1995**, *30*, 333–338.
19. Li, L. Y. T.; Campbell, D. A.; Bennett, P. K.; Henion, J. *Anal. Chem.* **1996**, *68*, 3397–3404.
20. Rodriguez, M.; Orescan, D. B. *Anal. Chem.* **1998**, *70*, 2710–2717.
21. Budde, W. L.; Eichelberger, J. W. In *Identification and Analysis of Organic Pollutants in Water*; Keith, L. H., Ed.; Ann Arbor Science Publishers: Ann Arbor, MI, 1976, pp 155–176.
22. Bellar, T. A.; Budde, W. L. *Anal. Chem.* **1988**, *60*, 2076–2083.
23. Betowski, L. D.; Kendall, D. S.; Pace, C. M.; Donnelly, J. R. *Environ. Sci. Technol.* **1996**, *30*, 3558–3564.
24. Cotter, R. J. *Anal. Chem.* **1980**, *52*, 1589A–1606A.
25. Winnik, W.; Brumley, W.; Betowski, L. *J. Mass Spectrom.* **1995**, *30*, 1574–1580.
26. Lattimer, R. P.; Schulten, H.-R. *Anal. Chem.* **1989**, *61*, 1201A–1215A.
27. Barber, M.; Bordoli, R. S.; Elliott, G. J.; Sedgwick, R. D.; Tyler, A. N. *Anal. Chem.* **1982**, *54*, 645A–657A.
28. Moseley, M. A.; Deterding, L. J.; Wit, J. S. M. de; Tomer, K. B.; Kennedy, R. T.; Bragg, N.; Jorgenson, J. W. *Anal. Chem.* **1989**, *61*, 1577–1584.
29. Caprioli, R. M. *Anal. Chem.* **1990**, *62*, 477A–485A.
30. Borgerding, A. J.; Hites, R. A. *Anal. Chem.* **1992**, *64*, 1449–1454.
31. Henry, C. *Anal. Chem.* **1997**, *69*, 625A–627A.
32. Aberth, W.; Straub, K. M.; Burlingame, A. L. *Anal. Chem.* **1982**, *54*, 2029–2034.
33. Falick, A. M.; Wang, G. H.; Walls, F. C. *Anal. Chem.* **1986**, *58*, 1308–1311.
34. Macfarlane, R. D. *Anal. Chem.* **1983**, *55*, 1247A–1264A.
35. Macfarlane, R. D.; Hu, Z.-H.; Song, S.; Pittenauer, E.; Schmid, E. R.; Allmaier, G.; Metzger, J. O.; Tuszynski, W. *Biol. Mass Spectrom.* **1994**, *23*, 117–130.
36. Hillenkamp, F.; Karas, M.; Beavis, R. C.; Chait, B. T. *Anal. Chem.* **1991**, *63*, 1193A–1203A.
37. Busch, K. L. *J. Mass Spectrom.* **1995**, *30*, 233–240.

38. Gibbs, R. A. *Anal. Chem.* **1990**, *62*, 1202–1214.
39. Newman, A. *Anal. Chem.* **1995**, *67*, 731A–734A.
40. Fenselau, C.; Ed. *Mass Spectrometry for the Characterization of Microorganisms*; American Chemical Society Symposium Series No. 541; American Chemical Society: Washington, DC, 1994.
41. Knapp, D. R., Ed. *Handbook of Analytical Derivatization Reactions*; John Wiley: New York, 1979.
42. Drozd, J.; Novak, J. P. *Chemical Derivatization in Gas Chromatography*; Elsevier Scientific: New York, 1981.
43. Frei, R. W.; Lawrence, J. F., Eds. *Chemical Derivatization in Analytical Chemistry*; Plenum Press: New York, 1981, Vol. I.
44. Blau, K.; Halket, J. H., Eds. *Handbook of Derivatives for Chromatography*; 2nd ed.; John Wiley: New York, 1994.
45. Dethlefs, F.; Gerhardt, K. O.; Stan, H.-J. *J. Mass Spectrom.* **1996**, *31*, 1163–1168.
46. Hawthorne, S. B.; Miller, D. J.; Nivens, D. E.; White, D. C. *Anal. Chem.* **1992**, *64*, 405–412.
47. Croft, M. Y.; Murby, E. J.; Wells, R. J. *Anal. Chem.* **1994**, *66*, 4459–4465.
48. Kawamura, K. *Anal. Chem.* **1993**, *65*, 3505–3511.
49. Seiber, J. N.; Woodrow, J. E. *Arch. Environm. Contam. Toxicol.* **1981**, *10*, 133–149.
50. Matthias, C. L.; Bellama, J. M.; Olson, G. J.; Brinckman, F. E. *Environ. Sci. Technol.* **1986**, *20*, 609–615.
51. Jiang, G. B.; Maxwell, P. S.; Siu, K. W. M.; Luong, V. T.; Berman, S. S. *Anal. Chem.* **1991**, *63*, 1506–1509.
52. Budzikiewicz, H.; Djerassi, C.; Williams, D. H. *Mass Spectrometry of Organic Compounds*; Holden-Day: San Francisco, CA, 1967, pp 471–477.
53. McCloskey, J. A.; Stillwell, R. N.; Lawson, A. M. *Anal. Chem.* **1968**, *40*, 233–236.
54. Risby, T. H.; Yergey, A. L. *Anal. Chem.* **1978**, *50*, 226A–333A.
55. Heeren, R. M. A.; Koster, C. G. de; Boon, J. J. *Anal. Chem.* **1995**, *67*, 3965–3970.
56. DeLuca, S. J.; Voorhees, K. J.; Sarver, E. W. *Anal. Chem.* **1986**, *58*, 2439–2442.
57. DeLuca, S.; Sarver, E. W.; Harrington, P. De B.; Voorhees, K. J.; *Anal. Chem.* **1990**, *62*, 1465–1472.
58. Ettre, L. S.; Horvath, C. *Anal. Chem.* **1975**, *47*, 422A–446A.
59. Novotny, M. *Anal. Chem.* **1981**, *53*, 1294A–1308A.
60. Dorschel, C. A.; Ekmanis, J. L.; Oberholtzer, J. E.; Warren, F. V.; Bidlingmeyer, B. A. *Anal. Chem.* **1989**, *61*, 951A–968A.
61. Dorsey, J. G.; Cooper, W. T. *Anal. Chem.* **1994**, *66*, 857A–867A.
62. Brown, P. R. *Anal. Chem.* **1990**, *62*, 995A–1008A.
63. Brown, P. R.; Hartwick, R. A., Eds. *High Performance Liquid Chromatography*; John Wiley: New York, 1989.
64. Bidlingmeyer, B. A. *Practical HPLC Methodology and Applications*; John Wiley: New York, 1993.
65. Snyder, L. R.; Kirkland, J. J.; Glajch, J. L. *Practical HPLC Method Development*; 2nd ed.; John Wiley: New York, 1997.
66. Smith, R. D.; Wright, B. W.; Yonker, C. R. *Anal. Chem.* **1988**, *60*, 1323A–1336A.
67. Hawthorne, S. B. *Anal. Chem.* **1990**, *62*, 633A–642A.
68. Kalinoski, H. T.; Wright, B. W.; Smith, R. D. *Biomed. Environ. Mass Spectrom.* **1986**, *13, 33–45.*
69. Smith, R. D.; Kalinoski, H. T.; Udseth, H. R. *Mass Spectrom. Rev.* **1987**, *6*, 445–496.
70. Via, J.; Taylor, L. T. *Anal. Chem.* **1994**, *66*, 1385–1395.

71. Erickson, B. *Anal. Chem.* **1997**, *69*, 683A–686A.
72. Pinkston, J. D.; Chester, T. L. *Anal. Chem.* **1995**, *67*, 650A–656A.
73. Jorgenson, J. W.; Lukacs, K. D. *Anal. Chem.* **1981**, *53*, 1298–1302 .
74. Baker, D. R. *Capillary Electrophoresis*; John Wiley: New York, 1995.
75. Moseley, M. A.; Jorgenson, J. W.; Shabanowitz, J.; Hunt, D. F.; Tomer, K. B. *J. Am. Soc. Mass Spectrom.* **1992**, *3*, 289–300.
76. Yu, Y.; Cole, R. B. *Environ. Sci. Technol.* **1997**, *31*, 3251–3257.
77. Colón, L. A.; Guo, Y.; Fermier, A. *Anal. Chem.* **1997**, *69*, 461A–467A.
78. Lee, E. D.; Mück, W.; Henion, J. D.; Covey, T. R. *Biomed. Environ. Mass Spectrom.* **1989**, 18, 253–257.
79. Song, X.; Budde, W. L. *J. Am. Soc. Mass Spectrom.* **1996**, *7*, 981–986.
80. Arpino, P. J.; Guiochon, G. *Anal. Chem.* **1979**, *51*, 682A–701A.
81. Lee, E. D.; Henion, J. D. *J. Chromatogr. Sci.* **1985**, *23*, 253–264.
82. Wit, J. S. M. de; Parker, C. E.; Tomer, K. B.; Jorgenson, J. W. *Anal. Chem.* **1987**, *59*, 2400–2404.
83. Parker, C. E.; Voyksner, R. D.; Tondeur, Y.; Henion, J. D.; Harvan, D. J.; Hass, J. R.; Yinon, J. *J. Forensic Sci.* **1982**, *27*, 495–505.
84. Voyksner, R. D.; Bursey, J. T. *Anal. Chem.* **1984**, *56*, 1582–1587.
85. Wit, J. S. M. de; Parker, C. E.; Tomer, K. B.; Jorgenson, J. W. *Biomed. Environ. Mass Spectrom.* **1988**, *17*, 47–53.
86. Pinkston, J. D.; Bowling, D. J. *Anal. Chem.* **1993**, *65*, 3534–3539.
87. Smith, R. D.; Udseth, H. R. *Anal. Chem.* **1987**, *59*, 13–22.
88. Kalinoski, H. T.; Smith, R. D. *Anal. Chem.* **1988**, *60*, 529–535.
89. McFadden, W. H.; Schwartz, H. L.; Evans, S. *J. Chromatogr.* **1976**, *122*, 389–396.
90. Scott, R. P. W.; Scott, C. G.; Munroe, M.; Hess Jr., J. *J. Chromatogr.* **1974**, *99*, 395.
91. Smith, R. D.; Johnson, A. L. *Anal. Chem.* **1981**, *53*, 739–740.
92. Hardin, E. D.; Fan, T. P.; Blakley, C. R.; Vestal, M. L. *Anal. Chem.* **1984**, *56*, 2–7.
93. Stroh, J. G.; Cook, J. C.; Milberg, R. M.; Brayton, L.; Kihara, T.; Huang, Z.; Rinehart Jr., K. L.; Lewis, I. A. S. *Anal. Chem.* **1985**, *57*, 985–991.
94. Hayes, M. J.; Schwartz, H. E.; Vouros, P.; Karger, B. L.; Thruston Jr., A. D.; McGuire, J. M. *Anal. Chem.* **1984**, *56*, 1229–1236.
95. Willoughby, R. C.; Browner, R. F. *Anal. Chem.* **1984**, *56*, 2626–2631.
96. Winkler, P. C.; Perkins, D. D.; Williams, W. K.; Browner, R. F. *Anal. Chem.* **1988**, *60*, 489–493.
97. Apffel, A.; Perry, M. L. *J. Chromatogr.* **1991**, *554*, 103–118.
98. Ligon Jr., W. V.; Dorn, S. B. *Anal. Chem.* **1990**, *62*, 2573–2580.
99. Vestal, M. L.; Winn, D. H.; Vestal, C. H. In *LC-MS: New Developments and Applications to Pesticide, Pharmaceutical and Environmental Analysis*; Brown, M. A., Ed.; ACS Symposium Series; American Chemical Society: Washington, DC, 1990, pp 215–231.
100. Cappiello, A.; Bruner, F. *Anal. Chem.* **1993**, *65*, 1281–1287.
101. Cappiello, A.; Famiglini, G. *Anal. Chem.* **1994**, *66*, 3970–3976.
102. Cappiello, A.; Famiglini, G. *Anal. Chem.* **1995**, *67*, 412–419.
103. Creaser, C. S.; Stygall, J. W. *Analyst* **1993**, *118*, 1467–1480.
104. Jones, T. L.; Betowski, L. D.; Lesnik, B.; Chiang, T. C.; Teberg, J. E. *Environ. Sci. Technol.* **1991**, *25*, 1880–1884.
105. Budde, W. L.; Behymer, T. D.; Bellar, T. A.; Ho, J. S. *J. Am. Water Works Assoc.* **1990**, *82*, 60–65.
106. Kim, S.; Sasinos, F. I.; Stephens, R. D.; Wang, J.; Brown, M. A. *Anal. Chem.* **1991**, *63*, 819–823.

107. Cappiello, A.; Famiglini, G.; Lombardozzi, A.; Massari, A.; Vadalá, G. G. *J. Am. Soc. Mass Spectrom.* **1996**, *7*, 753–758.
108. Ho, J. S.; Bellar, T. A.; Eichelberger, J. W.; Budde, W. L. *Environ. Sci. Technol.* **1990**, *24*, 1748–1751.
109. Slobodnik, J.; Jager, M. E.; Hoekstra-Oussoren, S. J. F.; Honing, M.; Baar, B. L. M. van; Brinkman, U. A. T. *J. Mass Spectrom.* **1997**, *32*, 43–54.
110. Jedrzejewski, P. T.; Taylor, L. T. *J. Chromatogr. A* **1994**, *677*, 365–376.
111. Edlund, P. O.; Henion, J. D. *J. Chromatogr. Sci.* **1989**, *27*, 274–282.
112. Doerge, D. R.; Miles, C. J. *Anal. Chem.* **1991**, *63*, 1999–2000.
113. Voyksner, R. D.; Straub, R.; Keever, J. T.; Freeman, H. S.; Hsu, W.-N. *Environ. Sci. Technol.* **1993**, *27*, 1665–1672.
114. Bellar, T. A.; Budde, W. L. Unpublished measurements, 1990.
115. Singh, R. P.; Brindle, I. D.; Jones, T. R. B.; Miller, J. M.; Chiba, M. *J. Am. Soc. Mass Spectrom.* **1993**, *4*, 898–905.
116. Ho, J. S.; Budde, W. L. *Anal. Chem.* **1994**, *66*, 3716–3722.
117. Hsu, J. *Anal. Chem.* **1992**, *64*, 434–443.
118. Kim, S.; Sasinos, F. I.; Stephens, R. D.; Brown, M. A. *Environ. Sci. Technol.* **1990**, *24*, 1832–1836.
119. Carpenter, J. C.; Cella, J. A.; Dorn, S. B. *Environ. Sci. Technol.* **1995**, *29*, 864–868.
120. Kim, S.; Sasinos, F. I.; Stephens, R. D.; Brown, M. A. *J. Agric. Food Chem.* **1990**, *38*, 1223–1226.
121. Clark, L. B.; Rosen, R. T.; Hartman, T. G.; Louis, J. B.; Rosen, J. D. *Int. J. Environ. Anal. Chem.* **1991**, *45*, 169–178.
122. Bellar, T. A.; Behymer, T. D.; Budde, W. L. *J. Am. Soc. Mass Spectrom.* **1990**, *1*, 92–98.
123. Jedrzejewski, P. T.; Taylor, L. T. *J. Chromatogr. A* **1995**, *703*, 489–501.
124. Cappiello, A.; Famiglini, G.; Bruner, F. *Anal. Chem.* **1994**, *66*, 1416–1423.
125. Cappiello, A.; Famiglini, G.; Palma, P.; Berloni, A.; Bruner, F. *Environ. Sci. Technol.* **1995**, *29*, 2295–2300.
126. Ho, J. S.; Behymer, T. D.; Budde, W. L.; Bellar, T. A. *J. Am. Soc. Mass Spectrom.* **1992**, *3*, 662–671.
127. Mattina, M. J. I. *J. Chromatogr.* **1991**, *554*, 385–395.
128. Apffel, A.; Perry, M. L. *J. Chromatogr.* **1991**, *554*, 103–118.
129. Wilkes, J. G.; Zarrin, F.; Lay Jr., J. O.; Vestal, M. L. *Rapid Commun. Mass Spectrom.* **1995**, *9*, 133–137.
130. Betowski, L. D.; Pace, C. M.; Roby, M. R. *J. Am. Soc. Mass Spectrom.* **1992**, *3*, 823–830.
131. Blakley, C. R.; Carmody, J. J.; Vestal, M. L. *J. Am. Chem. Soc.* **1980**, *102*, 5931–5933.
132. Blakley, C. R.; Carmody, J. J.; Vestal, M. L. *Anal. Chem.* **1980**, *52*, 1636–1641.
133. Blakley, C. R.; Vestal, M. L. *Anal. Chem.* **1983**, *55*, 750–754.
134. Vestal, M. L.; Fergusson, G. J. *Anal. Chem.* **1985**, *57*, 2373–2378.
135. Vestal, M. L. In *Mass Spectrometry*; McCloskey, J. A., Ed.; *Methods in Enzymology*; Vol. 193; Academic Press: New York, 1990, pp 107–130.
136. Vestal, M. L. In *Ion formation from organic solids: Proceedings of the Second International Conference*; Munster, Federal Republic of Germany, 7–9 September 1982; Benninghoven, A., Ed.; Springer Series in Chemical Physics, Vol. 25; Springer-Verlag: Berlin, New York, 1983, pp 246–263.
137. Alexander, A. J.; Kebarle, P. *Anal. Chem.* **1986**, *58*, 471–478.
138. Bencsath, F. A.; Field, F. H. *Anal. Chem.* **1988**, *60*, 1323–1329.

139. Abían, J.; Sánchez-Baeza, F.; Gelpí, E.; Barceló, D. *J. Am. Soc. Mass Spectrom.* **1994**, *5*, 186–193.
140. Voyksner, R. D.; Bursey, J. T.; Pellizzari, E. D. *Anal. Chem.* **1984**, *56*, 1507–1514.
141. Yergey, A.; Edmonds, C.; Lewis, I.; Vestal, M. *Liquid Chromatography/Mass Spectrometry, Techniques and Applications*; Plenum Press: New York, 1989.
142. Honing, M.; Barceló, D.; Baar, B. L. M. van; Brinkman, U. A. T. *J. Mass Spectrom.* **1996**, *31*, 527–536.
143. Voyksner, R. D. *Anal. Chem.* **1985**, *57*, 2600–2605.
144. Pleasance, S.; Anacleto, J. F.; Bailey, M. R.; North, D. H. *J. Am. Soc. Mass Spectrom.* **1992**, *3*, 378–397.
145. Volmer, D.; Levsen, K. *J. Am. Soc. Mass Spectrom.* **1994**, *5*, 655–675.
146. Evans, K. A.; Dubey, S. T.; Kravetz, L.; Dzidic, I.; Gumulka, J.; Mueller, R.; Stork, J. R. *Anal. Chem.* **1994**, *66*, 699–705.
147. Voyksner, R. D.; Haney, C. A. *Anal. Chem.* **1985**, *57*, 991–996.
148. Betowski, L. D.; Jones, T. L. *Environ. Sci. Technol.* **1988**, *22*, 1430–1434.
149. Betowski, L. D.; Ballard, J. M. *Anal. Chem.* **1984**, *56*, 2604–2607.
150. Puig, D.; Barceló, D.; Silgoner, I.; Grasserbauer, M. *J. Mass Spectrom.* **1996**, *31*, 1297–1307.
151. Vreeken, R. J.; Brinkman, U. A. T.; de Jong, G. J.; Barceló, D.; *Biomed. Environ. Mass Spectrom.* **1990**, *19*, 481–492.
152. Barceló, D.; Durand, G.; Vreeken, R. J.; Jong, G. J. de; Brinkman, U. A. T. *Anal. Chem.* **1990**, *62*, 1696–1700.
153. Honing, M.; Barceló, D.; van Baar, B. L. M.; Ghijsen, R. T.; Brinkman, U. A. T. *J. Am. Soc. Mass Spectrom.* **1994**, *5*, 913–927.
154. Volmer, D.; Levsen, K.; Honing, M.; Barceló, D.; Abian, J.; Gelpi, E.; van Baar, B. L. M.; Brinkman, U. A. T. *J. Am. Soc. Mass Spectrom.* **1995**, *6*, 656–667.
155. Horning, E. C.; Horning, M. G.; Carroll, D. I.; Dzidic, I.; Stillwell, R. N. *Anal. Chem.* **1973**, *45*, 936–943.
156. Carroll, D. I.; Dzidic, I.; Stillwell, R. N..; Horning, M. G.; Horning, E. C. *Anal. Chem.* **1974**, *46*, 706–710.
157. Thomson, B. A. *J. Am. Soc. Mass Spectrom.* **1998**, *9*, 187–193.
158. Covey, T. R.; Lee, E. D.; Bruins, A. P.; Henion, J. D. *Anal. Chem.* **1986**, *58*, 1451A–1461A.
159. Huang, E. C.; Wachs, T; Conboy, J. J.; Henion, J. D. *Anal. Chem.* **1990**, *62*, 713A–725A.
160. Bruins, A. P.; Weidolf, L. O. G.; Henion, J. D.; Budde, W. L. *Anal. Chem.* **1987**, *59*, 2647–2652.
161. Thomas, D.; Crain, S. M.; Sim, P. G.; Benoit, F. M. *J. Mass Spectrom.* **1995**, *30*, 1034–1040.
162. Lacorte, S.; Barceló, D. *Anal. Chem.* **1996**, *68*, 2464–2470.
163. Puig, D.; Silgoner, I.; Grasserbauer, M.; Barceló, D. *Anal. Chem.* **1997**, *69*, 2756–2761.
164. Ferrer, I.; Hennion, M.-C.; Barceló, D. *Anal. Chem.* **1997**, *69*, 4508–4514.
165. Castillo, M.; Alpendurada, M. F.; Barceló, D. *J. Mass Spectrom.* **1997**, *32*, 1100–1110.
166. Schaefer, W. H.; Dixon, F. *J. Am. Soc. Mass Spectrom.* **1996**, *7*, 1059–1069.
167. Fenn, J. B.; Mann, M.; Meng, C. K.; Wong, S. F.; Whitehouse, C. M. *Science* **1989**, *246*, 64–71.
168. Smith, R. D.; Loo, J. A.; Edmonds, C. G.; Barinaga, C. J.; Udseth, H. R. *Anal. Chem.* **1990**, *62*, 882–899.

169. Kebarle, P.; Tang, L. *Anal. Chem.* **1993**, *65*, 972A–986A.
170. Fenn, J. B. *J. Am. Soc. Mass Spectrom.* **1993**, *4*, 524–535.
171. Gaskell, S. J. *J. Mass Spectrom.* **1997**, *32*, 677–688.
172. Cole, R. B., Ed. *Electrospray Ionization Mass Spectrometry: Fundamentals, Instrumentation, and Applications*; Wiley-Interscience: New York, 1997.
173. Thomson, B. A.; Iribarne, J. V.; Dziedzic. P. J. *Anal. Chem.* **1982**, *54*, 2219–2224.
174. Whitehouse, C. M.; Dreyer, R. N.; Yamashita, M.; Fenn, J. B. *Anal. Chem.* **1985**, *57*, 675–679.
175. Banks, J. F.; Dresch, T. *Anal. Chem.* **1996**, *68*, 1480–1485.
176. Katta, V.; Chowdhury, S. K.; Chait, B. T. *Anal. Chem.* **1991**, *63*, 174–178.
177. Bruins, A. P.; Covey, T. R.; Henion, J. D. *Anal. Chem.* **1987**, *59*, 2642–2646.
178. Allen, M. H.; Vestal, M. L. *J. Am. Soc. Mass Spectrom.* **1992**, *3*, 18–26.
179. Van Berkel, G. J.; Glish, G. L.; McLuckey, S. A. *Anal. Chem.* **1990**, *62*, 1284–1295.
180. McLuckey, S. A.; Van Berkel, G. J.; Glish, G. L.; Huang, E. C.; Henion, J. D. *Anal. Chem.* **1991**, *63*, 375–383.
181. Van Berkel, G. J.; McLuckey, S. A.; Glish, G. L. *Anal. Chem.* **1992**, *64*, 1586–1593.
182. Kuhlmann, F. E.; Apffel, A.; Fischer, S. M.; Goldberg, G.; Goodley, P. C. *J. Am. Soc. Mass Spectrom.* **1995**, *6*, 1221–1225.
183. Van Berkel, G. J.; McLuckey, S. A.; Glish, G. L. *Anal. Chem.* **1991**, *63*, 2064–2068.
184. Van Berkel, G. J.; Asano, K. G. *Anal. Chem.* **1994**, *66*, 2096–2102.
185. Yang, Y.; Griffiths, W. J.; Sjövall, J.; Gustafsson, J.-Å.; Rafter, J. *J. Am. Soc. Mass Spectrom.* **1997**, *8*, 50–61.
186. Duffin, K. L.; Wachs, T.; Henion, J. D. *Anal. Chem.* **1992**, *64*, 61–68.
187. Hopfgartner, G.; Wachs, T.; Bean, K.; Henion, J. *Anal. Chem.* **1993**, *65*, 439–446.
188. Rule, G. S.; Mordehai, A. V.; Henion, J. *Anal. Chem.* **1994**, *66*, 230–235.
189. Lin, H.-Y.; Voyksner, R. D. *Anal. Chem.* **1993**, *65*, 451–456.
190. Ferrer, I.; Thurman, E. M.; Barceló, D. *Anal. Chem.* **1997**, *69*, 4547–4553.
191. Straub, R. F.; Voyksner, R. D. *J. Am. Soc. Mass Spectrom.* **1993**, *4*, 578–587.
192. Hughes, B. M.; McKenzie, D. E.; Duffin, K. L. *J. Am. Soc. Mass Spectrom.* **1993**, *4*, 604–610.
193. Chiron, S.; Papilloud, S.; Haerdi, W.; Barceló, D. *Anal. Chem.* **1995**, *67*, 1637–1643.
194. Crescenzi, C.; Di Corcia, A.; Marchese, S.; Samperi, R. *Anal. Chem.* **1995**, *67*, 1968–1975.
195. Molina, C.; Honing, M.; Barceló, D. *Anal. Chem.* **1994**, *66*, 4444–4449.
196. Edlund, P. O. E.; Lee, E. D.; Henion, J. D.; Budde, W. L. *Biomed. Environ. Mass Spectrom.* **1989**, *18*, 233–240.
197. Crescenzi, C.; Di Corcia, A.; Samperi, R.; Marcomini, A. *Anal. Chem.* **1995**, *67*, 1797–1804.
198. Crescenzi, C.; Di Corcia, A.; Marcomini, A.; Samperi, R. *Environ. Sci. Technol.* **1997**, *31*, 2679–2685.
199. Di Corcia, A.; Crescenzi, C.;Guerriero, E.; Samperi, R. *Environ. Sci. Technol.* **1997**, *31*, 1658–1663.
200. Corr, J. J.; Anacleto, J. F. *Anal. Chem.* **1996**, *68*, 2155–2163.
201. Charles, L.; Pépin, D. *Anal. Chem.* **1998**, *70*, 353–359.
202. Sheppard, R. L.; Henion, J. *Anal. Chem.* **1997**, *69*, 2901–2907.
203. Conboy, J. J.; Henion, J. D.; Martin, M. W.; Zweigenbaum, J. A. *Anal. Chem.* **1990**, *62*, 800–807.
204. Siu, K. W. M.; Gardner, G. J.; Berman, S. S. *Anal. Chem.* **1989**, *61*, 2320–2322.
205. Quirke, J. M. E.; Adams, C. L.; Van Berkel, G. J. *Anal. Chem.* **1994**, *66*, 1302–1315.
206. Smith, R. D.; Barinaga, C. J.; Udseth, H. R. *Anal. Chem.* **1988**, *60*, 1948–1952.

207. Lee, E. D.; Mück, W.; Henion, J. D.; Covey, T. R. *Biomed. Environ. Mass Spectrom.* **1989**, *18*, 844–850.
208. Smith, R. D.; Wahl, J. H.; Goodlett, D. R.; Hofstadler, S. A. *Anal. Chem.* **1993**, *65*, 574A–584A.
209. Severs, J. C.; Smith, R. D. *Anal. Chem.* **1997**, *69*, 2154–2158.
210. Siethoff, C.; Nigge, W.; Linscheid, M. *Anal. Chem.* **1998**, *70*, 1357–1361.
211. Song, X.; Budde, W. L. *J. Chromatogr. A* **1998**, *829*, 327–340.
212. Method 553, Revision 1.1, In *Methods for the Determination of Organic Compounds in Drinking Water*; Suppl. II, USEPA Report EPA/600/R-92/129, August 1992; URL http://www.epa.gov/nerlcwww/methmans.html.
213. *Test Methods for Evaluating Solid Waste, Physical/Chemical Methods*; USEPA Publication SW-846, 3rd ed. and Updates I, II, IIA, IIB, and III; URL http://www.epa.gov.epaoswer/hazwaste/test/8xxx.htm.

7

Elemental Measurements and Their Applications

The identification and measurement of the chemical elements is the original and one of the classical applications of mass spectrometry (MS). Elemental measurements by MS are widely used in the chemical and physical sciences and in chemical analysis. Some of the basic properties of most elements were measured with MS, and MS was the technique that enabled the discovery to be made of the naturally occurring isotopes of the elements. The relative abundances of some isotopes are routinely measured at the elemental level with MS. The atomic weights are derived from mass spectrometric measurements of the exact masses and relative abundances of the naturally occurring nuclides.

It is beyond the scope of this book to consider all applications of elemental measurements by MS. In this chapter, four important, or potentially important, strategic applications of elemental measurements are described. The inductively coupled plasma (ICP) technique is the source of elemental ions for the mass spectrometric methods considered in this chapter. The ICP/MS methods are widely used for elemental analyses in environmental science, government environmental regulatory programs, and in many other areas of chemical analysis. The four strategies are summarized below and described in more detail in subsequent sections of the chapter.

- The determination of the total amounts of target elements in a sample. The total amount of an element is the sum of the quantities in each of the elemental forms, compounds, ions, and oxidation states of the element. This measurement is often defined by the procedure used to prepare the sample for analysis.

- The determination of target analytes, after chromatographic separation, by measurements of target elements other than those commonly found in organic compounds.
- The broad spectrum analysis of a sample to determine the individual elemental forms, compounds, ions, and oxidation states of an element or elements in a sample.
- The determination of the elements, other than those commonly found in organic compounds, present in unknown substances separated by chromatography.

The measurement of the total amounts of target elements in a sample is important in environmental and other analyses and is used for compliance monitoring of some environmental regulations in the United States. Several standard USEPA mass spectrometric methods for total target elements in several types of environmental matrices are summarized. The determination of target analytes using elemental analysis by MS is a relatively new strategy, which is likely to become more important in the future. Several examples of this approach are presented. Elemental analysis can contribute to the broad spectrum analyses of samples and this strategy is briefly explored. The determination of ion composition with electron ionization and other ionization techniques (Chapter 5) is well established and elemental analysis can also provide valuable information about the composition of unknown substances.

Inductively Coupled Plasma/Mass Spectrometry

The ICP was developed as a vaporization, atomization, and excitation source for atomic emission spectrometry (AES) primarily by Velmer Fassel and associates at Iowa State University during the 1960s–1970s.[1] Elemental analysis with AES had been practiced since the mid-1930s using flames and arc or spark discharges to vaporize and atomize samples and excite the atoms for optical emission. The AES technique provided rapid simultaneous or sequential multielement determinations, but the flames and arc or spark discharges had significant limitations. Interferences from electrode and other sample components were not uncommon, elemental measurements were often imprecise, detection limits were not sufficiently low, and liquid samples were difficult to analyze.[1,2]

During the 1960s flame atomic absorption spectrometry (AAS) became the dominant technique for determinations of the elements in low concentrations in aqueous samples, and AES was abandoned by many analysts.[1] Aqueous sample aerosols could be injected directly into the flame and accurate, precise, selective, and sensitive determinations of many elements could be made. However, AAS was limited to measuring one element at a time and there was an increasing demand for rapid multielement determinations. In addition, instrument detection limits for some important elements, for example, As, Cr, Pb, Sb, Se, and Tl, in water and other environmental samples were $> 50\,\mu g/L$ and often much greater.[2] The high-temperature graphite furnace sample introduction system for AAS, which was developed during the 1970s, provided instrument detection limits of $< 5\,\mu g/L$ for most elements, and allowed automation of sample processing, but was still limited to measuring one element at a time.[2] Other AAS sample introduction techniques, especially chemical reduction and elemental Hg vaporization (cold vapor) and conversion of As and Se into volatile hydrides, provided similarly low detection limits. The graphite furnace and Hg cold vapor techniques are often referred to in the scientific literature as flameless AAS methods.

Spark source ionization MS was also employed for multielement analysis during the 1960s. However, this technique was generally limited to qualitative and semiquantitative analyses because of variable ion abundances. Expensive double-focusing electrostatic and magnetic sector mass spectrometers were required to compensate for the broad kinetic energy distributions of the ions produced in the spark source. Systems for the introduction of liquid samples were not well developed and measurements were subject to interferences from electrode materials and other sample components. Thermal emission ionization, in which ions are formed during the high temperature evaporation of substances from the surface of a heated solid, is a very sensitive ionization technique that is widely used for precise measurements of isotope abundances in inorganic solids. However, this technique is not suited to liquid samples.

The ICP was a revolutionary development of the late 1960s and early 1970s that allowed the direct injection of aqueous sample aerosols into a 5500–8000 K argon ion plasma.[1] The ICP technique provided very efficient vaporization, atomization, and excitation of elements with minimal or no interferences from background components or other analytes.[1,2] The combination of ICP and AES, with its rapid multielement measurement capabilities, was a major advance in elemental analytical capabilities. Argon is the standard support gas for ICP because, in part, it has few emission lines in the 200–400 nm wavelength region where many element emission lines occur.[3] The ICP/AES technique was commercialized by several manufacturers and became a very important analytical technique for rapid simultaneous or sequential multielement analyses of environmental and many other types of samples.[2]

During the late 1970s the ICP was investigated as an ion source for MS.[4] This combination quickly became one of the most useful and important techniques for rapid multielement analyses of gases, liquids, and solids.[3] The Ar ICP is a very efficient ion source that produces mainly singly charged ions. It is estimated that 54 elements, all metals, are ionized with 90% or greater efficiency. Only C, H, N, a few electronegative elements, and the noble gases are ionized with efficiencies less than 10%. A small number of elements, for example, As, B, Be, Hg, I, P, S, Se, and Te have estimated ionization efficiencies in the 10–90% range. Although Ar has a low ionization efficiency, it is present in the ion source in great quantity and gives a few significant ions at m/z 40 ($Ar^{+\cdot}$) and m/z 80 ($Ar_2^{+\cdot}$). Other background ions, for example, m/z 41 (ArH^+) and m/z 56 ($ArO^{+\cdot}$), are formed from sample components, usually water, and mineral acids that are used to ensure dissolution of some analytes. These ions obscure ions from some elements, but techniques are available to circumvent most of these interferences.

Figure 7.1 is a diagram of an ICP/MS system that has several types of sample introduction systems. This diagram, which is not drawn to scale, shows the general features of the ICP ion source, but does not include valves used to isolate parts of the system and other similar components. It is beyond the scope of this chapter to describe the details of construction of commercial ICP/MS systems, which often have some proprietary features that may contribute to performance characteristics and other specifications. Manufacturer's literature and other reference books should be consulted for details of instrument designs and performance specifications.[3, 5] Samples are introduced into ICP/MS instruments either as vapors or as aerosols generated by nebulization of a flowing liquid with Ar. Solid materials are usually dissolved and introduced in solution, but laser ablation and thermal vaporization techniques are also used with ICP/MS.

Gases are injected directly into the ICP torch as shown in Figure 7.1 and sources of vapors include the effluents from gas chromatography (GC) or supercritical fluid chromatography (SFC) and the volatile products of chemical reactions, for example, the volatile hydrides of As or Se. Liquid samples are commonly introduced by flow injection (Figure 7.1) or as effluents from liquid chromatography (LC), ion exchange chromatography (IC), and capillary electrophoresis (CE) systems. Brief descriptions of these separation techniques are in Chapters 2 (GC) and 6 (LC, IC, SFC, and CE). Liquids are injected first into a nebulizer, which produces an aerosol spray into a spray chamber. Large droplets and condensate are removed in the spray chamber, and the fine aerosol particles are transported by the flowing Ar gas into the ICP torch. The spray chamber is similar to the desolvation chamber used with the particle beam LC interface described in Chapter 6. However, the efficiencies of transport of analytes from the sample to the ICP torch are only in the 1–5% range. If a low flow rate separation column is used, for example, a microbore LC column or CE column,

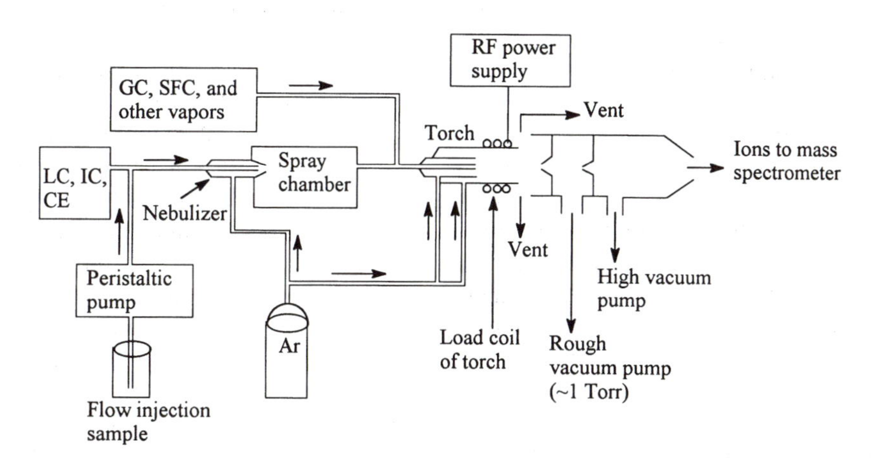

Figure 7.1 Diagram of an ICP/MS system with several types of sample introduction systems.

a supplemental flow of liquid into the nebulizer can be used or the spray chamber can be by-passed and the nebulizer spray injected directly into the ICP torch.

Vaporization of the sample, atomization of molecules and ions, and ionization occurs in the ICP torch at atmospheric pressure. Ions are transported through a cooled sampling orifice and a smaller skimmer orifice into the high vacuum of the mass spectrometer. The Ar and other neutral vapors from the sample rapidly diffuse in the plasma sampling region and are largely removed from the interface by several or more stages of vacuum pumping. This pressure reduction system is similar to those used in several LC/MS interfaces described in Chapter 6. Like some of the atmospheric pressure ionization (API) LC/MS systems, ions are formed in the interface between the sample introduction system and the mass spectrometer. Ions are then largely separated from the neutral gases in the interface, and mainly ions are injected through a series of focusing lenses into the high vacuum of the mass spectrometer. The efficiencies of commercial interfaces in transporting ions into the mass spectrometer has improved substantially since the first interface designs of the late 1970s.

The ICP/MS was commercialized in the early 1980s and even early models had element detection limits much lower than those of ICP/AES.[3] While ICP/MS does not simultaneously detect and measure ions from various elements, it does provide very fast sequential data acquisition using the continuous measurement of spectra (CMS) and selected ion monitoring (SIM) techniques described in Chapter 2 ("Data Acquisition Strategies"). Mass spectrometry provides many additional benefits including ease of interpretation of masses and charges that are familiar to chemists instead of obscure atomic emission or absorption wavelengths that are not familiar to most chemists. Mass spectra of mixtures of elements contain few ions compared to the potentially very large number of lines in atomic emission spectra. Mass spectrometry also provides the abundances of the isotopes of the elements that are used to ensure correct identifications. When isotopes are present, as they are for many elements, MS provides several choices of quantitation ions and allows quantitation, using the isotope dilution method of internal standardization (Chapter 2, "Quantitative Analysis"). High resolution and tandem MS are used with ICP ionization to provide enhanced selectivity, exact m/z measurements, and structural information (Chapter 5).

Competition among a number of manufacturers for a share of a growing market during the 1990s led to the development of bench-top instruments and instruments with lower prices than earlier models.[6] Commercial ICP/MS systems are automated and include standard operating system and applications software analogous to that supplied with conventional GC/MS and other mass spectrometer systems. Instrument detection limits are three orders of magnitude less than those of ICP/AES for most elements, which allows not only very low concentration measurements, but also permits analyses of very small or diluted samples.[7] Dilution with deionized water minimizes the quantities of dissolved solids introduced into the interface. Dissolved solids have a tendency to clog and corrode the sampling orifice and skimmer, which increases the need for maintenance. It is desirable to keep dissolved solids in samples to less than about 0.1–0.5% by weight. Detection limits in ICP/MS are not generally limited by instrument sensitivity, but by background elements introduced during sampling, sample shipment, and sample preparation. Clean laboratory hoods devoid of metal fixtures and continuously purged with filtered air are required to utilize the full sensitivity of contemporary ICP/MS instruments.

Total Amounts of Elements in a Sample

The total amount of an element in a sample is defined as the sum of the quantities in each of the elemental forms, compounds, ions, and oxidation states of the element. Some of these substances, or species, may also be present in different physical phases within the sample. For example, the total chlorine in a water sample may be divided among Cl_2, dissolved Cl^- and ClO^- ions, the volatile compound ClO_2, volatile organic compounds, and insoluble semivolatile organic compounds adsorbed on sediment particles. Similar diversity of chemical combination, water solubility, oxidation state, and so on, occurs with many elements. Therefore, the physical state of the sample, the sampling procedure, and the procedure used to prepare the sample for analysis are very important variables that can have a significant impact on the results of measurement. Many metals, for example, Cr, Hg, Fe, or Mn, are routinely measured in various samples as the total elements without regard to chemical combination, charge, physical phase, structure, or oxidation state.

The analytical information needed from an elemental analysis should be specified in the objectives of the scientific study, industrial project, or government regulatory program. The information needed from a geochemical or minerals exploration project may be much different than the information needed from an environmental study. Geochemical investigations may specify the complete dissolution or digestion of soil or rock samples so that the true total amounts of some or all elements in the sample can be determined. This may require aggressive treatment of the sample with a mixture of concentrated hydrochloric and nitric acids (aqua regia), perchloric acid, some other strong reagent, or a thermal fusion process. The processing of environmental samples can vary widely from the strong acid dissolution of inhalable air particulates to the gentle extraction of a soil sample using a procedure that simulates natural leaching by rainwater. This section is focused on analytical methods for total elements in environmental air, water, soil, sediment, and some tissue samples. Information on elemental analyses in geochemical studies, minerals exploration, semiconductor production, and other areas of specialization is available in reviews and reference books.[5,7] Regardless of the specific area of application, the scope of the determination of the total amounts of elements in a sample is explicitly or implicitly contained in the description of the complete analytical method.

Environmental Elemental Analyses

The determination of certain total elements in several kinds of environmental media were significant factors in the development of public concern for the rapidly degrading state of the environment in the United States during the late 1960s and 1970s (Chapter 1). Measurements of total Pb in house dust and children's blood were major factors in the development of environmental regulations that limited the use of Pb pigments in household paint. Measurements of total Hg in fish, soils, and sediments resulted in the banning of Hg-containing pesticides and the phase out of chemical production processes that used Hg electrodes and catalysts. The elements and all compounds of Ag, As, Be, Cd, Cr, Cu, Hg, Ni, Pb, Sb, Se, Tl, and Zn were declared priority pollutants (PP in Table 7.1) in the 1976 consent decree (Chapter 3, "Target Analytes—The Priority Pollutants").[8] More recently the determination of total Pb in automobile tailpipe emissions, ambient air particulates, and roadside soils contributed to the banning of Pb antiknock additives in gasoline. Measurements of total Pb in drinking water led to the banning of Pb solder in drinking water supply pipes. The Clean Air Act Amendments of 1990 designated the com-

Table 7.1 Elements that are consent decree priority pollutants (PPs) and/or hazardous air pollutants (HAPs) listed in the Clean Air Act Amendments of 1990 and/or have maximum contaminant levels (MCLs) or monitoring requirements in drinking water or are candidates for regulation

Elements	Symbol	PP[8]	HAP	MCL[9] (μg/L)	ICP/MS methods
Aluminum	Al			[a]	200.8, 6020
Antimony	Sb	+	+	6	200.8, 6020
Arsenic	As	+	+	50[b]	200.8, 6020
Barium	Ba			2000	200.8, 6020
Beryllium	Be	+	+	4	200.8, 6020
Cadmium	Cd	+	+	5	200.8, 6020
Chromium	Cr	+	+	100	200.8, 6020
Cobalt	Co		+		200.8, 6020
Copper	Cu	+		[c]	200.8, 6020
Lead	Pb	+	+	[c]	200.8, 6020
Manganese	Mn		+	[a]	200.8, 6020
Mercury	Hg	+	+	2	200.8 (SIM only)
Molybdenum	Mo				200.8
Nickel	Ni	+	+	[d]	200.8, 6020
Selenium	Se	+	+	50	200.8
Silver	Ag	+		[a]	200.8, 6020
Thallium	Tl	+		2	200.8, 6020
Thorium	Th				200.8
Uranium	U				200.8
Vanadium	V				200.8
Zinc	Zn	+		[a]	200.8, 6020

[a]Nonenforcable secondary drinking water regulation and monitoring recommendation.
[b]Interim value.
[c]MCL is not promulgated. Concentration controlled by source water monitoring, water treatment regulations, and the ban on Pb-based solder in drinking water systems. Action levels for Pb and Cu in drinking water are 10 and 1300 μg/L, respectively.
[d]Drinking water monitoring required by the Code of Federal Regulations.

pounds of 11 elements as hazardous air pollutants (HAPs) and these elements are listed in Table 7.1.

For the 1976 consent decree the USEPA initiated a substantial analytical program to determine the 13 priority pollutant elements in wastewater effluents from industries in 21 categories.[8] The consent decree required not only the determination of the 13 elements, but also all compounds of these elements. This would have been a very costly and difficult task, and the requirement was addressed by measuring the total amounts of the 13 elements in the wastewater samples. This required an analytical method designed to produce sample aliquots that were representative of all physical phases in the samples and all chemical forms of the elements. Strong mineral acids were used for dissolution of the solids dispersed in the wastewater samples. Most measurements were made with ICP/AES, which effectively vaporized and atomized all chemical entities containing the elements. The ICP/AES method provided instrument detection limits of $< 10\,\mu g/L$ for many elements, and where detection limits were insufficient, for example, for As, Hg, Sb, Se, and Tl, the flameless AAS methods provided instrument detection limits $< 5\,\mu g/L$.[2,8]

On 3 December 1979 the USEPA proposed, as a part of wastewater regulations, an analytical method entitled *Inductively Coupled Plasma (ICP) Optical Emission Spectrometric Method for Trace Element Analysis of Water and Wastes.*[10] This method included options for either vigorous digestion or moderate treatment of unfiltered samples with concentrated nitric and hydrochloric acids. Analyses of the treated samples by ICP/AES gave either total elements from the vigorous acid digestion, or total recoverable elements from the moderate acid treatment. Another option in the method was filtration of the sample, acidification of the filtrate with nitric acid to pH < 2, and separate analyses of the filtrate and filtered solids to give dissolved total elements and suspended total elements, respectively.

The public comments to the proposed wastewater regulations included objections to the utilization of the rather new ICP/AES technology in a government regulatory program. These objections were not unlike those received in response to the proposals of GC/MS methods for organic pollutants in wastewater (Chapters 3 and 4). For nearly 5 years the ICP/AES method was evaluated in multilaboratory studies with a variety of wastewater samples to prove the validity of the method. The ICP/AES method was promulgated as USEPA Method 200.7 in final wastewater regulations in the Federal Register of 26 October 1984.[11] Method 200.7 and the flame and flameless AAS methods were used for essentially all the total element determinations during the golden era of environmental analytical chemistry (Chapter 1). A similar ICP/AES method was used for the determination of total elements in air particulates.[12] In this method, samples are collected on glass or quartz fiber filters and dissolved in a mixture of concentrated nitric and hydrochloric acids with ultrasonic agitation and moderate heating.

While Method 200.7 and similar ICP/AES methods provided adequate detection limits and sample processing productivity for most elements in wastewater and air particulates, a multielement method with lower detection limits was needed for ambient and drinking water samples. The flameless AAS methods had the needed instrument detection limits, but were limited to measurements of one element at a time. With the commercialization of ICP/MS instruments during the early 1980s, this technique was explored for multielement analyses of aqueous environmental samples. The sample preparation procedure in method 200.7 for dissolved elements in water was used with ICP/MS for the determination of 49 elements in over 250 samples from lakes in the eastern United States.[13] This approach provided submicrogram per liter detection limits for nearly all elements, and the USEPA began using ICP/MS during the late 1980s for analyses of surface water, ground water, and soil leachate samples from existing and abandoned hazardous waste sites.

Method 6020. Method 6020, and a derivative called 6020M-CLP, were developed for the determination of total elements in samples from existing and abandoned solid and hazardous waste sites.[14] This method provides procedures for the analysis of acidic aqueous samples by ICP/MS, but refers to other methods for dissolution and digestion procedures for various solid and liquid samples. The 15 analytes listed in Method 6020 are indicated by the number 6020 in the last column of Table 7.1. Method 6020 specifies internal standards for quantitative analysis to compensate for variabilities in instrument responses caused by sample matrices, instrument and interface conditions, and operating parameters (Chapter 2, "Quantitative Analysis"). The recommended internal standards, for example, ^{209}Bi and ^{103}Rh, are elements that are not likely to be present, except in extremely low concentrations, in environmental samples and that are not subject to interferences from other elements. Quantitation ions are also recommended and adjustments to ion abundances are specified to correct for interferences with some quantitation ions. The selection of quantitation ions and corrections for interferences are discussed in a subsequent section.

Method 200.8. Method 200.8 was developed primarily for application in USEPA drinking water and wastewater regulatory and compliance monitoring programs.[15] It is a complete analytical method for the determination of the 21 elements listed in Table 7.1 by ICP/MS. Method 200.8 is applicable to the determination of the total amounts of the dissolved elements in ground water, surface water, and drinking water and to the total amounts of recoverable elements in these waters and in wastewater, sludge, and soil after appropriate sample preparation. The sample preparation procedures in Method 200.8 are the same or similar to those described for ICP/AES Method 200.7. Like Method 6020, Method 200.8 specifies internal standards to compensate for variabilities in instrument responses caused by sample matrices, instrument and interface conditions, and operating parameters. Estimated instrument detection limits are in the range 0.02–0.9 μg/L (5 μg/L for Se) with continuous measurement of spectra (CMS) and 0.002–0.07 μg/L (1.3 for Se) with selected ion monitoring (SIM). These estimates are for pneumatic nebulization sample introduction, and some of them may be significantly lower with contemporary commercial ICP/MS systems. Method detection limits (Chapter 2, "Detection Limits") are usually somewhat higher, but still generally at the submicrogram per liter level, and depend on the sample matrix and other operating conditions.

Quantitation Ions and Corrections for Interferences. Methods 6020 and 200.8 cite recommended quantitation ions, which are selected to avoid, as much as possible, interferences from other analytes and background ions. Table 7.2 lists the nuclides of selected elements of environmental significance and their integer

Table 7.2 Integer masses, exact masses, and natural abundances of some environmentally significant nuclides and some low-resolution interferences

Nuclide	Integer mass (Da)	Exact mass (Da)	Natural abundance (%)	Quantitation ion (Method 200.8)	Some low-resolution interferences
Ar	40	39.962384	99.60		
As	75	74.921594	100	+	$^{40}Ar^{35}Cl^+$
Be	9	9.012182	100	+	
Cd	106	105.906461	1.25		$^{90}Zr^{16}O^+$ (51.45%)
Cd	108	107.904176	0.89		$^{92}Mo^{16}O^+$, $^{92}Zr^{16}O^+$
Cd	110	109.903005	12.49		$^{94}MO^{16}O^+$, $^{94}Zr^{16}O^+$
Cd	111	110.904182	12.80	+	$^{95}Mo^{16}O^+$ (15.92%)
Cd	112	111.902758	24.13		$^{94}MO^{16}O^+$, $^{94}Zr^{16}O^+$
Cd	113	112.904400	12.22		$^{97}Mo^{16}O^+$ (9.55%)
Cd	114	113.903357	28.73		$^{98}Mo^{16}O^+$ (24.13%)
Cd	116	115.904754	7.49		$^{100}Mo^{16}O^+$ (9.63%)
Cr	50	49.946046	4.35		$^{50}Ti^+$ (5.4%), $^{35}Cl^{15}N^+$
Cr	52	51.940510	83.79	+	$^{35}Cl^{16}OH^+$, $^{40}Ar^{12}C^+$
Cr	53	52.940651	9.50		$^{37}ClO^+$
Cr	54	53.938883	2.37		$^{37}ClOH^+$, $^{40}Ar^{14}N^+$
Hg	196	195.965807	0.15		
Hg	198	197.966743	9.97		
Hg	199	198.968254	16.87		
Hg	200	199.968300	23.10		
Hg	201	200.970277	13.18		
Hg	202	201.970617	29.86	+	
Hg	204	203.973467	6.87		$^{204}Pb^+$
Pb	204	203.973020	1.4		$^{204}Hg^+$
Pb	206	205.974440	24.1	+	
Pb	207	206.975872	22.1	+	
Pb	208	207.976627	52.4	+	
Sb	121	120.903821	57.36		$^{40}Ar^{81}Br^+$
Sb	123	122.904216	42.64	+	
Se	74	73.922475	0.89		
Se	76	75.919220	9.36		$^{40}Ar^{36}Ar^+$
Se	77	76.919913	7.63		$^{40}Ar^{37}Cl^+$
Se	78	77.917308	23.78		$^{40}Ar^{38}Ar^+$
Se	80	79.916520	49.61		$^{40}Ar_2{}^{+\cdot}$
Se	82	81.916698	8.73	+	$^{82}Kr^{+\cdot}$ (11.6%), $^{81}BrH^+$
Tl	203	202.972320	29.52		
Tl	205	204.974401	70.48	+	

Source: Data from Ref. 16.

masses, exact masses, natural abundances, quantitation ions, and some ions that are interferences at low resolution. Some elements, for example, As and Be (Table 7.2), have just one naturally occurring nuclide and only one possible quantitation ion. If an interference is present for this ion, as there is for $^{75}As^+$ from the background ion $^{40}Ar^{35}Cl^+$, a strategy is needed to avoid the interference and measure the correct $^{75}As^+$ ion abundance.

If the sample has a very low Cl content, and HCl is avoided in the sample preparation process, the contribution of $^{40}Ar^{35}Cl^+$ to m/z 75 may be insignificant. This can be tested by observing the abundance at m/z 77 due to $^{40}Ar^{37}Cl^+$, which should be very low if the Cl content of the sample is low and if the $^{77}Se^+$ ion is not present. If the m/z 77 ion is significant, there are several possible strategies. The usual approach, which is specified in Methods 6020 and 200.8, is to adjust the measured m/z 75 ion abundance using the m/z 77 ion abundance and the natural abundances of the Cl isotopes, which are 75.77% for ^{35}Cl and 24.23% for ^{37}Cl (Table 5.2). The m/z 75 abundance is reduced by 75.77/24.23 = 3.13 times the m/z 77 abundance. If Se is present in the sample, the m/z 77 abundance is also adjusted by using the Se natural abundance values. Commercial ICP/MS data systems usually have software to support ion abundance adjustments, but software calculations should be checked manually at least once to ensure that the correct algorithms are used by the software. A strategy that avoids ion abundance corrections is high-resolution MS to separate the $^{75}As^+$ and $^{40}Ar^{35}Cl^+$ ions. A resolving power, $m/\Delta m_{10\%v}$, of 7778 is required and ICP/MS instruments with this capability, although expensive, have been developed and are commercially available.[7,17,18]

If an element has several or more naturally occurring isotopes, for example, most of those in Table 7.2, selection of the isotope with the largest natural abundance for the quantitation ion can provide the lowest detection limits, but may not be the best choice. The ^{52}Cr, ^{202}Hg, and ^{205}Tl isotopes with 83.79, 29.86, and 70.48% natural abundances, respectively, are the best choices because the probability of interferences with these ions is very low. The sum of the abundances of three Pb isotope ions, which account for 98.6% of all Pb, are used for measurement of Pb because of the variabilities of the natural abundances of Pb from various deposits in the world. However, for some other elements, for example, Cd, Sb, and Se, the isotopes with the highest natural abundances also have significant interferences, and even lower abundance isotopes are not always completely free of interferences (Table 7.2).

The quantitation ion specified for Cd is m/z 111, which has a 12.80% natural abundance. This ion is selected to minimize interferences with Mo and Zr oxide ions that form in the interface and have strong bonds (Table 7.2). An ion abundance adjustment similar to that described for As is appropriate at m/z 111 in samples containing Mo. A resolving power > 32,000 is required to separate $^{111}Cd^+$ from $^{95}Mo^{16}O^+$. The quantitation ion specified for Se is m/z 82, which has only a 8.73% natural abundance, but is selected to eliminate interferences with Ar ions (Table 7.2). Interferences from the $^{82}Kr^{+\cdot}$ ion are eliminated by using very pure Ar in the ICP. The ^{80}Se isotope has a 49.61% natural abundance and would provide lower detection limits, but it has a significant interference from the $^{40}Ar_2^{+\cdot}$ ion (Table 7.2). A resolving power of 9699 is required to separate these ions. All potential interferences must be assessed with all samples to determine if ion abundance corrections are required.

Regulatory Adoption and Multilaboratory Validation. Method 200.8 was approved in 1994 by the USEPA for monitoring the total amounts of 16 of the elements listed in Table 7.1 in drinking water.[19] The 16 elements are those that have either the maximum contaminant levels (MCLs) shown in Table 7.1 or other monitoring requirements or recommendations as indicated in footnotes "a", "b", and "c". Method 200.8 also was proposed in 1995 by the USEPA for monitoring the total amounts of 18 of the elements listed in Table 7.1 in wastewater.[20] The 18 elements are Al, Ba, Co, Mn, Mo, V, and all those designated as priority pollutants in Table 7.1 except Hg. Method 200.8 was evaluated in a 13-laboratory study of reagent water, drinking water, ground water, and wastewater fortified with all the analytes in Table 7.1 except Hg.[21] The analytes were in the 0.8–200 μg/L concentration range and recoveries were generally in the 95–105% range with between-laboratory relative standard deviations (RSDs) of about 4–8% for all samples except wastewater. Recoveries from wastewater averaged 100% and RSDs averaged 8%.

Determination of Specific Chemical Entities

Speciation is defined as the identification and measurement of the individual elemental forms, compounds, ions, and oxidation states of an element that contribute to the total amount of the element in a sample. These chemical entities are often called *species* and the term *speciation* is often used in inorganic analytical chemistry and elemental analysis.[22] Although *speciation* is rarely used in organic chemical analysis, the concept is widely employed. The analytical methods for individual organic compounds described in Chapters 3 and 4 are methods for the speciation of the compounds that contribute to the total organic carbon in a sample. Speciation is a broad concept that is divided into two general strategies and one special strategy in this chapter.

Two general analytical strategies are defined in Chapter 2 ("General Analytical Strategies and Samples") and both of these are applicable to the determination of specific chemical entities with ICP/MS. In the target analyte (TA) strategy the goal is to determine specific substances which are either known or thought to be in the sample and which must be determined to meet the project goals. The TA strategy is widely used in analytical chemistry, dominates much of environmental analysis, and is illustrated by many analytical methods for organic compounds described in Chapters 3 and 4. If a TA, or analytes, contains an element, or elements, that can be measured with high sensitivity and selectivity by ICP/MS, this technique may be the best choice for the determination of the TA or analytes. However, separation of the TAs from each other and from other entities containing the element is required before measurement. This is usually accomplished with an on-line chromatographic separation that can provide the additional benefit of separation of substances that cause interferences in

ICP/MS determinations of some analytes. For example, if all Cl-containing entities are separated from a TA containing. ^{75}As, the interference from $^{40}Ar^{35}Cl^{+}$ may be eliminated.

The other general analytical strategy defined in Chapter 2 is the broad spectrum (BS) analysis of a sample to determine what substances are present. The BS strategy is far more general, complex, and difficult than the TA strategy, but is needed to discover unexpected sample components and to solve real environmental and other problems. No single analytical method can reasonably be expected to identify and measure all the chemical substances in a sample. The BS analytical methods described in Chapters 3 and 4 are designed to encompass a wide variety of compounds with generally similar properties, for example, vapor pressures and water solubilities. A similar approach is necessary to determine all the elemental forms, compounds, ions, and oxidation states of all the elements in a sample.

A special application of elemental analysis by ICP/MS is the determination of the elements present in an unknown substance separated by chromatography. During the analyses of samples using the GC/MS methods in Chapters 3 and 4 or the techniques for compounds not amenable to GC in Chapter 6, unknown substances are often discovered and inadequate information is available to identify the substance (Chapter 2, "Identification Criteria"). Elemental analysis of these unidentified substances can provide valuable information to aid in the identification of the substance.

Target Analytes

Substances containing elements other than those commonly found in organic compounds are suitable for the target analyte (TA) strategy with on-line chromatographic separation and ICP/MS (Figure 7.1). The elements in Table 7.1, and some others, are contained in a variety of industrial compounds and natural products that are dispersed in the environment. Compounds of these elements are also determined after reactions with various reagents to give derivatives that are amenable to chromatographic separations.

Some of these analytes are organometallic, that is, they have a metal-to-carbon bond, and are either neutral compounds or ions. Some environmentally important organometallic compounds are described in Chapter 6 in the section entitled "Classes of Compounds and their Structures and Properties". Structures of six representative and important organometallic compounds are shown in the first two rows of Chart 7.1. Other compounds containing the elements in Table 7.1 are chelates in which the element is part of a ring structure and typically attached to the compound through bonds to N, O, and/or S. Chelates are either neutral compounds or ions, and the structures of six representative chelates are shown in the last two rows of Chart 7.1. Coordination complexes of many elements are well known and contain a central metal atom surrounded by ligands that are often small molecules or ions such as ammonia, chloride, CO, and alkyl- or aryl-phosphines. Coordination complexes can be either neutral compounds or ions. Carboxylic or inorganic acid salts of many elements are ionic but often exist as ion pairs in water. In general an enormous number of compounds containing the elements in Table 7.1, and other elements, could be present in a sample and many are, or could be, of interest as TAs.

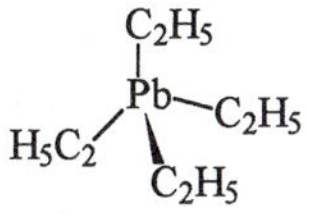

Tetraethyllead

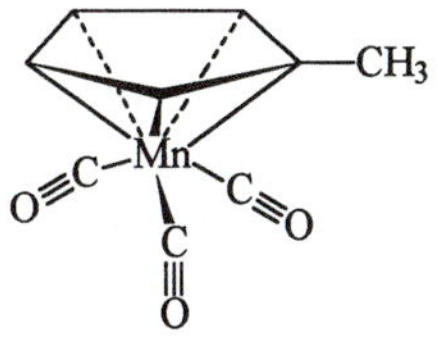

Methylcyclopentadienyl manganese tricarbonyl

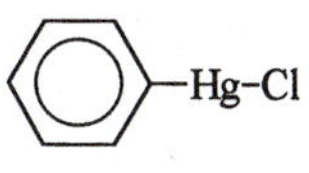

Phenylmercuric chloride

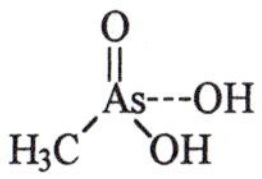

Methylarsonic acid

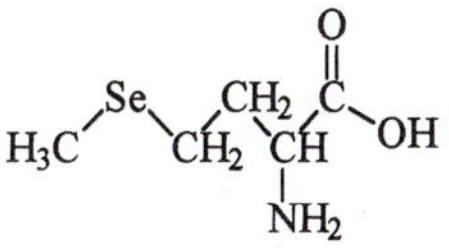

Selenomethionine

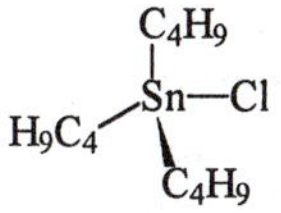

Tributyltin chloride

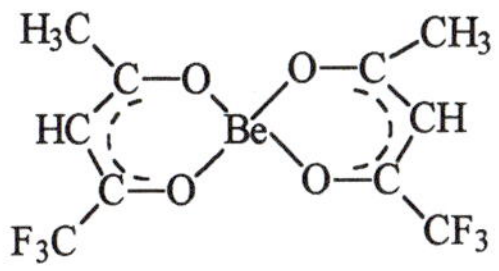

Beryllium bis(trifluoroacetylacetonate)

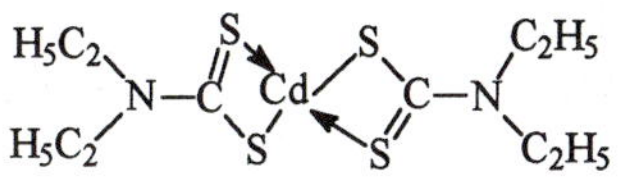

Cadmium bis(N,N'-diethyldithiocarbamate)

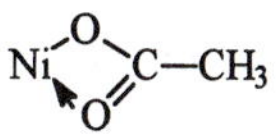

Nickel (II) acetate ion (+1)

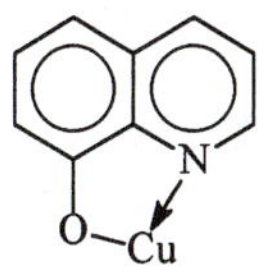

Copper (II) quinoline-8-oxide ion (+1)

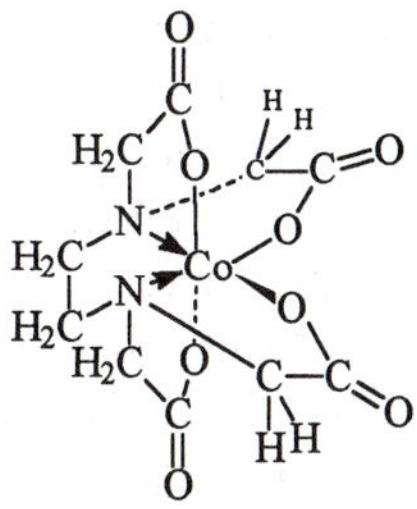

Cobalt (III) ethylenediamine tetraacetate ion (-1)

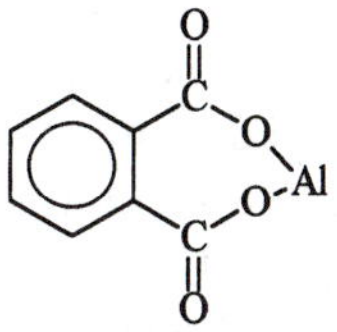

Aluminum phthalate ion (+1)

Chart 7.1

Compounds Amenable to GC. The effluent from a GC column is readily introduced into the ICP torch (Figure 7.1) through a heated transfer line, and this connection provides high analyte transport efficiency.[22] With narrow-bore GC columns, which operate with low carrier gas flow rates, an auxiliary gas flow is merged with the GC effluent to ensure efficient transfer of analytes into the ICP torch.

A small group of neutral analytes have sufficient thermal stabilities and vapor pressures at typical GC column oven temperatures to allow separation by GC. Compounds amenable to GC are mainly those in which the element is bonded exclusively to small nonpolar organic groups, typically methyl or ethyl groups. Elements that form these compounds include As, Hg, Pb, Sb, Se, Sn, and Te. An example from this class of compounds is tetraethyllead (left end of first row in Chart 7.1). Larger alkyl or aryl groups, for example, *n*-butyl or phenyl, bonded to these elements generally reduce vapor pressures and make these compounds less amenable to GC. Most other organometallic compounds containing simple σ bonds to organic groups are highly reactive and are subject to rapid oxidation and hydrolysis in air and water. Some cyclopentadienyl derivatives, for example, the sandwich compound bis(cyclopentadienyl)iron (ferrocene), and related compounds, may be amenable to separation by GC.

The compounds tetramethyllead, tetraethyllead (Chart 7.1), tetramethyltin, tetraethyltin, dimethylmercury, diethylmercury, and dimethylselenide were collected from air by cryogenic trapping (Chapter 3, "Volatile Organic Compounds in Air"), separated by packed-column GC, and measured by ICP/MS.[23] Several of the symmetrical Pb compounds and mixed methylethyllead compounds were detected in ambient air samples in areas where leaded gasoline was still in use. Tetramethyltin and dimethylselenide were also detected in some samples.

Compounds in which the element is bonded to one or more small organic groups and a carboxylate anion, halide, or some other inorganic anion are typically not amenable to GC. These compounds include phenylmercuric chloride (right end of first row in Chart 7.1), dialkyllead acetates, and tri-*n*-butyltin chloride (right end of second row in Chart 7.1). This class of compounds can be converted into derivatives amenable to GC with various alkylating agents or, for some compounds, reducing agents (Chapter 6, "Chemical Derivatization"). The reagent sodium tetraethylborate was used in aqueous solution to replace the Cl in *n*-butyltin trichloride, di-*n*-butyltin dichloride, tri-*n*-butyltin chloride, methylmercuric chloride, and trimethyllead chloride with an ethyl group, which allowed separation of the derivatives by GC and detection by ICP/MS.[24]

A derivatization technique that produces neutral compounds amenable to GC is the reaction of metal ions with fluorinated β-diketones to give thermally stable chelates. These derivatives were developed during the 1960s and early 1970s for the determination of small quantities of some metals in geochemical materials including lunar samples. An example is beryllium bis(trifluoroacetylacetonate) whose structure is shown at the left end of the third row in Chart 7.1. Chelates of Al, Be, Cr, the lanthanides, Zr, and other elements have been prepared and subjected to either GC separation or direct insertion probe EI MS.[25–28]

Compounds and Ions Not Amenable to GC. Most organometallic compounds, metal chelates, coordination complexes, and metal salts are ions or exist as ion pairs in water or contain very polar bonds and are not amenable to separation by GC. Although chemical reagents can be used to produce volatile derivatives, this technique has many limitations and disadvantages (Chapter 6, "Chemical Derivatization"). An alternative strategy is utilization of the separation techniques briefly described in Chapter 6, that is, supercritical fluid chromatography (SFC), normal- and reverse-phase liquid chromatography (LC), ion exchange chromatography (IC), and capillary electrophoresis (CE). Several of these are more compatible with ICP/MS than others, but most have favorable properties for some analytes.

The effluent from SFC is injected directly into the ICP torch through a heated transfer line (Figure 7.1) containing a restrictor which maintains supercritical pressure in the column until decompression.[22] The total vaporized effluent is injected into the torch and this provides high analyte transport efficiency. This separation technique is best suited to nonvolatile or thermally unstable analytes that are not ions and are not so polar to preclude solubility in the usual supercritical CO_2 or modified supercritical CO_2 mobile phases. Compounds in this category include tetra-*n*-butyltin, tri-*n*-butyltin chloride, tetraphenyltin, and triphenyltin chloride, which have been separated by SFC and detected with ICP/MS.[22] However, the high levels of CO_2 in the torch produces significant background ions including $^{40}Ar^{12}C^+$, which interferes with the preferred $^{52}Cr^+$ quantitation ion.

Condensed-phase LC and IC effluents are usually injected into the nebulizer of the spray chamber (Figure 7.1) and this is a limitation because analyte transport efficiencies with this technique are typically in the 1–5% range.[22] Narrow-bore LC and IC columns, which utilize lower flow rates, are connected to direct injection nebulizers and much higher analyte transport efficience can be obtained. Some neutral organometallic and other compounds that are not suitable for GC can be separated by normal-phase LC, which uses a polar stationary phase and a nonpolar or low polarity mobile phase. However, typical normal-phase LC solvents, which include hydrocarbons, methylene chloride, and similar solvents, are not well tolerated in the ICP interface. Carbon deposits form in the vicinity of sampling orifice and the smaller skimmer orifice and cause major maintenance problems. Some carbon-containing background ions can also be troublesome. Normal-phase LC is the least useful of the condensed-phase chromatographic separation techniques that have been combined with ICP/MS.

Reverse-phase LC is the workhorse of the condensed-phase separation techniques and is appropriate for a wide range of slightly to moderately polar compounds. Gradient elutions are widely used and these typically begin with a high percentage of water and end with a high percentage of a miscible organic solvent. Analytes are retained on the nonpolar or slightly polar stationary phase and eluted as the mobile phase gradually becomes less polar. Methanol is a widely used solvent with reverse-phase LC and is more compatible with ICP/MS than the equally widely used acetonitrile, which produces more severe carbon deposits and more background ions. Solvents containing oxygen are generally preferred with ICP/MS because they produce fewer carbon deposits. Some organo As, Hg, and Pb compounds of the type discussed in this section have been separated by reverse-phase LC and detected with ICP/MS.[22]

Perhaps the most ICP-compatible condensed-phase chromatographic separation technique is IC. Various aqueous buffer solutions are standard mobile phases used to separate ions on anion or cation exchange resins. The ion suppression techniques commonly used with conductivity detectors are unnecessary with ICP, but dissolved solids should be maintained below about 0.1–0.5% by weight . Since many metals and other elements exist in water as ions, IC combined with ICP/MS is a valuable technique that has been used to determine a variety of TAs.[22] For example, BrO_3^- is a drinking water disinfection by-product produced when source water containing Br^- is disinfected with ozone. Bromate has a MCL of 10 μg/L in drinking water in the United States and can be determined by combined IC/ICP/MS.[29,30] Another technique designed for the separation of ions in solution is CE, which provides resolving power and has other advantages (Chapter 6). However, CE flow rates are exceedingly low, typically in the low nanoliter per minute range, and an auxiliary sheath liquid is used to provide a stable electrical ground and sufficient mobile phase flow to ensure efficient transfer of analytes into the ICP torch. Combined CE/ICP/MS has been evaluated with a variety of metal ions and has potential for many applications.[31,32]

Broad Spectrum Analysis

The broad spectrum (BS) analytical strategy was defined in Chapter 2 in the section entitled "General Analytical Strategies and Samples". Broad spectrum analytical methods used for general classes of organic compounds are described in Chapters 3 and 4. The BS approach is also considered in Chapter 6 in the section entitled "Strategies for Broad Spectrum Analyses". A complete BS analysis of a sample has the objective of determining all the chemical entities present including the organic and inorganic compounds, the neutral compounds and ions, and the low and high molecular weight substances. Clearly no single analytical method is capable of such a comprehensive analysis, but a combination of methods can reveal a wealth of information about the sample.

An ICP/MS method that includes an appropriate sample preparation and one or more chromatographic separations could provide BS information about the compounds and ions in a sample containing elements that are readily identified and measured with this technique. This strategy is not well developed probably because of past and current emphasis on measurements of the total amounts of elements in a sample and the determination of TAs. However, the concept of *speciation*, which was briefly defined at the beginning of the section "Determination of Specific Chemical Entities", implies not just the determination of TAs, but the determination of all species or entities containing an element or elements.

A BS analytical method could consist of a sample preparation procedure, a generalized anion exchange IC or CE separation, and ICP/MS. This method could be tested with a broad variety of anions including AsO_3^{3-}, AsO_4^{3-}, Br^-,

BrO^-, BrO_2^-, BrO_3^-, BrO_4^-, Cl^-, ClO^-, ClO_2^-, ClO_3^-, ClO_4^-, $Cr_2O_7^{2-}$, PO_3^{3-}, PO_4^{3-}, SO_3^{2-}, SO_4^{2-}, and others including various organic anions containing sulfonic or phosphonic acid groups. Identification criteria based on retention or migration times and elemental analyses, including isotope distributions, would then be established. Real environmental or other aqueous samples, or aqueous extracts of solid samples, could then be analyzed and the anions present determined in a BS approach. A similar approach for cations is a reasonable possibility.

Determination of the Elements Present in Unknown Substances

During either TA or BS analyses of real environmental or other samples, using the methods or techniques in Chapters 3, 4, or 6, substances are found that cannot be identified by using standard identification criteria or other techniques for the identification of nontarget analytes (section of Chapter 2 entitled "Identification Criteria", and Chapter 5). When this occurs, transfer of the separation technique to the ICP/MS system can provide valuable information about the elements that are present in the unknown substance. This approach is likely to be more useful with the condensed-phase separation techniques in Chapter 6 than with the GC separation techniques in Chapters 3 and 4 because elements easily detected by ICP/MS are more likely to be in nonvolatile or thermally unstable analytes.

This strategy has been applied is some environmental studies. In one example, several unknown substances were detected in the total ion chromatogram from the IC/particle beam/MS analysis of aqueous leachates from a hazardous waste site.[33] The anion exchange separation was transferred to the ICP/MS system, and the presence of sulfur and chlorine in the unknowns was confirmed. This information was used to develop a hypothesis and to identify chlorobenzene sulfonic acids in these samples.

Summary

Four strategies using elemental analysis by ICP/MS are described in this chapter. By far the most widely used is the determination of the total amounts of target elements in a sample. A strategy for TA determination is a chromatographic or other separation combined with ICP/MS. This is most effective when the TAs contain elements, other than those commonly found in organic compounds, that are efficiently measured with ICP/MS. The strategy of BS analysis by ICP/MS is poorly developed, but ICP/MS can be employed to determine which elements, other than those commonly found in organic compounds, are present in unknown substances separated by chromatography or other separation techniques.

REFERENCES

1. Fassel, V. A. *Anal. Chem.* **1979**, *51*, 1290A–1308A.
2. Kopp, J. F.; Martin, T. D. *ASTM Standardization News*, February 1983, pp 18–21.
3. Houk, R. S. *Anal. Chem.* **1986**, *58*, 97A–105A.
4. Houk, R. S.; Fassel, V. A.; Flesch, G. D.; Svec, H. J.; Gray, A. L.; Taylor, C. E. *Anal. Chem.* **1980**, *52*, 2283–2289.
5. Montaser, A., Ed. *Inductively Coupled Plasma Mass Spectrometry*; Wiley-VCH Verlag: Weinheim, Germany, 1998.
6. Newman, A. *Anal. Chem.* **1996**, *68*, 46A–51A.
7. Olesik, J. W. *Anal. Chem.* **1991**, *63*, 12A–21A.
8. Keith, L. H.; Telliard, W. A. *Environ. Sci. Technol.* **1979**, *13*, 416–423.
9. Title 40 Code of Federal Regulations, Part 141.62.
10. *Guidelines Establishing Test Procedures for the Analysis of Pollutants; Proposed Regulations, Federal Register*; 3 December 1979, Vol. 44, Appendix IV, 69559–69564.
11. *Guidelines Establishing Test Procedures for the Analysis of Pollutants Under the Clean Water Act; Final Rule and Interim Final Rule and Proposed Rule, Federal Register*; 26 October 1984, Vol. 49, 43433–43438; Title 40, Code of Federal Regulations, Part 136, Appendix C.
12. Harper, S. L.; Walling, J. F.; Holland, D. M.; Pranger, L. J. *Anal. Chem.* **1983**, *55*, 1553–1557.
13. Henshaw, J. M.; Heithmar, E. M.; Hinners, T. A. *Anal. Chem.* **1989**, *61*, 335–342.
14. *Test Methods for Evaluating Solid Waste, Physical/Chemical Methods*, USEPA Publication SW-846, 3rd ed. and Updates I, II, IIA, IIB, and III; URL http://www.epa.gov.epaoswer/hazwaste/test/8xxx.htm.
15. Method 200.8, Revision 5.4 in *Methods for the Determination of Metals in Environmental Samples*; Suppl. I, USEPA Report EPA/600/R-94/111, May 1994; URL http://www.epa.gov/nerlcwww/methmans.html.
16. Lide, D. R., Ed.-in-Chief *CRC Handbook of Chemistry and Physics*, 78th ed.; CRC Press: Boca Raton, FL, 1997/1998.
17. Stuewer, D.; Jakubowski, N. *J. Mass Spectrom.* **1998**, *33*, 579–590.
18. Moens, L.; Jakubowski, N. *Anal. Chem.* **1998**, *70*, 251A–256A.
19. *Analytical Methods for Regulated Drinking Water Contaminants; Final Rule, Federal Register*; 5 December 1994, Vol. 59, 62456–62471; Title 40 Code of Federal Regulations, Parts 141 and 143.
20. *Guidelines Establishing Test Procedures for the Analysis of Pollutants: New Methods; Proposed Rule, Federal Register*; 18 October 1995, Vol. 60, 53988–54006.
21. Longbottom, J. E.; Martin, T. D.; Edgell, K. W.; Long, S. E.; Plantz, M. R.; Warden, B. E. *J. AOAC Int.* **1994**, *77*, 1004–1023.
22. Vela, N. P.; Olson, L. K.; Caruso, J. A. *Anal. Chem.* **1993**, *65*, 585A–597A.
23. Pecheyran, C.; Quetel, C. R.; Lecuyer, F. M. M.; Donard, O. F. X. *Anal. Chem.* **1998**, *70*, 2639–2645.
24. Moens, L.; De Smaele, T.; Dams, R.; Van Den Broeck, P.; Sandra, P. *Anal. Chem.* **1997**, *69*, 1604–1611.
25. Eisentraut, K. J.; Griest, D. J.; Sievers, R. E. *Anal. Chem.* **1971**, *43*, 2003–2007.
26. Frew, N. M.; Leary, J. J.; Isenhour, T. L. *Anal. Chem.* **1972**, *44*, 665–671.
27. Sieck, R. F.; Banks, C. V. *Anal. Chem.* **1972**, *44*, 2307–2312.
28. Tsuge, S.; Leary, J. J.; Isenhour, T. L. *Anal. Chem.* **1973**, *45*, 198–200.

29. *National Primary Drinking Water Regulations: Disinfection Byproducts; Final Rule, Federal Register*; 16 December 1998, Vol. 63, 69390–69476; Title 40 Code of Federal Regulations, Parts 9, 141, and 142.
30. Creed, J. T.; Brockhoff, C. A. *Anal. Chem.* **1999**, *71*, 722–726.
31. Liu, Y.; Lopez-Avila, V.; Zhu, J. J.; Wiederin, D. R.; Beckert, W. F. *Anal. Chem.* **1995**, *67*, 2020–2025.
32. Kinzer, J. A.; Olesik, J. W.; Olesik, S, V. *Anal. Chem.* **1996**, *68*, 3250–3257.
33. Brown, M. A.; Kim, S.; Roehl, R.; Sasinos, F. I.; Stephens, R. D. *Chemosphere* **1989**, *19*, 1921–1927.

8

Measurements in the Field or a Processing Plant

It is sometimes advantageous to identify and measure analytes in the field or in a processing plant.[1–4] A major benefit of field or in-plant measurements is the greatly reduced time required to obtain results and utilize the information compared to the time required to deliver samples to a remote laboratory and wait for results. Another significant benefit is the cost savings projected from measurements in the field or in a processing plant. However, the economic and other benefits of field measurements should be carefully assessed for each specific situation. Some authors and advisors, in their enthusiasm to promote field analytical methods, may overstate the benefits of field measurements and overstate the deficiencies of shipping samples to a remote permanent laboratory. For some applications, such as utilization of measurements for feedback control of a chemical or physical process, the analytical results are often needed within a period that is so short that the only practical option is an in-plant measurement.

Like most strategies that greatly reduce the time and cost required to accomplish an objective, some reduction in completeness, flexibility, and quality can be expected from field and in-plant measurements. A major benefit of using mass spectrometry (MS) for these analyses is that a higher quality of information can be obtained compared to most other field measurement techniques. While the cost of field MS equipment may be higher than that of other types of instrumentation, this is often a small price to pay for the more reliable information. There are many different levels of field or in-plant measurements, which differ in both response time and the quality of information that can be provided.

A field laboratory can be equipped with instrumentation and other equipment and provide analyses that are essentially the same as those provided in a permanent laboratory. The field laboratory, which might be located at an abandoned hazardous waste site, is usually installed in a movable building, a truck trailer, or a recreational vehicle with an auxiliary power supply to provide standard electrical services and air conditioning. While precautions are taken to prevent damage to instruments and other equipment during movement of the laboratory, the operations of a field laboratory are not unlike a fixed laboratory. Standard analytical methods, such as those described in Chapters 3 and 4, can be implemented and, with appropriate quality control, the results should be similar to those obtained in a conventional laboratory.

One important limitation of a field laboratory is that usually only a few small or bench-top instruments can be accommodated, which limits the scope of the analyses that can be implemented. The economics of a field laboratory depend on the volume of samples analyzed over a number of months or years and the savings derived by producing results quickly, for example, within a few hours, to provide direction to construction equipment operators, field sampling teams, or well drillers. The analytical methods employed in a field laboratory are essentially the same as those used in a permanent laboratory and, therefore, field laboratory operations are not considered further in this chapter.

A significantly different and lower cost strategy is a highly mobile or portable analytical instrument deployed in the field in a small van or a sport utility vehicle, or moved with a hand cart, or carried by a person. This type of instrument is designed for field applications and is ruggedly constructed to prevent damage from vibration, dust, heat, humidity, and other environmental stressors. Typically quadrupole or ion-trap mass spectrometers have been used for this application and they are modified to operate with a portable generator or on battery power and packaged to reduce weight and size. Mass spectrometers have been developed which can be carried in a backpack to a field sampling station.

The analytical methods used with portable mass spectrometers are generally greatly simplified compared to laboratory methods.[4] Often the information needed in the field is qualitative or an estimate of whether certain analytes are above or below an action level concentration. While gas chromatography (GC) is used for some analyses, this capability is frequently not included in field instruments. When analytes are not separated by GC, spectra of mixtures are often encountered and interferences are probable from compounds that give ions that have the same *m*/*z* as analyte ions. Under these circumstances an ion-trap mass spectrometer with tandem capabilities is recommended so that individual analyte ions can be isolated in the trap, fragmented, and measured to enhance the reliability of the determination (Chapter 5, "Tandem Mass Spectrometry"). A tandem mass spectrometer cannot, however, distinguish all possible analytes especially isomers and components of very complex mixtures. With this type of portable instrument, discrete samples are often analyzed and results can be available within 5–30 min. Sample introduction systems are also available for short- or long-term continuous measurements of some analytes. The economics of this approach depend on whether the analytical results are of sufficient quality and are available within the time required to provide direction to construction equipment operators, field sampling teams, well drillers, or others whose work depends on the information.

Continuous measurements in the field are needed when the substances present and their concentrations are changing rapidly. The determination of the source of fugitive emissions that are rapidly dispersed in the atmosphere or in a flowing stream requires continuous and sometimes mobile continuous monitoring. Continuous measurements are also needed when results are required within a time that is much shorter than that required to analyze discrete samples in a field laboratory or with a portable field instrument. Long-term continuous monitoring is often required to determine changing concentrations in a processing plant and provide rapid feedback of results that are used in process control strategies. Equipment used for continuous monitoring in a processing plant may have fewer constraints in regard to size and weight, but other factors and sampling techniques are similar to those of field operations. Standard chromatographic separation techniques are generally too slow for continuous measurements although GC separations, especially fast GC, are used for process control.

Techniques for continuous measurements have been developed primarily for volatile substances, as defined in Chapter 3, in air and water, and emitted from soils and other solids. These techniques are also applicable to discrete samples obtained using various sampling procedures including those described in Chapter 3. Techniques for continuous measurements of semivolatile substances (Chapter 4) and compounds not amenable to GC (Chapter 6) in condensed-phase samples are generally not developed, but considerable potential exists for advances in this area.

Continuous Measurements of Volatile Substances in Air

One of the first applications of atmospheric pressure chemical ionization (APCI) MS during the early 1980s was continuous monitoring of ambient air and emissions from stationary and mobile air pollution sources.[5] Ambient or source air is continuously drawn or pumped into the ionization chamber where a corona discharge needle at 2–6 kV produces electrons and reagent ions from the atmospheric components water, nitrogen, and oxygen (Chapter 5, "Chemical and Related Ionization Techniques"). Reagent ions and thermal electrons react with volatile analytes to give analyte ions that enter the mass spectrometer through a small sampling orifice. Figure 6.8 is a diagram of an APCI ionization chamber showing the corona discharge needle and the ion sampling orifice (Chapter 6, "Mass Spectrometer Interfaces"). For atmospheric sampling the liquid chromatography effluent inlet and nebulizer shown in Figure 6.8 are replaced by an air inlet system.

A triple quadrupole mass spectrometer was used with APCI to measure collision induced dissociation spectra of analyte ions selected in the first-stage analyzer (Chapter 5, "Tandem Mass Spectrometry"). This allows determination of many specific air components without interferences from other ions with the same m/z. The APCI/MS/MS technique provides rapid responses to the presence and varying concentrations of volatile substances. However, APCI is a selective ionization process and while detection limits for some analytes are in the sub parts per 10^9 by volume range, analytes with low proton or electron affinities and weak acids such as phenol are not efficiently ionized with APCI.[6] Many of the volatile organic compounds listed in Table 3.1 have low proton or electron affinities or are very weak acids. Some of the APCI–triple quadrupole instruments were installed in large vehicles and used to detect fugitive emissions and to monitor air pollution while the vehicle was stationary or being driven on a highway. A significant limitation of this approach is the high cost of acquisition and operation of a triple quadrupole spectrometer in a vehicle or in the field. However, the commercial availability of APCI sources on compact ion-trap mass spectrometers with tandem capabilities provides a viable alternative to the triple quadrupole instrument.

Glow discharge ionization with both quadruple and ion-trap mass spectrometers was investigated for continuous sampling of air to detect explosives.[7,8] Nitroaromatic compounds (Chapter 4, "Classes of Semivolatile Organic Compounds"), nitric acid esters, and *N*-nitro compounds (Chapter 6, "Classes of Compounds and their Structures and Properties") readily form negative ions by electron capture in a glow discharge. These compounds are detected at sub-picogram or low parts per 10^{12} by volume concentrations in air with an ion-trap mass spectrometer. This technique is well suited to field measurements particularly with a compact ion-trap mass spectrometer with tandem capabilities (Chapter 5, "Tandem Mass Spectrometry"). However, detection of explosives is a specialized application and this ion source is not often used for general investigations.

Most portable mass spectrometers are quadrupole or ion-trap instruments that use conventional electron ionization. These instruments are designed for field or in-plant operations, powered by portable generators or batteries, and have minimum size and weight. Direct air sampling, which was used with the APCI/MS/MS and glow discharge ionization techniques, is also used with these systems. In one design the air sample is continuously drawn with a sampling pump into a manifold where it is mixed with dry nitrogen to lower the relative humidity or dry helium, which is needed in an ion trap, and with known amounts of internal standards for quantitative analysis.[4] The mixture is then drawn into the mass spectrometer through a capillary restrictor and transfer line. An important limitation of direct air sampling is that water vapor and atmospheric gases are introduced into the mass spectrometer and these can cause interferences and undesirable ion–molecule reactions.

Membrane Introduction

Cross-linked poly(dimethylsilicone) membranes in the form of sheets or hollow fibers are widely used to partition small and nonpolar volatile organic compounds (VOCs) from water and introduce them into a mass spectrometer.[3,9–11] This technique, which is known as membrane introduction mass spectrometry (MIMS), is also useful for air sampling and has the advantage over direct air sampling of minimizing the transport of water vapor and other polar substances into the mass spectrometer. The MIMS technique is simple and rugged and is increasingly used for continuous monitoring of ambient air, emissions from stationary and mobile sources of air pollution, soil gases, and process streams in manufacturing plants.

Figure 8.1 is a diagram of one design of a continuous sampling semipermeable membrane introduction system.[12] Helium from a flow controller, which is not shown in the figure, passes at an appropriate rate through a silicone hollow-fiber membrane located in a sampling manifold. Air is pumped into the manifold and flows coaxially over the membrane in a direction counter current to the flow of helium carrier gas inside the membrane. Low molecular weight and generally less polar compounds are absorbed by the membrane, diffuse through the polymer matrix, evaporate in the stream of helium, and are transported into the mass spectrometer. A jet separator interface (Figure 2.3) can be used before the mass spectrometer to enrich the gas stream in analytes with respect to helium and to reduce the volume of gas that enters the spectrometer.[10, 13] Many of the VOCs listed in Table 3.1 are amenable to introduction with this technique although membrane transport efficiencies are lower with the larger and/or more polar compounds. A small number of the compounds in Table 4.1 are probably amenable to membrane introduction.

A membrane introduction system similar to that depicted in Figure 8.1 and equipped with a jet separator was used to determine benzene, carbon tetrachloride, 1,1,1-trichloroethane, and toluene in air in the range 0.1–187 parts per 10^9 by volume.[13] A sheet membrane inlet was used to study several organosulfur compounds including two dialkylsulfides, a dimethyl disulfide, and ethanethiol in air in the low microgram per cubic meter concentration range.[14] In laboratory experiments the biogenic hydrocarbon isoprene was measured by GC/MS after membrane sampling of air with a system similar to that shown in Figure 8.1 and cryogenic trapping before thermal desorption into the GC system.[15]

Continuous Measurements of Volatile Substances in Water

The continuous measurement of VOCs in water requires an efficient method of partitioning the analytes from the bulk sample matrix. One approach is similar

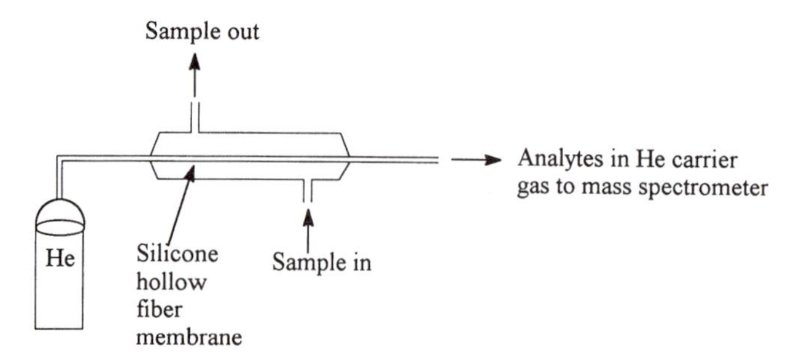

Figure 8.1 Diagram of a continuous sampling semipermeable membrane introduction system. (Redrawn from Ref. 12, published in 1991 by the American Chemical Society.)

to the nebulizer–spray chamber technique used to produce aerosols for elemental analysis (Figure 7.1). Water is pumped through a nozzle into a spray chamber where vaporized analytes are carried by an independently injected stream of helium into a gas stream splitter and by a restricted transfer line into the mass spectrometer.[4] A slightly different technique is a nebulizer similar to that shown in Figure 7.1 used in the spray and trap technique for volatiles in water.[16] Just as with direct sampling of air, an important limitation of the spray chamber vaporization technique is that water vapor and atmospheric gases are introduced into the mass spectrometer and these can cause interferences and undesirable ion–molecule reactions.

Membrane Introduction

The MIMS technique, which was briefly described in the previous section, has been widely used in several different configurations to determine VOCs in a variety of aqueous samples.[3,9–12] The design shown in Figure 8.1 is often called the *flow over* configuration because the air or water sample flows over the hollow-fiber membrane.[17] Another option is the *flow through* technique in which the aqueous sample is pumped through a loop of hollow-fiber membrane at the end of a direct insertion probe where the exterior of the membrane is continuously exposed to the vacuum of the mass spectrometer.[18] A similar design uses a flat sheet membrane at the end of the probe, and the aqueous sample is continuously pumped across the inside surface of the membrane while the outside surface is exposed to the vacuum of the mass spectrometer.[9] The latter two designs have the advantage of low detection limits and of giving very rapid responses to the appearance of analytes in the sample or to changes in their concentrations. However, they also have a higher risk than the *flow over* configuration in Figure 8.1 of a catastrophic accident in the event of a leak or a ruptured membrane. The effects of hollow-fiber dimensions, sample flow rates, membrane configuration, and temperature on permeation rates from air and water samples have been studied using some of the compounds in Table 3.1, but the helium purged design in Figure 8.1 was not included in this study.[17]

Submicrogram per liter detection limits were demonstrated for benzene, carbon tetrachloride, 1,1-dichloroethene, trichloroethene, and vinyl chloride in water using the MIMS configuration in Figure 8.1 with the continuous helium purge of the interior of the membrane.[12] Similar detection limits were demonstrated for 59 of the VOCs in Table 3.1 with the *flow through* technique, using a direct insertion probe with a loop of hollow-fiber membrane close to the entrance of an ion-trap mass spectrometer.[19] Substantialy improved detection limits of 500 pg/L–1 ng/L were demonstrated for toluene and *trans*-1,2-dichloroethene in water with a similar probe and a selective ion-trap storage technique that rejected unwanted ions and allowed the accumulation of analyte ions in the trap.[20] Ground water samples were analyzed in the field and benzene, toluene, the xylenes, and trichloroethene were determined by using the MIMS configuration shown in Figure 8.1 with a continuous helium purge of the membrane and two standard quadrupole mass spectrometers.[21] Multivariate calibration was used for the determination of several mixtures of benzene, ethylbenzene, toluene, and the xylenes by using a direct insertion sheet membrane probe.[22] This calibration technique allows the determination of compounds that have quantitation ions of the same *m*/*z*.

A potential process monitoring application of MIMS is the continuous measurement of volatile disinfection by-products (DBP) that are formed when water containing natural orgnic compounds is disinfected with chlorine, ozone, and other strong oxidizing agents. Continuous monitoring of DBPs in the treatment plant would provide real-time concentrations of DBPs for feedback and control of the treatment process parameters to minimize the formation of DBPs. Another potential application is to assess the temporal variations in the concentrations of DBPs at the point of use, which would allow estimates of actual human exposure and risk assessment.

Discrete Samples

The techniques described in the preceding sections for continuous measurements of VOCs in air or water are also used to measure the same analytes in discrete samples. However, with discrete samples, chromatographic separations, particularly GC, are feasible in the field but are not always used in order to keep the instrumentation as simple as possible and produce more timely results.[4] When chromatographic separations are not used the probability of obtaining mass spectra of mixtures is high and interferences are probable from compounds that give ions with the same *m*/*z* as the analyte ions. Under these circumstances an ion-trap mass spectrometer with tandem capabilities is recommended so that individual analyte ions can be isolated in the trap, fragmented, and measured to enhance the reliability of the determination (Chapter 5, "Tandem Mass Spectrometry"). A tandem mass spectrometer cannot, however, distinguish all possible analytes especially isomers and components of very complex mixtures.

The sampling and sample preparation techniques for discrete samples described in Chapters 3 and 4 are often used in the field with some simplifications. For example, the purge and trap technique for VOCs in water, which is described in Chapter 3 (under "Volatile Organic Compounds in Water"), is adapted to field measurements by replacement of the trap with a gas stream splitter and a restricted transfer line to the mass spectrometer.[4] An important limitation of this technique is that water vapor and atmospheric gases are introduced into the mass spectrometer and these can cause interferences and undesirable ion–molecule reactions.

The sampling and sample preparation techniques for VOCs in air described in Chapter 3 (under "Volatile Organic Compounds in Air") are also applicable to field measurements. An adsorbent trap similar to those described in Chapter 3 was used with an automated sample introduction system and an ion-trap GC/MS system in the field to determine chlorofluorocarbons and hydrochlorofluorocarbons in ambient air with sub parts per 10^{12} by volume detection limits.[23]

Near-continuous sampling and detection of VOCs in air was achieved by using an adsorbent trap and a very fast thermal desorption system.[24]

A portable mass spectrometer operated in the field with battery power was used to determine polychlorinated biphenyls in soils and organic solvent extracts with a thermal desorption probe and a very short temperature programmed fused silica capillary GC column.[25] Thermal desorption GC/MS was briefly described in Chapter 4 (under "Semivolatile Organic Compounds in Air, Sediments, Soil, Solid Wastes, and Tissue").

Summary

Techniques are described for the introduction of analytes from continuous flows and discrete samples into a mass spectrometer in the field or in a processing plant. These sample introduction and analysis techniques are applied to environmental, process control, and other measurements. Chromatographic separations, which are usually time consuming, require additional equipment, and provide less timely results, are not well suited to field operations. Compact ion-trap mass spectrometers with tandem capabilities can be used in the field to overcome most of the limitations of not having a chromatographic separation.

REFERENCES

1. McDonald, W. C.; Erickson, M. D.; Abraham, B. M.; Robbat, A. *Environ. Sci. Technol.* **1994**, *28*, 336A–343A.
2. Poppiti, J. *Environ. Sci. Technol.* **1994**, *28*, 536A–539A.
3. Kotiaho, T. *J. Mass Spectrom.* **1996**, *31*, 1–15.
4. Wise, M. B.; Guerin, M. R. *Anal. Chem.* **1997**, *69*, 26A–32A.
5. Mitchum, R. K.; Korfmacher, W. A. *Anal. Chem.* **1983**, *55*, 1485A–1499A.
6. Sunner, J.; Nicol, G.; Kebarle, P. *Anal. Chem.* **1988**, *60*, 1300–1307.
7. McLuckey, S. A.; Glish, G. L.; Asano, K. G.; Grant, B. C. *Anal. Chem.* **1988**, *60*, 2220–2227.
8. McLuckey, S. A.; Van Berkel, G. J.; Goeringer, D. E.; Glish, G. L. *Anal. Chem.* **1994**, *66*, 737A–743A.
9. Kotiaho, T.; Lauritsen, F. R.; Choudhury, T. K.; Cooks, R. G.; Tsao, G. T. *Anal. Chem.* **1991**, *63*, 875A–883A.
10. Wong, P. S. H.; Cooks, R. G.; Cisper, M. E.; Hemberger, P. H. *Environ. Sci. Technol.* **1995**, *29*, 215A–218A.
11. Srinivasan, N.; Johnson, R. C.; Kasthurikrishnan, N.; Wong, P.; Cooks, R. G. *Anal. Chim. Acta* **1997**, *350*, 257–271.
12. Slivon, L. E.; Bauer, M. R.; Ho, J. S.; Budde, W. L. *Anal. Chem.* **1991**, *63*, 1335–1340.
13. Cisper, M. E; Gill, C. G.; Townsend, L. E.; Hemberger, P. H. *Anal. Chem.* **1995**, *67*, 1413–1417.
14. Ketola, R. A.; Mansikka, T.; Ojala, M.; Kotiaho, T.; Kostiainen, R. *Anal. Chem.* **1997**, *69*, 4536–4539.
15. Colorado, A.; Barket, D. J.; Hurst, J. M.; Shepson, P. B. *Anal. Chem.* **1998**, *70*, 5129–5135.
16. Matz, G.; Kesners, P. *Anal. Chem.* **1993**, *65*, 2366–2371.

17. LaPack, M. A.; Tou, J. C.; Enke, C. G. *Anal. Chem.* **1990**, *62*, 1265–1271.
18. Bier, M. E.; Cooks, R. G. *Anal. Chem.* **1987**, *59*, 597–601.
19. Bauer, S.; Solyom, D. *Anal. Chem.* **1994**, *66*, 4422–4431.
20. Soni, M.; Bauer, S; Amy, J. W.; Wong, P. Cooks, R. G. *Anal. Chem.* **1995**, *67*, 1409–1412.
21. Virkki, V. T.; Ketola, R. A.; Ojala, M.; Kotiaho, T.; Komppa, V.; Grove, A.; Facchetti, S. *Anal. Chem.* **1995**, *67*, 1421–1425.
22. Ohorodnik, S. K.; Shaffer, R. E.; Callahan, J. H.; Rose-Pehrsson, S. L. *Anal. Chem.* **1997**, *69*, 4721–4727.
23. Simmonds, P. G.; O'Doherty, S.; Nickless, G.; Sturrock, G. A.; Swaby, R.; Knight, P.; Ricketts, J.; Woffendin, G.; Smith, R. *Anal. Chem.* **1995**, *67*, 717–723.
24. Mitra, S.; Feng, C.; Zhang, L.; Ho, W.; McAllister, G. *J. Mass Spectrom.* **1999**, *34*, 478–485.
25. Robbat, A.; Liu, T.-Y.; Abraham, B. M. *Anal. Chem.* **1992**, *64*, 358–364.

INDEX

In this index, *f* refers to a figure; *t* refers to a table.